Lecture Notes in Mechanical Engineering

Lecture Notes in Mechanical Engineering (LNME) publishes the latest developments in Mechanical Engineering—quickly, informally and with high quality. Original research reported in proceedings and post-proceedings represents the core of LNME. Volumes published in LNME embrace all aspects, subfields and new challenges of mechanical engineering. Topics in the series include:

- Engineering Design
- Machinery and Machine Elements
- Mechanical Structures and Stress Analysis
- Automotive Engineering
- Engine Technology
- Aerospace Technology and Astronautics
- Nanotechnology and Microengineering
- Control, Robotics, Mechatronics
- MEMS
- Theoretical and Applied Mechanics
- Dynamical Systems, Control
- Fluid Mechanics
- Engineering Thermodynamics, Heat and Mass Transfer
- Manufacturing
- Precision Engineering, Instrumentation, Measurement
- Materials Engineering
- Tribology and Surface Technology

More information about this series at http://www.springer.com/series/11236

A. K. Lakshminarayanan · Sridhar Idapalapati
M. Vasudevan

Editors

Advances in Materials and Metallurgy

Select Proceedings of ICEMMM 2018

 Springer

Editors
A. K. Lakshminarayanan
Department of Mechanical Engineering
SSN College of Engineering
Chennai, India

M. Vasudevan
Indira Gandhi Centre for Atomic Research
Kalpakkam, India

Sridhar Idapalapati
Nanyang Technological University
Singapore, Singapore

ISSN 2195-4356   ISSN 2195-4364 (electronic)
Lecture Notes in Mechanical Engineering
ISBN 978-981-13-1779-8   ISBN 978-981-13-1780-4 (eBook)
https://doi.org/10.1007/978-981-13-1780-4

Library of Congress Control Number: 2018949339

This Springer imprint is published by the registered company Springer Nature Singapore Pte Ltd.
The registered company address is: 152 Beach Road, #21-01/04 Gateway East, Singapore 189721, Singapore

Preface

The Two-Day International Conference on "Engineering Materials, Metallurgy and Manufacturing" (ICEMMM 2018) is organized by the Department of Mechanical Engineering, SSN College of Engineering, Kalavakkam, Chennai. The advent of the digital age is seeing a robust growth of research in smart materials and technologies, while the manufacturing processes ensure making these new age products and prototypes. The need for energy savings, environment protection, and green-rated processes are further accelerators in the right direction. We are therefore happy to organize the International Conference on Engineering Materials, Metallurgy and Manufacturing—ICEMMM 2018, on February 15–16, 2018, in our esteemed institution. The aim and scope of the conference is to provide a robust platform for academicians, researchers, scientists, and students to present their ongoing research work in cutting-edge areas of materials, metallurgy, and manufacturing. It is also the endeavor of this conference to bring to the forefront the seminal research works in these key areas by the eminent scholars worldwide and provide them an opportunity to share their success stories with an earnest audience. This conference hopes to provide enough food for thought on the challenges faced by the industry and eagerly awaits the amalgamation of ideas and solutions to address the current and future requirements. We sincerely hope all the delegates will cherish the exchange of new ideas and reap considerable benefits from the face-to-face interaction and the potential future networking.

Chennai, India
Singapore
Kalpakkam, India

A. K. Lakshminarayanan
Sridhar Idapalapati
M. Vasudevan

ICEMMM 2018—Core Organizing Committee

Conveners

Dr. K. S. Vijay Sekar, Associate Professor
Dr. A. K. Lakshminarayanan, Associate Professor

Co-conveners

Dr. L. Poovazhagan, Associate Professor
Dr. M. Dhananchezian, Associate Professor
Dr. K. Jayakumar, Associate Professor

Acknowledgements

We express our gratitude to the management of SSN Institutions for giving us the opportunity to organize this important conference and also thank all the delegates for making this event a grand success. We thank the keynote speakers; conference chairs; advisory, program, and technical committees; reviewers; colleagues; the college administration; our friends; students; and well-wishers for making this event a memorable one. We thank Springer for having consented to publish our papers through two book volumes. We thank one and all.

Contents

About the Editors

Dr. A. K. Lakshminarayanan holds a Ph.D. in manufacturing engineering from Annamalai University, in the area of welding of ferritic stainless steel. He received the Best Faculty Award from the Department of Mechanical Engineering, SSN college of Engineering for the years 2013–14, 2014–15, and 2015–16. His areas of specialization include materials joining, surface engineering, advanced metal-forming processes, fatigue, and stress corrosion cracking of welds. He has published more than 50 research papers in international peer-reviewed journals, and his research work has been presented at over 30 international and national conferences. He is also a reviewer for various international journals. He has successfully completed a sponsored external project on "Developing FSW tools to weld stainless steels," funded under the Government of India's DST Young Scientist Scheme.

Dr. Sridhar Idapalapati is currently Associate Professor at the School of Mechanical and Aerospace Engineering, Nanyang Technological University (NTU), Singapore. He is Cambridge Nehru Scholar and holds a Ph.D. in mechanics of solids from the University of Cambridge and a first-class honors (HMT Gold Medal recipient) degree in mechanical engineering from Jawaharlal Nehru Technological University (JNTU), India. Prior to joining NTU, he worked as Visiting Scientist at the National Institute of Standards and Technology (NIST), USA. His research interests include design with sandwich structures, composite materials, powder consolidation, biomaterials, and hard tissue implants. Recently, he has also begun working on wind energy. He has published more than 70 papers in international journals and seven book chapters.

Dr. M. Vasudevan received his Ph.D. in metallurgical engineering from IIT Madras, India. He currently holds the position of Section Head, Advanced Welding Processes and Modeling Section, Materials Technology Division, Indira Gandhi Centre for Atomic Research (IGCAR), Kalpakkam. He is also Professor in the Department of Atomic Energy, Homi Bhabha National Institute. He is an expert in the areas of welding nuclear structural materials and modeling weld phenomena. He has published more than 120 research papers in international journals and received the Metallurgist of the Year Award from the Indian Institute of Metals in 2016.

Sliding Wear Response of Typical Bearing Materials

Ram Naresh, Ankit Aman, Anuj Gupta and S. Sivarajan

Abstract Bearing performance is an important factor to improve efficiency of pumps. Depending on specific pump requirements, pump producers have various alternatives for selection of bearing materials. Little friction, slightest to no wear, low-power utilization, and extended life are important features for pump bearings. Serious attention must be given to the tribological evaluation of typical bearing materials. In this work, a pin-on-disk wear testing instrument is employed to study the sliding wear response of phosphor bronze, nylon 6/6, brass, DLC-coated steel, and WC/C coated steel. The tests were performed with an applied load of 10 N and sliding velocity of 0.6 m/s. EN 31 steel heat treated to 55 HRC was selected as disk material. The experiments are conducted as per ASTM G99 standards. The low value of coefficient of friction was noticed in nylon 6/6 and DLC-coated steel.

Keywords Bearing material · Friction · Wear · Nylon · Coated steel
Brass

1 Introduction

Pump performance is remarkably affected by the life and performance of bearings. Different materials have been attempted for bearing parts in pumps, automobile engines, general industrial use, marine diesel engines, hydraulic turbine, and electrical generator shafts [1, 2]. Most of the bearing materials are still copper based. Lead- and tin-based materials are also familiar [3]. Bearing materials with polymer base surpassed conventional bearing materials. The more and more difficult operating environment, which recent designs of pumps and engines have dictated on bearing material, have led to the use of unconventional and novel bearing materials. Low friction, wear resistance, cost, conformability, and long life

R. Naresh · A. Aman · A. Gupta · S. Sivarajan (✉)
School of Mechanical and Building Sciences, Vellore Institute of Technology,
Chennai, India
e-mail: sivarajan.s@vit.ac.in

© Springer Nature Singapore Pte Ltd. 2019
A. K. Lakshminarayanan et al. (eds.), *Advances in Materials and Metallurgy*,
Lecture Notes in Mechanical Engineering,
https://doi.org/10.1007/978-981-13-1780-4_1

are the major factors that go with the choice of bearing materials. Traditional bearing materials are quite expensive, and this requires a quest for low-priced bearing materials. Steel and gray irons are less costly and easily available. Nitriding treatment is applied to plain carbon steel and gray iron [4]. This treated steel performs adequately in relation to traditional bearing materials.

Polymer materials with inherent lubricity are to a greater extent substituting metallic components in bearings where dry, corrosive, and abrasive environment develops [5]. Polymers are able to take in impact loading and bending due to its elastic properties. In comparison to metals, the frictional coefficient of polymers is reliant on the load, reducing with increase in load. Surface adhesion and deformation are the factors that contribute to friction in polymers. Friction and wear properties of bearing materials were tested in oil and gas lubricated condition by several researchers [6–10]. In the present paper, the frictional behaviors of phosphor bronze, brass, nylon 6/6, DLC-coated steel, and WC/C coated steel, in dry condition have been determined using a pin-on-disk tribometer with a uniform load of 10 N and uniform sliding speed of 0.6 m/s.

2 Experimental Work

Friction and wear tests were conducted as per ASTM G99 standard [11]. For the pin-on-disk wear test, two specimens are used. One, a pin with a radiused tip, is arranged perpendicular to the flat circular disk. The testing apparatus brings about the disk specimen to revolve about the disk center. The sliding track is a circle on the disk surface. The plane of the disk was located horizontally. The pin specimen is pushed against the disk at a definite load by means of an arm and attached weights. The present study is conducted in order to compare the frictional and wear under dry environment for phosphor bronze, brass, nylon 66, DLC-coated steel, and WC/C coated steel, sliding on an EN31 (hardness 55 HRC).

Wear results were described as volume loss for the pin and the disk separately. The amount of wear was evaluated by measuring linear dimensions of both specimens before and after the test. The length change of the pin and the depth change of the disk wear track are determined by stylus profiling. Linear wear values were changed to wear volume using suitable geometric relations. Wear results were acquired by performing a test for a chosen sliding distance, load, and speed. The wear will depend upon factors such as the applied load, machine features, sliding speed, sliding distance, environment, and the material properties. The system has a load cell that allows the coefficient of friction to be determined. The pin specimen is cylinder with a radiused tip. The circular disk has a diameter of 55 mm and thickness of 10 mm. A surface roughness of 0.8 μm is maintained. Attention must be taken in surface preparedness to prevent subsurface damage that modifies the material significantly. Friction and wear tests were conducted with an applied load of 10 N and sliding speed of 0.6 m/s at room temperature.

Before measuring or weighing, the specimens were cleaned and dried. Attention is taken to remove all dirt from the specimens. Non-chlorinated cleaning agents and solvents are used for cleaning. The disk was inserted securely in the holding device so that the disk is fixed perpendicular to the axis of the revolution. The pin specimen was attached safely in its holder and adjusted so that the specimen is perpendicular to the disk surface when in contact. The mass was added to the system lever to establish the chosen force pressing the pin against the disk. The motor was started, and the speed was adjusted to the wanted value while holding the pin specimen out of contact with the disk. The motor was stopped. The revolution counter was fixed to the required number of revolutions. The test was started with the pin and disk in contact under load. The test is stopped when the required number of revolutions was carried out. The specimens were removed, and loose wear debris is cleaned. The pin and disk wear track dimensions were re-measured accurately. The test was repeated with extra specimen to acquire sufficient data for statistically important results. Figure 1 shows the photo of pin-on-disk tribometer. Table 1 gives the properties of bearing material. The wear parameters were selected as per previous literature studies.

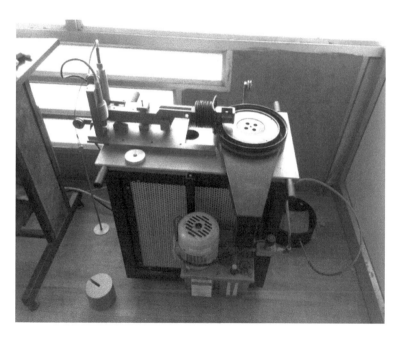

Fig. 1 Photo of pin-on-disk tribometer

Table 1 Material specification

S. no.	Material	Composition (%wt.)	Hardness	Density (g/cu cm)
1.	Phosphor bronze (UNS C51000)	Cu 98, Sn 1.2, P 0.30, Zn 0.35, Fe 0.10, Pb, 0.05	78 (BHN)	8.8
2.	Brass (UNS C46400)	Cu 64.5, Zn 35.3, Fe 0.05, Pb 0.15	58 BHN	8.55
3.	Nylon 66	Aliphatic polyamide	80 shore D	1.15
4.	EN-31(hardened)	Fe 96.32, Cr 1.6, C.90, Mn 0.75, Si 0.35, S 0.040, P 0.040	60 HRC	8.86

3 Results and Discussion

3.1 Wear Results

It is advantageous for a bearing material to display a low frictional coefficient and low specific wear rate over a range of functional regime. The results of wear of phosphor bronze sample depicted two contrasting regions as shown in Fig. 2. At the beginning, the wear rate increased up to a sliding distance of 400 m as the original surface asperities are removed. Between 400 and 1000 m of sliding distance, the plastic deformation formed along the wear test causes delamination and the wear rate is escalated. The results of the wear of brass sample show three contrasting regions as shown in Fig. 2. In the initial runoff period, limited wear rate is observed as the original surface asperities are taken off. The elimination of surface asperities is followed by a polished surface when the sliding distance is between 400 and 600 after a sliding distance of 600 m; there is a steady-state wear. The results of wear of nylon sample showed negligible wear for a sliding distance of 1000 m.

The plot of the DLC-coated specimen (Fig. 3) although starts at zero and continues to possess negative values for a time period of 400 s. After a transient state for 15 s, steady-state interaction is achieved. This negative sign of wear implies that the disk with which the pin is interacting is being worn off by the coating. This is a demonstration of abrasive wear mechanism where the pin is abrading the disk. After a time of 400 s, the coating has worn off and the substrate material of the pin begins interacting with the disk. This phenomenon can be avoided by providing an interlayer of nitride coating as suggested in [8]. The wear track displays the presence of foreign material which corresponds to the wear particles from the pin. This presence indicates that during the interaction of the surfaces, adhesive wear has taken place during the steady-state wear phase which was followed by abrasive wear between the specimens after the coating was extinguished. The wear rate was calculated to be 1.4893×10^{-6} m^2/N.

The plot of WC-C (Fig. 3) shows similar behavior to DLC as it starts at zero and continues to possess a negative value up to a time elapse of 200 s. The steady-state

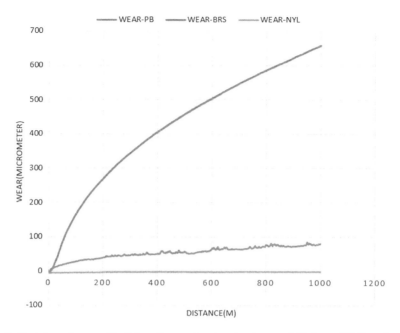

Fig. 2 Variation of wear with sliding distance for phosphor bronze (PB), brass (BRS), and nylon 6/6 (Nyl)

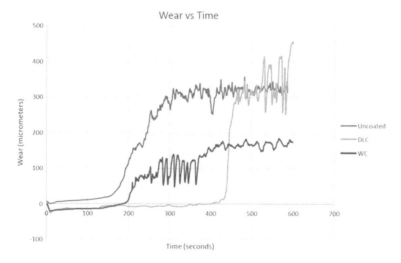

Fig. 3 Variation of wear with time for DLC-coated steel and WC/C coated steel

operation culminates at 200 s which indicates the exhaustion of the coating. The interaction of the pin substrate material and disk begins at this point and reaches steady-state phase after a short period of operation in the transient phase. The wear track formed on the disk by this specimen illustrates material transfer from the pin. This is similar to the case of the DLC pin which shows that the material transfer occurred during the steady-state interaction between the coating and disk. This is followed by the interaction between the pin substrate material and disk. Wear rate was calculated to be 1.7937×10^{-6} m^2/N.

3.2 Friction Results

The fluctuation of coefficient of friction with sliding distance is presented in Fig. 4. For phosphor bronze, the coefficient of friction is 0.3 at the beginning of the test and then varied in the range of 0.9–1.2 for the remaining sliding distance. The friction coefficient of brass was around 0.3 for most of the sliding distance tested. In case of brass, transfer of material due to adhesion at the initial stage of sliding has controlled the coefficient of friction. When brass pin slides on steel, friction coefficient is governed by adhesion through shearing of brass pin asperities. Subsequently, the wear debris produced in the wear process is formed of brass only. No traces of steel are present in either the worn surface of brass pin or the debris. The friction coefficient for nylon is 0.28 for most of the sliding distance tested.

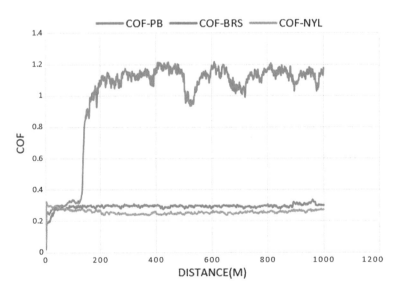

Fig. 4 Variation of coefficient of friction with sliding distance for phosphor bronze (PB), brass (BRS), and nylon 6/6 (Nyl)

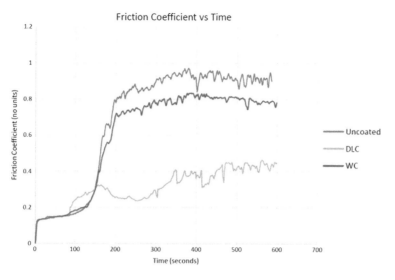

Fig. 5 Variation of coefficient of friction with time for DLC-coated steel and WC/C coated steel

The fluctuations of coefficient of friction with time for DLC-coated steel and WC/C coated steel are shown in Fig. 5. The plot of DLC varies indeterminately for 350 s which is followed by the attainment of steady state. The mean value of friction coefficient is 0.387. The plot of WC-C varies in the transient state for 200 s after which it reaches steady state. The average value of friction coefficient is 0.803. The wear volume loss of pin and disk offers another perspective on analyzing the performance of the coatings. As a general rule of thumb, the greater the volume loss of disk, the more abrasive the pin is and the more the pin can endure. The more the pin volume loss, the more the susceptible it is to wear out soon and less it can endure. The pin and disk wear results are shown in Table 2.

Table 2 Pin and disk wear for phosphor bronze, brass, and nylon 6/6

S. No.	Material	Load (in N)	Sliding distance (m)	Wear volume (10^{-3} cm^3)	Pin wear volume (10^{-3} mm^3)	Disk wear volume (mm^3)	Wear rate (10^{-8} cm^2)	Wear resistance (10^8 cm^{-2})
1.	Phosphor bronze (UNS C51000)	10	1000	2.04545	92.94	4.385	2.0454	0.4888
2.	Brass(UNS C46400)	10	1000	6.7836	1020	5.089	6.7836	0.1474
3.	Nylon	10	1000	1.7391	12.27	2.6	1.7391	0.5750

4 Conclusions

Friction and wear of different bearing materials under dry conditions were examined utilizing a pin-on-disk tribometer. The following conclusions were drawn.

1. The maximum value of friction coefficient was observed in phosphor bronze bearing while the minimum value of friction coefficient was observed in nylon 66 bearing without lubrication.
2. From wear rate results, maximum values were recorded for brass and the minimum values were recorded for nylon 66.
3. DLC-coated steel can be an excellent choice as bearing material under dry conditions since it recorded an average coefficient of 0.383, while it is inadvisable to use WC/C coated steel in dry condition for bearings.

Acknowledgements The authors are grateful to the Management of VIT, Chennai for their support to carry out this research work.

References

1. Nagata M, Fujita M, Yamada M, Kitahara T (2012) Evaluation of tribological properties of bearing materials for marine diesel engines utilizing acoustic emission technique. Tribol Intl 46:183–189
2. Gebretsadik DW, Hardell J, Prakash B (2015) Tribological performance of tin-based overlay plated engine bearing materials. Tribol Intl 92:281–289
3. Zeren A (2007) Embeddability behaviour of tin-based bearing material in dry sliding. Mater Des 28:2344–2350
4. Rae AR, Marappan R, Mohanram PV (1992) Performance studies on treated steels as substitutes for conventional bearing materials. Wear 155:15–29
5. Thorp JM (1982) Friction of some commercial polymer-based bearing materials against steel. Tribol Intl 69–74
6. Abdollaha MFB, Mazlan MAA, Amiruddina H, Tamaldin N (2013) Frictional behavior of bearing material under gas lubricated conditions. Proc Eng 68:688–693
7. Goudarzi MM, Jahromi SAJ, Nazarboland A (2009) Investigation of characteristics of tin-based white metals as a bearing material. Mater Des 30:2283–2388
8. Feyzullahoglu E, Sakiroglu N (2010) The wear of aluminium-based journal bearing materials under lubrication. Mater Des 31:2532–2539
9. Miyajima T, Tanaka Y, Iwai Y, Kagohara Y, Haneda S, Takayanagi S, Katsuki H (2013) Friction and wear properties of lead-free aluminum alloy bearing material with molybdenum disulfide layer by a reciprocating test. Tribol Intl 59:17–22
10. Kuanyshev M, Nuralin B, Salimov B, Kaukarov A, Murzagaliev A, Narikov K, Shakeshev B (2017) The improvement of friction bearing manufacturing technology by using copper alloy. Int J Adv Manuf Technol 88:317–324
11. ASTM G99 (2010) Standard test method for wear testing with a Pin-on-Disk Apparatus. ASTM International, West Conshohocken, PA. www.astm.org

Investigating and Enhancing the Properties of Biodegradable Polymer Cassava Starch with Acrylonitrile Butadiene Styrene by the Addition of Cardanol Oil

T. Panneerselvam, T. Prasanna Vengatesh and S. Raghuraman

Abstract Biodegradable polymers are produced in a way to decay by the natural cause of micro-living organisms. Cassava starch which is a biodegradable polymer used for biomedical, packaging items. But the molecular weight is less than other biodegradable polymers. The aim of this research work is to investigate the properties of biodegradable polymer manufactured from cassava starch containing the combination of acrylonitrile butadiene styrene (ABS), a non-biodegradable polymer with the cardanol oil, a bio-phenolic compound which is employed as the compatibilizer. They are blended with melt blending process using injection moulding. The use of cardanol oil in the mixture resulted in increase in the level of compatibility evidenced by increase in the intermolecular weight, and the results have shown the substantial plasticizing effect, increase of mechanical properties (tensile strength, impact strength and flexural strength) and increase in the intermolecular weight which will make from biodegradable property of 'disposables' to applications for 'durables'.

Keyword Biodegradable polymers · Acrylonitrile butadiene styrene Cassava starch · Cardanol oil · Mechanical properties

T. Panneerselvam · T. P. Vengatesh (✉) · S. Raghuraman
School of Mechanical Engineering, SASTRA University,
Thanjavur 613401, Tamilnadu, India
e-mail: yohanprasanna1994@gmail.com

T. Panneerselvam
e-mail: tpansel@mech.sastra.edu

S. Raghuraman
e-mail: raghu@mech.sastra.edu

© Springer Nature Singapore Pte Ltd. 2019
A. K. Lakshminarayanan et al. (eds.), *Advances in Materials and Metallurgy*,
Lecture Notes in Mechanical Engineering,
https://doi.org/10.1007/978-981-13-1780-4_2

1 Introduction

The polymers used nowadays are mostly non-biodegradable type based on their chemical and physical properties. The degradation is mainly based on three categories, namely, photodegradation, thermal degradation and biodegradation. Thermal and photodegradation are done by the natural effect of heat and light from the sun. The biodegradation is caused by the microorganisms. The biodegradation starts from the fragmentation and ends with mineralization. The fragmentation is the mechanical breaking of the substance, and mineralization is the chemical disintegration of the substance. Generally, the non-biodegradable materials do not undergo biodegradation [1]. These non-biodegradable polymers produce lots of wastes which are hazard to the nature and cannot be recycled. Hence, biodegradable polymer is required.

The biodegradation can be carried out in different ways as per the standard, ASTM D883 [2]. Biodegradable plastic, called as a polymer composite, degrades by the nature of the living microorganisms (e.g. fungi and bacteria). The biodegradation process is happened by two different mechanisms, called biofragmentation and biomineralization. Biofragmentation, a phenomenon in which a polymeric material goes to macroscopic disintegration due to geological reasons over the period of time or changed to fossil fuels in due course of time by the action of microorganisms [3].

Acrylonitrile butadiene styrene (ABS), thermoplastic polymer, has wider applications in the engineering field. ABS consists of styrene, butadiene and acrylonitrile. The ABS composition attributes to many characteristics like ease of processing, toughness, rigidity, impact strength, etc. ABS is mainly used in automotive sector for manufacturing headlight housings, radiator grills, interior trim parts, consoles and instrument panels [4].

Cassava, starchy tuberous root comes under the Euphorbiaceous family and it is largely cultivated in tropical regions of Asia, America and Africa. Cassava starch is a thermoplastic material by polymerization with plasticizer, and it has compatibility to produce as well as develop biodegradable polymers. The researchers have reported in their work about glycerol and water which can be used as a plasticizer in preparing starch. But the use of starch in biodegradable polymers preparation for industrial applications is scarce due to its low stability, lesser strength and poor processability. Even if it has limitations, the cassava starch could be used as a matrix material in composites preparation by the injection moulding process with suitable reinforcements [5].

However, a polymer composite preparation requires an utmost care in selecting a suitable blending material. The blending material actually increases the miscibility in order to attain homogeneity in the composite. Cardanol oil is precipitation of liquid from cashew nut shell, and it is a type of phenol. It has a good compatible capacity which increases the intermolecular weight of the component [6]. The process is done by the addition of hydrochloric acid, glycerine and sodium hydroxide. The hydrochloric acid is used to accelerate the reaction. The glycerine is used to improve the glass transition temperature of the cassava starch. The sodium hydroxide is used to neutralize the solution.

2 Experimental Works

2.1 Materials

ABS polymer was provided by Noor trading Mart, Trichy in form of black pellets. At the received condition, the density of ABS was measured to 1.04 g/cc and a melt flow index was found to be 23 g/10 min. Cassava starch (18 wt% amylose and 82 wt% amylopectin) was supplied by Linga Starch Products, Salem. The cashew nut shell liquid (CNSL) was bought from Golden Chemicals Pvt. Ltd., Pondicherry. The glycerine, HCl and NaOH are received from Meenakshi Chemicals Pvt. Ltd., Pondicherry.

2.2 Processing

Initially, the starch and the ABS pallets were dried at 50 °C for 24 h. After this process is over, injection moulding was done. In the injection moulding, the starch and the ABS were initially introduced for melting and then they were blended. The starch/ABS mixing ratio was kept as constant at the rate of 80/20 wt% in order to get a good balance as well as to improve toughness property of a final polymer product. The other sample is prepared with the same ratio of starch/ABS with 5% of cardanol oil. At the given ratio, the glycerine, NaOH and HCl were also added to the polymer mixture. The blended starch/ABS was moulded by using a hand-operated injection moulding machine. Temperature of the mould was set to 65 °C, and then pellets were poured into the cylinder for melting, nearly 5 min and finally injected. For tensile testing, flexural testing and impact testing, the mould was prepared to 120 mm × 30 mm × 4 mm to meet the specimen preparation size needed as per the ASTM standards [7].

2.3 Characterization

2.3.1 Scanning Electron Microscopy (SEM)

Scanning electron microscope (SEM) was used to study the microstructure of starch/ABS blends. The surface is gold coated to avoid any sample degradation. The prepared surfaces were observed under various magnifications (500X 1000X,

2000X, 5000X and 10,000X). Testing was done at TESCAN Vega 3 SEM CENTAB, SASTRA Deemed University.

2.3.2 Impact Testing

Izod impact test was carried out on Starch/ABS specimens to assess its impact strength using an Izod impact testing machine at Metmech engineering and Analytical laboratory, Chennai. The sample was cut to 65.5 mm × 12.5 mm × 2 mm with the 45° V notch in the middle of length as per the ASTM D256 [8]. The sample was cut to the standard size by the waterjet cutting at ALIND-waterjet cutting, Chennai.

2.3.3 Flexural Testing

The flexural strength test was performed to starch/ABS blends by the flexural strength testing machine at Omega Inspection and Analytical Laboratory, Chennai. The specimen size was 50 mm × 20 mm × 4 mm as per the ASTM D790 [9].

2.3.4 Tensile Testing

The tensile tests of elaborated starch/ABS blends were assessed by Universal Testing Machine at Metmech Engineering and Analytical Laboratory, Chennai. The specimen was made to dog bone shape with 100 mm × 20 mm × 4 mm (6 mm breadth in middle for 38 mm) as per the ASTM D638 [10]. The sample was cut to the standard size by the waterjet cutting at ALIND-waterjet cutting, Chennai.

3 Results and Discussion

3.1 Morphology Observation

Morphological observations of starch/ABS composites were performed through SEM to reveal the dispersion states of starch. Figure 1a shows the blends of starch/ABS (80/20 wt%) without cardanol oil and Fig. 1b shows the blends with cardanol, 5 wt%. The image was taken in both 5000X magnification and 10,000X magnification. In this mixture, the glycerol increases the plasticizing capacity of starch in preheating and mixing with ABS. But ABS is immiscible with starch because it is not homogenous in nature. When the cardanol oil is mixed with the starch/ABS blends, it increases the blending of starch/ABS due to its compatible process.

The dots in the image are the starch which was not completely miscible state in the image, Fig. 1a. But as shown in Fig. 1b the dots were gradually reduced due to

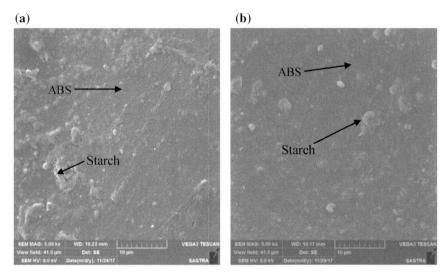

Fig. 1 Scanning electron micrographs 5000X of starch/ABS (80/20 wt%) blends **a** without and **b** with 5 wt% cardanol oil

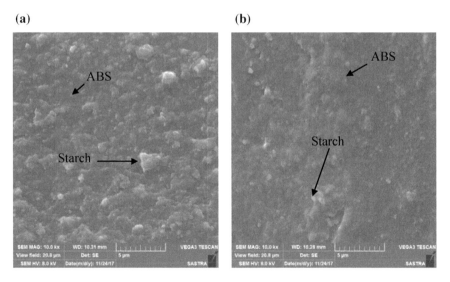

Fig. 2 Scanning electron micrographs 10,000X of starch/ABS (80/20 wt%) blends **c** without and **d** with 5 wt% cardanol oil

the addition of cardanol oil. The blending of starch is high with ABS in the image, Fig. 1b. This is because of the compatible capacity of the cardanol oil. Figure 2c, d shows the deep investigation of the blending and the huge difference in blending. The smoothness of the blends is increased due to the addition of cardanol oil.

The smooth surface of starch/ABS blends reveals the good compatibility after being mixed with cardanol oil. From Figs. 1 and 2, it is inferred that the starch and ABS are immiscible but the cardanol oil has increased the compatible property of the starch and ABS. So the blending of starch and ABS is high when cardanol oil is mixed.

3.2 Mechanical Properties Analysis

3.2.1 Impact Testing

Table 1 shows the impact test results of starch/ABS blends without and with cardanol oil. Compared to the impact strength of ABS, the results of blends are very low. But a small increase in impact strength is observed when the cardanol oil is mixed with starch/ABS blends. This result shows that the addition of cardanol oil gives more impact strength to the polymer material. The starch/ABS blends are not fully miscible as shown in Fig. 1, but the bond between them is increased when the cardanol oil is mixed. This shows the advantage of compatible oil mixed in the blending process.

3.2.2 Flexural Testing

The flexural test results of starch/ABS blends are shown in Table 2 which shows the bending strength and load-bearing capacity of the material. It is found that the load-bearing capacity is better for the material without mixing the oil, but the bending capacity of the material is high when the oil is mixed. This shows the flexibility of the material. Polymer requires more flexibility when compared to other material. So the cardanol oil increases the bending property of the material, but its brittleness is high when it is compared to normal blend, and the breakage at load given shows the brittleness of the material. Due to its high compatibility, it increases the brittleness of the material.

Table 1 Impact test results

Sample no.	Material	Impact load (KJ/m^2)
Sample 1	ABS + starch	0.22
Sample 2	ABS + starch + cardanol oil	0.24

Table 2 Flexural test results

Sample No.	Material	Flexural load (KN)	Stroke (mm)
Sample 1	ABS + starch	0.124	4.75
Sample 2	ABS + starch + cardanol oil	0.118	5.65

3.2.3 Tensile Testing

The tensile test shows huge variation of the blends without (Sample 1) and with (Sample 2) oil mixing. Figure 3 shows the tensile properties of the starch/ABS without cardanol oil (Sample 1).

The result shows that tensile strength of sample 1 is 16.99 MPa. The load is 0.42 kN when the breakage is happened. The maximum displacement is observed to 2.62 mm, and the percent elongation is 2.75%. The strain rate is 6.55. So starch/ABS (80/20 wt%) blends cannot be used for the application, when the starch is mixed directly to the ABS.

Figure 4 shows the tensile test results of starch/ABS blends with cardanol oil (Sample 2). The load is 0.56 kN when breakage is happened, and this is comparatively good to Sample 1. The displacement is 4.76 mm which is higher than the

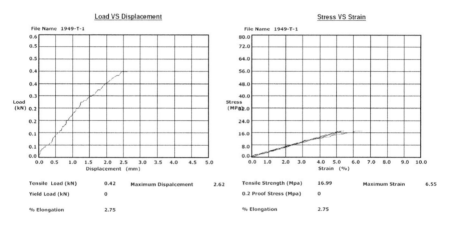

Fig. 3 The UTM tensile test report on starch/ABS (80/20 wt%) blends without cardanol oil

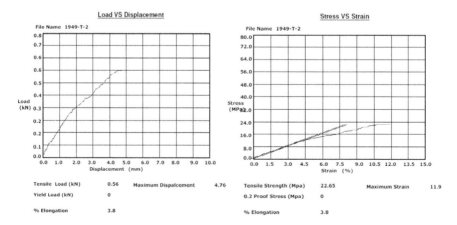

Fig. 4 The UTM tensile test report on starch/ABS (80/20 wt%) blends with cardanol oil

displacement observed in Sample 1. The percent elongation is 3.8%, higher than the Sample 1 and the strain rate is 11.9 compared to starch/ABS blends without cardanol oil. The ultimate tensile strength of the starch/ABS blend is high (22.65 kN) when the oil is mixed with the blending. But the tensile strength of Sample 2 is moderately increased and Sample 2 gives good results when the cardanol oil is mixed with the starch/ABS blends. This is due to the compatible property of the cardanol oil which increases the miscibility of the ABS and starch and increases the mechanical properties of the material.

Normally, when the starch is mixed with the synthetic polymer, it reduces its mechanical property which makes it not suitable for application. But the cardanol oil increases its property which is comparatively good, suitable for application, and the durability of the polymer is increased. Figures 1 and 2 also show that the miscibility of the ABS and starch is increased when the cardanol oil is mixed. This shows that the starch/ABS blends with cardanol oil could be used for commercial application to achieve the benefit of biodegradability.

4 Conclusions

The cardanol oil effects on the properties of biodegradable polymer studied in this work are summarized as follows:

- The SEM images show the immiscibility/miscibility of the starch and ABS without/with cardanol oil and it is observed that cardanol oil has increased the blending of the material, starch/ABS.
- The impact test result has shown that the cardanol oil mixed with starch/ABS blends has increased the impact strength of the material.
- The flexural test results reveal that the load-bearing capacity is better for the material, starch/ABS blends without mixing of the cardanol oil, but the bending capacity of starch/ABS blends is high when the cardanol oil is mixed with the blends.
- The tensile strength of starch/ABS (80/20 wt%) blends with cardanol oil is found to be higher than the tensile strength of starch/ABS (80/20 wt%) blends without cardanol oil.

However, further investigation would be needed to improve the properties of biodegradable polymer in order to compete with the properties of ABS and its applications.

Acknowledgements Authors are very much grateful to Prof. R. Sethuraman, The Vice Chancellor, SASTRA Deemed University, Thanjavur—613 401 for the permission granted to publish this work.

References

1. Krzan A (2011) Bio-degradable polymers and plastics. Central Europe programme
2. Hemjinda S, Miertus S, Corti A, Chiellini E (2006) Standardization and certification in the area of environmentally degradable plastics, vol 91
3. Laycock B (2017) Lifetime prediction of bio-degradable polymers. Polym Sci 71:189
4. Ben Difallah B, Kharrat M, Dammak M, Monteil G (2012) Mechanical and tribological response of ABS polymer matrix filled with graphite powder. Mater Des 34:782–787
5. Navia DP, Villada HS (2005) Thermoplastic cassava flour
6. Alexsandra M, Rios DS, Mazzetto SE (2009) Cashew nut shell liquid (cnsl) as source of eco-friendly antioxidants for lubricants. In: The 13th International electronic conference on synthetic organic chemistry
7. Rigoussen A, Verge P, Raquez JM, Habibi Y, Dubois P (2017) In-depth investigation on the effect and role of cardanol in the compatibilization of PLA/ABS immiscible blends by reactive extrusion. Eur Polym J 93:272–283
8. Hung TA (2006) Izod impact tester. Hung Ta^{TM} Izod Impact Tester Model HT8041B (ASTM D256 Methods A, C, Dand E and ASTM E23)
9. Specimens P (2002) Flexural properties of unreinforced and reinforced plastics and electrical insulating materials. ASTM Des D 790–02(14):146–154
10. ASTM Handbook (2004) Introduction to tensile testing ASTM D638. Tensile testing, 2nd edn., pp 1–13

Investigation on the Microstructure and Mechanical Properties of AA6082/TiO₂ Surface Composites Produced by Friction Stir Processing

Deepak Sandar, Rishika Chatterjee, R. Jayendra Bharathi and Peddavarapu Sreehari

Abstract AA6082–TiO_2 surface composites are prepared by friction stir processing (FSP). Fabricating defect-free surface composite with uniform particle distribution by FSP is a challenging task. In this study, TiO_2 particle-reinforced AA6082 alloy surface composites were fabricated using different FSP parameters to find the optimum tool rotation speeds and traverse speeds. This investigation was carried out with optical microscopy and was found that TiO_2 reinforcement particles are disintegrated and distributed homogenously by the stirring action of the tool, which also forges the plasticized material. The hardness profile was obtained using Vickers microhardness test across the thermomechanically affected zone and welded zone and found that the retreating side had more hardness. The tensile test was also performed to identify the change in the strength due to the surface composite.

Keywords Alloy · Composite · Tool rotation · Welding speed Reinforcements

1 Introduction

Manufacturers and industrialists always look for better performing materials and innovative performances. Aluminium alloys never settle without satisfying all needs at all times. With their extreme lightness, very good thermal and electrical conductivity, corrosion resistance, diversity of alloys and intermediates, ease of use and recycling properties they came to forefront in the usage [1–3]. Their common

D. Sandar · R. Chatterjee · R. J. Bharathi · P. Sreehari (✉)
School of Mechanical Engineering, SASTRA Deemed University,
Thanjavur, India
e-mail: sreehari@mech.sastra.edu

R. J. Bharathi
e-mail: jayendrabharathi@mech.sastra.edu

© Springer Nature Singapore Pte Ltd. 2019
A. K. Lakshminarayanan et al. (eds.), *Advances in Materials and Metallurgy*,
Lecture Notes in Mechanical Engineering,
https://doi.org/10.1007/978-981-13-1780-4_3

applications include light aircraft structures, automotive chassis, bodies, engine blocks, bridges and superstructures like passenger ships and merchant ships [4–6].

The 6xxx series alloys of aluminium are commonly used for automotive applications. The major alloying elements are magnesium and silicon which provides the sufficient hardness and formability property at molten state to the alloy [6–8]. With these properties, casting can be done easily to have intricate products for various applications as listed above. But the main defect appearing during 6xxx series production is the hot brittleness. Excess percentage of magnesium content (around 2%) in the system increases the tendency to hot brittleness of the alloy, and the silicon introduction leads to the recrystallization cracks. Such defects call for additional reinforcements like titanium and its compounds (Titanium dioxide and titanium boride) to eliminate hot brittleness. Titanium alloys are one of the toughest materials in the world [8]. They are even used in cryogenic applications because of their impressive toughness. Titanium embrittles rapidly when it is allowed to oxidize. The absorption of oxygen by titanium when the material is heated causes an increase in hardness of the surface, thus counteracting the embrittlement characteristics of aluminium alloys at elevated temperatures [7]. So, in the current work, titanium dioxide particles are chosen to reinforce the aluminium 6082 matrix.

Conventional metal matrix preparation methods like stir casting, pressure infiltration, diffusion bonding, reactive processing, and physical vapour deposition use elevated temperatures to fabricate MMCs. Later, when they are cooled to ambient temperatures after diffusion bonding occurs, residual stresses are generated in the composite [7, 10]. This mechanism significantly affects the mechanical properties of MMC in all loading conditions. But friction stir processing adopts the principle of friction stir welding which in turn is dynamic plastic deformation to fill the welds [5]. Since friction stir processing is a solid-state processing technique and as it is performed only on the surface, it has lots of advantages like avoiding diffusion, creation of homogenous fine and equiaxed grains, and thus the development of superplastic properties [6–10]. However, the microstructure and mechanical properties of the processed material can be enhanced by changing the processing parameters, and thus, on the whole, the composite production becomes very easy to handle [4, 5].

FSP of various Al–Mg alloys has resulted in improvement in their mechanical properties [11]. However, since the technique is relatively new, there are many outstanding issues which need better scientific understanding. The development of microstructure during the process is one such issue which needs particular attention [12]. FSP has been developed as a generic tool for materials processing and modification. In this technique, specially designed non-consumable cylindrical tool rotating at high speed is plunged into the alloy till the shoulder contacts the surface and is traversed along a particular length at a desired traversed speed. The most significant feature of the friction stir processed material was that even after this significant improvement in strength there was little loss of ductility [8].

AA6082–T6 alloy plates of 8 mm thickness were used for this study. Two grooves of 5 mm deep each and 0.8 and 1.6 mm width were made along the centre line of the plate using milling and compacted with TiO_2 powder. The average size of TiO$_2$ particles used in this work was 25 μm. A cylindrical pin profile tool made of high-speed steel was used to perform FSP. Necessary dimensions [5], shoulder diameter 12 mm and pin diameter 6 mm with height 4.5 mm, are chosen for the tool profile. Surface composites are a category of composites in which reinforcement of particles is limited to the surface layer of 1–5 mm while material beneath the surface maintains the original structure. The surface composites differ from metal matrix composites (MMCs)/functionally graded composites (FGCs). Particles are in-forced throughout the whole volume in MMCs, and gradual transition in volume percentage of reinforcement exists in FGCs [13–17].

The various strategies to incorporate the reinforcement particles are FSP with a uniform layer of reinforcement particles placed on the workpiece before processing, a groove in the workpiece filled with reinforcement particles and tool comprising a consumable rod of aluminium drilled with holes to accommodate reinforcement particles. Of these, the second method of a groove in the workpiece filled with reinforcement particles was the most convenient and effective [18]. Thus, finding the optimum tool rotation and feed to for uniform distribution of reinforcements on the surface of composites is a challenging task. In the present work, TiO$_2$ particle-reinforced AA6082 alloy surface composite has been fabricated using different tool rotation speeds and traverse speeds. Optical microscopy is used for the investigation. Vickers hardness test is used to plot the hardness profile across the FSP zone for the selected specimen along with the uniaxial tensile testing.

2 Experimental Procedure

The composition of the commercially procured AA6082–T6 alloy is shown in Table 1. To perform the FSP, two grooves of 0.8 and 1.6 mm width with 5 mm depth were made using wire cut electrical discharge machining (EDM). After the

Table 1 Composition of aluminium 6082 T6 alloy

	Chemical element	Amount present in (%)
1.	Manganese (Mn)	0.4–1.00
2.	Iron (Fe)	0.0–0.50
3.	Magnesium (Mg)	0.60–1.20
4.	Silicon (Si)	0.70–1.30
5.	Copper (Cu)	0.0–0.10
6.	Zinc (Zn)	0.0–0.20
7.	Titanium (Ti)	0.0–0.10
8.	Chromium (Cr)	0.0–0.25
9.	Others	0.0–0.05
10.	Aluminium (Al)	Remaining

grooves were made, the grooves were cleaned with acetone to remove the unnecessary oxide layers formed on the surface. Then, the grooves were compacted with TiO_2 powder.

2.1 Friction Stir Welding Procedure

Friction stir processing (FSP) is a solid-state technique that does not melt the material, rather it plastically deforms the material.

A non-consumable cylindrical tool was used which is brought in contact with the workpiece while it is rotating at the desired speed, and is plunged into the alloy until the shoulder of the tool touches the surface. After the tool makes contact, it is made to travel along the groove which is compacted with TiO_2 powder. As the tool is continuously rotating, there is a large amount of heat that is generated along with the severe plastic deformation and mixes the TiO_2 particles with the base alloy around the tool without a melt. As tool traverses, the mix is travelled around the tool and left behind the wake of the tool which in turn forged further produce the surface composite. This severe plastic deformation yields a refined equiaxed grain structure in the FSP zone [5–9]. Table 2 shows the various experiments conducted at the different possible combinations of speeds and feeds.

3 Results and Discussion

Initially, the FSP is carried out with two different groove widths, 1.6 and 0.8 mm which were least possible with the in-house facility, which necessitates the finding of the optimum width at which the groove gets proper compaction to produce the quality surface composite. When the FSP was performed on the wider 1.6 mm width, tunnel defect was spotted macroscopically on the cross section of the weld sample (Fig. 1b). Whereas with the thinner 0.8-mm-width groove (Fig. 1a),

Table 2 Experiments carried out at the speed and feed combinations

NO	Speed (rpm)	Feed (mm/min)	No	Speed (rpm)	Feed (mm/min)
01	560	40	09	900	40
02	560	50	10	900	50
03	560	63	11	900	63
04	560	80	12	900	80
05	710	40	13	1120	40
06	710	50	14	1120	50
07	710	63	15	1120	63
08	710	80	16	1120	80

(a)   (b)

Fig. 1 Samples depicting the cross section after FSP, **a** with 0.8 mm groove, **b** with 1.6 mm groove, tunnel defect shown with an arrow

(a)   (b)

(c)   (d)

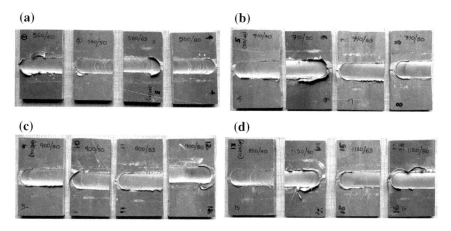

Fig. 2 Samples depicting the friction stir processed specimens at various speeds/feeds mentioned. **a** With 560 rpm speed and feed rates of 40, 50, 63 and 80 mm. **b** With 710 rpm speed and feed rates of 40, 50, 63 and 80 mm. **c** With 9000 rpm speed and feed rates of 40, 50, 63 and 80 mm. **d** With 11200 rpm speed and feed rates of 40, 50, 63 and 80 mm

this defect was not found. The reasons are bifold: insufficient compaction and the thermal energy resulted by plastic strain induced by the high speed and the ideal traversal speed were not sufficient, resulted in the non-uniform distribution of the reinforcement with leaving an unfilled gap behind. Thus, the further investigation was carried out with 0.8-mm-width groove.

Totally, 16 combinations with different speeds and feeds (Table 2) are friction stir processed and the specimens are shown in Fig. 2. Figure 2a shows the samples with 560 rpm as speed and with feed rates of 40, 50, 63 and 80 mm. Figure 2b shows samples with 710 rpm as speed and with feed rates of 40, 50, 63 and 80 mm. Figure 2c shows samples with 900 rpm as speed and with feed rates of 40, 50, 63 and 80 mm. Figure 2d shows samples with 900 rpm as speed and with feed rates of 40, 50, 63 and 80 mm.

From Fig. 2, it can be observed that the specimens performed at a lower speed and feed rate, scratches present in the weld puddle and the weld was not smooth. But a good finish is observed when higher speed and feed rate were used. Because low speeds do not produce enough amount of heat to maintain the aluminium at a specific temperature so that it can be moldable and can solidify on cooling. By increasing the rotation speed, the amount of heat generation increases and this maintains the aluminium at the state of thermoplasticity. This concludes that 1120 rpm and 80 mm/min were observed to be the best suitable speed and feed combination which furnished a good quality weld. When the optimum speed and feed are chosen, the quantity of material that is extruded per rotation increases. The more the material, the more will be the mixing and the stronger will be the extrusion. This increases the strength of the metal. As the extrusion is stronger, it leads to finer grain growth.

From Fig. 2, it can be identified that the speed and feed combinations 1, 6, 11 and 16 have fine weld puddle. This says that the heat energy produced by mechanical stirring has correctly synchronized with the traversal speed which resulted in quality weld [3]. But anyhow we can observe the lateral flashes in the above-mentioned weld samples which are gradually decreasing as we reach the higher speed and feed. However, the quality of weld seems good with very less lateral flashes and we can still observe the slightly dot structured puddle which says that the high heat generated has made the top layer of the material to cling on to the forging shoulder of the tool, which when removed results in such disturbed structure.

From the optical micrographs (Fig. 3), fine grains are observed to be produced at (a) and (d) sample. In the case of Fig. 3a even though the speed is low, the feed was equally less enough to build the thermal input required by slow movement, thereby leading to fine grain formation. This concludes that the speed and feed combination 16 proved to be the best among chosen combinations. Thus, the further analysis of microhardness test and uniaxial tensile testing was carried out on the combination 16.

3.1 Mechanical Tests

3.1.1 Hardness Test

A detailed Vickers hardness test was performed on the weld sample 16 and shown in Fig. 4. From the variation of hardness from retreating side (on left side) TMAZ zone on one side to weld zone and then on the TMAZ zone on the advancing side (on right side), it is observed that the composite formed also has high hardness value on the retreating side, less value on the advancing side and even more lower value on the weld zone. The weld zone consists of nuggets which are dynamically recrystallized after attaining high temperatures due to constant high-speed stirring. So the weld zone essentially consists of fine grains which are softened and the

(a)   **(b)**

(c)   **(d)**

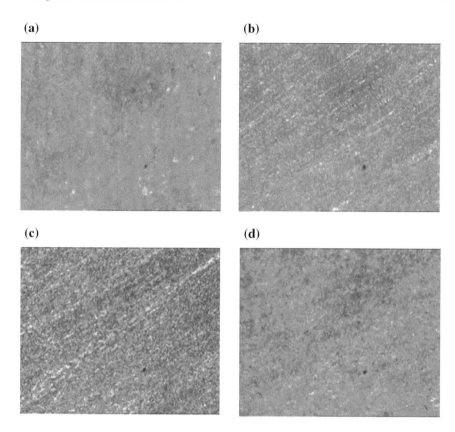

Fig. 3 Microstructure at 100X, Clockwise from top left, **a** 560/40, **b** 710/50, **c** 900/63, **d** 1120/80

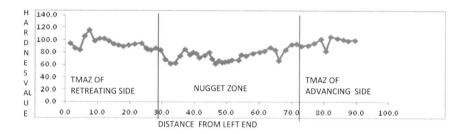

Fig. 4 Hardness profile shows the Vickers hardness values across various zones for sample 16

hardness profile is mainly affected by the precipitates (T6) that get dispersed. Moreover, the very low hardness value of nugget zone is also due to the high heat generated due to high-speed rotation of the tool which has led to grain refinement. These fine grains reduce the hardness of the base metal compared to the neighbour heat-affected one and TMAZ. The heat-affected zone consists of elongated grain

Table 3 Observed values for test parameters

Sl no.	Test parameters	Observed values
01	Gauge thickness (mm)	7.81
02	Gauge width (mm)	6.30
03	Original cross-sectional area (mm^2)	49.20
04	Ultimate tensile load (kN)	6.21
05	Ultimate tensile strength (N/mm^2 or Mpa)	126.00
06	Fracture location	Weld

structure which is artificially aged by T6 heat treatment. When such zone comes in contact with partial heat, it increases the base hardness. From Fig. 4 again, we can confirm that the retreating side has more hardness than weld and advancing side. But here another important observation to be noted is the reduced gap between the hardness value of TMAZ and weld zone.

3.1.2 Tensile Test

Table 3 shows the observed values for test parameters. From the tensile test results (Fig. 5), we were able to identify a huge drop in the ultimate tensile strength value. Figure 5b, c shows the location of the fracture. On inspection, it was found out that tunnelling defect had initiated the crack at the bottom part of the retreating side which could be due to improper material flow that was caused by higher rpm of tool

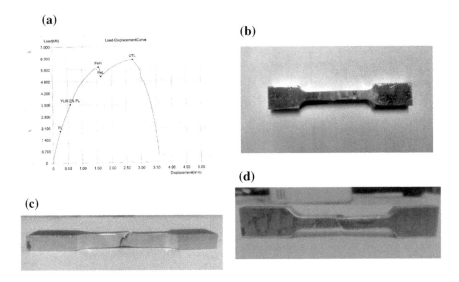

Fig. 5 From clockwise left to right. **a** Load versus displacement plot, **b**, **c** and **d** show the fracture of the tensile test specimen

and high feed. So, the distribution of the material did not happen uniformly around the tool. Even though the crack has initiated on the retreating side, the crack has tried to propagate into the nugget zone, but the presence of fine grains has restricted the crack's mobility and has thus helped to retain the tendency of ductility. We can also see the cup and cone structure formation on the top side of the weld which shows that it has undergone ductile fracture.

4 Conclusion

The surface composite was prepared by FSP and micro-tested by optical microscopy and Vickers hardness. A tensile test was performed on the sample 16 and the following conclusions were drawn:

1. Defect-free weld and fine grains with uniform spread reinforcement particle were observed in the speed and feed combination 16 on thin 0.8 mm groove.
2. The retreating side had more hardness due to major grain dislocations caused by opposite direction of tool speed and feed.
3. The reinforcement particle has improved the hardness of the weld zone and has reduced the difference between the hardness values of TMAZ and weld zone.
4. Tunnelling defect was found at the bottom part of the retreating side which led to low tensile strength.

References

1. Mishraa RS, Ma ZY (2005) Friction stir welding and processing. Mater Sci Eng R 50:1–78
2. Salih OS, Ou H, Sun W, McCartney DG (2015) A review of friction stir welding of aluminium matrix composites. Mater Des 86:61–71
3. Ma ZY (2008) Friction stir processing technology: a review. Metall Mater Trans A 39 (3):642–658
4. Jana A, Siddhalingeshwar IG, Mitra R (2013) Effect of homogeneity of particle distribution on tensile crack propagation in mushy state rolled in situ Al–4.5Cu–5TiB$_2$ particulate composite. Mater Sci Eng, A 575:104–110
5. Peddavarapu S, Raghuraman S, Bharathi RJ et al (2017) Micro structural investigation on friction stir welded Al–4.5Cu–5TiB$_2$ Composite. Trans Indian Inst Met 70:703
6. Herbert MA, Sarkar C, Mitra R, Chakraborty M (2007) Microstructural evolution, hardness, and alligatoring in the mushy state rolled cast Al-4.5Cu alloy and In-Situ Al4.5Cu-5TiB$_2$ Composite. Metall Mater Trans A 38(9):2110–2126
7. Sundaram NS, Murugan N (2010) Tensile behavior of dissimilar friction stir welded joints of aluminium alloys. Mater Des 31:4184–4193
8. Palanivel R, Laubscher RF, Dinaharan I, Murugan N (2016) Tensile strength prediction of dissimilar friction stir-welded AA6351–AA5083 using artificial neural network technique. J Brazilian Soc Mech Sci Eng 38(6):1647–1657
9. Cavaliere P (2013) Friction stir welding of Al alloys: analysis of processing parameters affecting mechanical behavior. Procedia CIRP 11:139–144

10. Gandraa J, Pereirab D, Mirandab RM, Vilaçac P (2013) Influence of process parameters in the friction surfacing of AA 6082-T6 over AA 2024-T3. Procedia CIRP 7:341–346
11. Adamowski J, Gambaro C, Lertora E, Ponte M, Szkodo M (2007) Analysis of FSW welds made of aluminium alloy AW6082-T6. Arch Mater Sci Eng 28(8):453–460
12. Raja AR, Yusufzai MZK, Vashista M (2016) Characterization of advancing and retreating weld of friction stir welding of aluminium. ICAMM
13. Zheng Q, Feng X, Shen Y, Huang G, Zhao P (2016) Dissimilar friction stir welding of 6061 Al to 316 stainless steel using Zn as a filler metal. J Alloy Compd 686:693–701
14. Huang Y, Wang Y, Wan L, Liu H, Shen J, dos Santos JF, Zhou L, Feng J (2016) Material-flow behavior during friction-stir welding of 6082-T6 aluminum alloy. Int J Adv Manuf Technol 87:1115–1123
15. Fonda R, Reynolds A, Feng CR, Knipling K, Rowenhorst D (2013) Material flow in friction stir welds. Metall Mater Trans A 44(1):337–344
16. Adamowski J, Szkodo M (2007) Friction stir welds (FSW) of aluminium alloy AW6082-T6. J Achievements Mater Manufacturing Eng 20
17. Asensio-Lozano J, Suárez-Peña B, Voort GFV (2014) Effect of processing steps on the mechanical properties and surface appearance of 6063 aluminium extruded products. Materials 7:4224–4242
18. Sharma V, Prakash U, Kumar BVM (2015) Surface composites by friction stir processing: a review. J Mater Process Technol 224:117–134

Microstructure Evolution and Mechanical Properties of Al 1050/Al 5083 Laminate Composites Produced by Accumulative Roll Bonding Process

L. Poovazhagan, P. Ruthran, S. Sreyas, A. Thamizharasan and S. Thejas

Abstract Al 1050/Al 5083 multilayer laminate composites were produced by accumulative roll bonding (ARB) process up to three cycles at room temperature. Tensile strength along rolling direction and hardness was evaluated for base alloys and laminate composites. As compared to the base alloys, tensile strength and hardness of Al laminate composites increase significantly after every ARB cycle. The third cycle ARBed sample exhibited the tensile strength of 202 MPa, which is 29% more than the first cycle ARBed sample. Similarly, the BHN (Brinell hardness number) of the third cycle ARBed sample is 73, which is 46% more than the first cycle ARBed sample. ARBed samples show very limited elongation. Microstructural evolutions of ARBed sheets were analyzed by scanning electron microscopy (SEM). Analysis revealed that grains of Al were refined significantly after every ARB cycle.

Keywords Accumulative roll bonding · Laminate composites · Tensile strength Hardness · Grain refinement

1 Introduction

Grain size can be considered as an important microstructural factor affecting almost all the physical and mechanical properties of polycrystalline materials [1]. Severe plastic deformations (SPD) are family of forming processes by which ultrafine-grained materials can be produced [2]. The significant property of SPD materials is improved strength-to-weight ratio. Equal channel angular pressing (ECAP), high-pressure torsion (HPT), and accumulative roll bonding (ARB) are the important SPD methods applied widely to produce bulk ultrafine-grained materials. Compared to ARB, the efficiency of ECAP and HPT are much lower [3]. Both ECAP and HPT required special equipments to produce ultrafine-grained

L. Poovazhagan (✉) · P. Ruthran · S. Sreyas · A. Thamizharasan · S. Thejas
Department of Mechanical Engineering, SSN College of Engineering, Chennai, India
e-mail: poovazhaganl@ssn.edu.in

© Springer Nature Singapore Pte Ltd. 2019
A. K. Lakshminarayanan et al. (eds.), *Advances in Materials and Metallurgy*,
Lecture Notes in Mechanical Engineering,
https://doi.org/10.1007/978-981-13-1780-4_4

29

structure in materials. The ARB process discovered three decades ago is a promising forming method by which severe plastic deformation can be induced in materials [4].

ARB is a process that induces ultra-large plastic strain in a material by rolling the material in between two rollers of equal diameter running at the same speed using a common gear train. As the name suggests, the process apart from strengthening a material also joins two metals together [5–8]. This type of bonding technique requires a good surface treatment of the two sheets to be rolled for a strong bond to be produced between them. The bonding that occurs between the two metals does not occur due to the presence of external heating unlike in the case of welding; the bonding here is because of the force applied on the two sheets of metals by the two rollers. The major technical advantage of ARB is that it makes use of conventional rolling setup [4]. ARB process can fabricate bulk ultrafine-grained samples without any pores compared to powder metallurgy route in which pores are ceased to exist [9]. ARB can produce bulk samples in the form of thin sheets, which in turn directly used for secondary forming operations.

The uniqueness of ARB is apart from bonding and strengthening the metals; this process can be done infinitely on the material obtained from the previous cycle [10–14]. That is, for the first cycle when two metal sheets are bonded, it is taken care that the bonded material coming out of the rolling machine is of the same thickness of one sheet of parent material. Now, this bonded material is cut into two pieces and sent into the rolling machine after proper stacking. So the material coming out of the rolling machine will again be of the thickness of the parent metal. This allows us to do this process infinitely as the thickness of the sample at the end of every cycle is a constant. In this research work, two dissimilar aluminum alloys were roll boned using ARB process up to three cycles and results were presented.

2 Experimental Procedure

The materials used in this work were two different commercial aluminum alloy sheets, namely, aluminum 1050 and aluminum 5083. The average mechanical properties of these alloys are given in Table 1. Sheets of Al 1050 and Al 5083 of dimensions 150 mm × 50 mm × 0.4 mm were degreased in acetone bath and grounded using hand grinding machine. The two sheets were stacked, and the corners of strips were tied using aluminum wire and then roll bonding was carried out. Attention was paid to get the proper alignment of two sheets before roll bonding. Laboratory rolling mill with a loading capacity of 20 tons was used for roll bonding. The roll bonding process was attempted in dry condition. The roll bonding procedure was repeated up to three cycles, and the reduction amount equals to 50% after every cycle. The schematic of roll bonding process is illustrated in Fig. 1. The rolling machine used in this work is shown in Fig. 2. Rollers and screws are used to maintain the gap between rollers.

Table 1 Mechanical properties of Al 1050 and Al 5083

Material	Yield strength (MPa)	Tensile strength (MPa)	% elongation	Brinell hardness number (BHN)
Al 1050	85	105	12	34
Al 5083	115	270	12	75

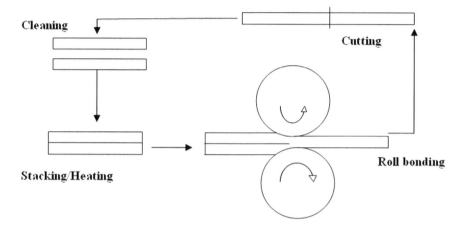

Fig. 1 Schematic of ARB process

Fig. 2 Two high rolling machines

Surface cleaning is a preliminary step in the surface treatment process. Purpose of cleaning the surface is to remove the grease and other foreign particles from the surface of the metal. The presence of foreign particles restricts the adequate bond formation. Foreign particles found on the surface of the metal are mostly oil, grease, and sometimes dust particles. Surface treatment is the main process without which bond formation is not possible. Aluminum is a reactive metal and quickly reacts with atmospheric oxygen to form oxide layer. The aim of surface treatment was to eliminate the oxide layer from the surface and to expose the pure aluminum. This was done using a simple hand grinding machine. The acetone-treated aluminum samples were ground completely throughout the surface. Upon doing so, pure aluminum was exposed. Apart from removal of the oxide coating, the grinding process on aluminum increased the roughness of the sample. Increased roughness implies larger surface area exposure for bonding. So grinding has a double functionality in case of ARB.

After grinding the two pieces, the samples were stacked one over the other. Stacking not only involves the mounting of two pieces one over the other but also includes tying them together. The two pieces have to be tied because, in the process of rolling, if the two pieces were not tied, the pieces would never bond together but will be coming out as two different sheets misaligned in the vertical direction due to the excessive compressive force exerted at the rollers. Sometimes, there are also chances that the two sheets may have a horizontal misalignment if it were not tied (these misalignments were observed while doing the experiment without tying the 2 sheets). So after the two sheets are stacked such that the ground faces of the samples face each other, tying is done using aluminum wire. The material of wire used is also important because, if a wire of a harder material were used, it can cause severe indentation on the aluminum surface which is softer. Sometimes, the hardwire material can also cut the aluminum sheet while it is being roll bonded.

After proper stacking of the two sheets, the sheets were fed into the rolling machine. Roll bonding has a standard procedure. The initial thickness of the sample after stacking is measured (0.8 mm). The stacked sample is placed between the two rollers of the rolling machine, and the screw on the machine is adjusted so that the two rollers touch either face of the sample. The sample is removed from the rollers. Now the screw is adjusted such that the gap between the rollers is lesser than half the thickness of the stacked sample (stacked sample = 0.8 mm; gap between rollers <0.4 mm for 50% reduction). Now the sample was fed into the machine while the rollers were rotating. The sample coming out of the other end of the machine was a bonded sample. The final thickness of the sample was measured. Cut the sample obtained at the end of first cycle into two pieces of equal length. The two pieces were cleaned using acetone. The cleaned pieces were ground using a hand grinder. The two pieces were stacked using proper aluminum wire. The procedure was repeated hereafter. The ARBed samples after each cycle are represented in Fig. 3. After each ARB cycle, specimens required for experimental work were cut.

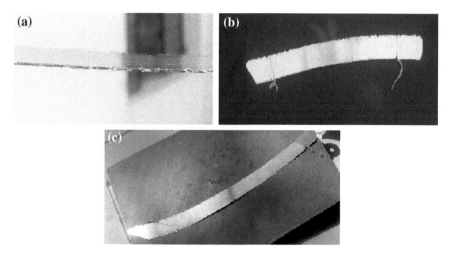

Fig. 3 ARBed aluminum laminate samples after **a** cycle 1, **b** cycle 2, **c** cycle 3

3 Results and Discussion

3.1 Tensile Strength

To find out the change in the tensile strength after each ARB cycle, tensile specimens were prepared from Al 1050, Al 5083 and ARBed samples. The specimen thickness, % reduction, strain-induced, and accumulative strain are given in Table 2. Yield strength, ultimate strength, % elongation, and Brinell hardness number (BHN) after each ARB cycle are indicated in Table 3. The ARBed sample properties are compared with Al 1050. Sample specimens for tensile strength are shown in Fig. 4. After each cycle, strength of the laminate composites was increased significantly when compared to Al 1050 and correspondingly % elongation reduced. Maximum improvement in strength was observed after third ARB cycle. For example, the commercial pure Al 1050 sheets possess the tensile strength of 85 MPa. The laminate composites after third ARB cycle possess the tensile strength of 202 MPa. The improvement in strength of laminate composites after third cycle is around 137% when compared to commercial pure Al 1050. Consequently, the ductility of the laminate composites reduces marginally.

Table 2 Accumulative strain of ARBed laminate composites

Cycle no.	Initial thickness (mm)	Final thickness (mm)	Reduction (%)	Strain-induced	Accumulative strain	No. of layers
1	0.8	0.4	50	0.5	0.5	2
2	0.8	0.4	50	0.5	1.0	4
3	0.8	0.4	50	0.5	1.5	8

Table 3 Mechanical properties of ARBed laminate composites

Cycle no.	Yield strength (MPa)	Ultimate strength (MPa)	% elongation	Hardness (BHN)
1	110	157	10	50
2	124	185	9	65
3	146	202	8	73

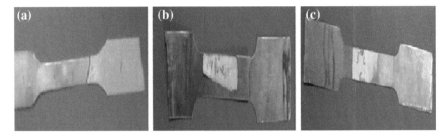

Fig. 4 Tensile test specimens prepared from ARBed aluminum laminates after **a** cycle 1, **b** cycle 2, **c** cycle 3

The reason for improvement in strength in laminate composites can be understood from the following discussion. It is a well-known fact that plastic deformations generally occur with the movement of dislocations (slip). The process of ARB, in essence, is a plastic deformation process. But what it achieves is this; it causes the dislocations to move rapidly and entangle with each other [15]. In this process, dislocations may cancel out each other or may form jogs, thereby limiting their own movement. With each successive cycle, this carries on. So if the dislocations cannot move easily, then plastic deformation will also not occur that easily. That is why the material strain hardens and its strength increases. After each ARB cycle, grain refinement in material took place. Hence, the mean grain size of the laminate composites reduces significantly. According to the Hall–Petch equation, the finer the grains, the more will be the strength of material [16].

3.2 Hardness Test

The effect of ARB on the hardness (BHN) behavior of aluminum laminate composites is presented in Fig. 5. After each ARB cycle, BHN gradually increases. For comparison, aluminum alloy Al 1050 has a hardness value of 34 BHN and the aluminum laminate composites after third cycle has a hardness value of 73 BHN. This is equal to more than 114% increase in hardness value. After each ARB cycle, grain size reduces significantly and the bonding between two aluminum alloys improved. These two are the primary reason for the increase in hardness of the laminate composites.

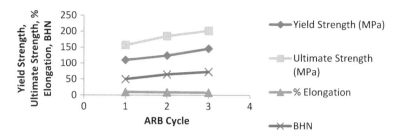

Fig. 5 ARB cycle versus mechanical properties

3.3 Microstructure

The microstructure was analyzed using a scanning electron microscope (SEM) which can operate at 30 kV and can magnify up to 25000X. SEM was conducted using Carl Zeiss.

MA15/EVO 18. Figure 6 indicates the ARBed sample SEM images after cycles 1, 2, and 3. It was clearly observed that grain size of laminate composites reduces after each ARB cycle. To quantify the grain size, high-resolution SEM was taken on the ARBed samples, which is represented in Fig. 7. Figure 7a represents the high-resolution SEM image of two-cycled ARB sample. The average grain size

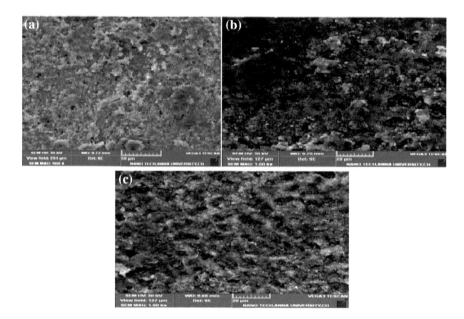

Fig. 6 Scanning electron microscopic images of ARBed samples after **a** cycle 1, **b** cycle 2, **c** cycle 3

Fig. 7 SEM images of ARBed samples after **a** cycle 2, **b** cycle 3

observed is around 0.11 μm. Figure 7b represents the high-resolution SEM image of three-cycled ARB sample. The average grain size observed is around 0.09 micrometers. Reduction in grain size is due to the increased number of ARB cycles.

4 Conclusion

- ARB process was successfully used to fabricate Al 1050/Al 5083 laminate composites up to three cycles. The properties of laminate composites were compared with the properties of commercial pure Al 1050.
- Results indicate that the yield strength, tensile strength, and BHN value increase with increase in number of ARB cycle. The maximum tensile strength of 202 MPa observed in three-cycled ARBed sample which is 92% more than the UTS of Al 1050. Similarly, maximum BHN of 73 was observed in three-cycled ARBed sample which is 115% more than BHN of Al 1050.
- Results also show that the ductility of ARBed samples reduces marginally after each cycle.
- Microstructure reveals that the average grain size reduces with increase in ARB cycle.

References

1. An J, Lu Y, Xu DW, Liu YB, Sun DR, Yang B (2001) Hot-roll bonding of Al-Pb bearing alloy strips and hot dip aluminized steel sheets. J Mater Eng Perform 10(2):131
2. Verlinden B (2002/2004) Severe plastic deformation of metals. In: Zhu YT et al (eds) The second and third international conference on ultrafine grained materials, TMS
3. Azushima A, Kopp R, Korhonen A, Yang DY, Micari F, Lahoti GD, Groche P, Yanagimoto J, Tsuji N, Rosochowski A, Yanagida A (2008) Severe plastic deformation (SPD) processes for metals. CIRP Ann Manuf Technol 57:716–735
4. Tsuji N, Saito Y, Lee SH, Minamino Y (2003) ARB and other new techniques to produce bulk ultrafine grained materials. Adv Eng Mater 5(5):338–344
5. Eizadjou M, Talachi AK, Manesh HD, Shahabi HS, Janghorban K (2008) Investigation of structure and mechanical properties of multi-layered Al/Cu composite produced by accumulative roll bonding (ARB) process. Compos Sci Technol 68(2008):2003–2009
6. Toroghinejad MR, Jamaati R, Dutkiewicz J, Szpunar JA (2013) Investigation of nanostructured aluminum/copper composite produced, by accumulative roll bonding and folding process. Mater Des 51(2013):274–279
7. Hsieh C-C, Chen M-C, Wu W (2013) Mechanical property and fracture behavior of Al/Mg composite produced by accumulative roll bonding technique. J Compos 2013(748273):8 pages
8. Salimi S, Izadi H, Gerlich AP (2011) Fabrication of an aluminum–carbon nanotube metalmatrix composite by accumulative roll-bonding. J Mater Sci 46:409–415
9. Yu HL, Lu C, Tieu AK, Kong C (2014) Fabrication of nanostructured aluminum sheets using four-layer accumulative roll bonding. Mater Manuf Process 29(4)
10. Jamaati R, Toroghinejad MR, Amirkhanlou S, Edris H (2015) On the achievement of nanostructured interstitial free steel by four-layer accumulative roll bonding process at room temperature. Metall Mater Transactions A 46A:4013–4019
11. Lahiri D, Bakshi SR, Keshri AK, Liu Y, Agarwal A (2009) Dual strengthening mechanisms induced by carbon nanotubes in roll bonded aluminum composites. Mater Sci Eng A 523:263–270
12. Bachmaier A, Pippan R (2013) Generation of metallic nanocomposites by severe plastic deformation. Int Mater Rev 58(1):41–62
13. Amirkhanlou S, Ketabchi M, Parvin N, Khorsand S, Bahram R (2013) Accumulative press bonding; a novel manufacturing process of nanostructured metal matrix composites. Mater Des 51:367–374
14. Kim HS, Estrin Y Bush MB (2000) Plastic deformation behaviour of fine-grained materials. Acta Mater 48(2000):493–504
15. Nieh TG, Wadsworth J (1991) Hall-petch relation in nanocrystalline solids. Scr Metall Mater 25(4):955–958

Synthesis and Characterization of Nano-Glass Particles Reinforced AZ91D Magnesium Alloy Composites

G. Anbuchezhiyan, B. Mohan and T. Muthuramalingam

Abstract In the present study, a considerable importance is being given to find a novel material for marine application. The synthesis of nano-particles reinforced with matrix alloy using conventional casting offers low wettability between reinforcement and the molten metal. Hence, an effort has been made to develop nano-glass particles reinforced magnesium alloy matrix to enhance the wettability between the matrix and the reinforcement. The die cast magnesium unified with various mass fractions percentage of nano-glass particles of particle size 30 μm has been synthesized using stir casting method. The influence of nano-reinforcement on micro structure and mechanical properties of developed nano-composites have been analyzed. From the experimental results, it has been inferred that synthesized composites have homogeneous grain structure, low density, minimal porosity, and excellent matrix-reinforcement interfacial integrity. The main interfacial phase has been identified as $Mg_{17}Al_{12}$ between the reinforcement and matrix alloy.

Keywords Compressive properties · Density · Microstructure
Nano-composites · Stir casting · Tensile strength

G. Anbuchezhiyan (✉)
Department of Mechanical Engineering, Valliammai Engineering College,
Kattankulathur, India
e-mail: tsgaaa1981@gmail.com

B. Mohan
Department of Mechanical Engineering, Anna University
CEG Campus, CEG Campus, Chennai, India
e-mail: mohan@mitindia.edu

T. Muthuramalingam
Department of Mechatronics Engineering, SRM Institute of Science
and Technology, Kattankulathur, India
e-mail: muthu1060@gmail.com

© Springer Nature Singapore Pte Ltd. 2019
A. K. Lakshminarayanan et al. (eds.), *Advances in Materials and Metallurgy*,
Lecture Notes in Mechanical Engineering,
https://doi.org/10.1007/978-981-13-1780-4_5

39

1 Introduction

Owing to its low density and higher strength to weight ratio magnesium-based alloys are primarily used in marine industries especially for subsea application. However, the exploitation of magnesium alloys is insufficient due to their low strength, poor heat resistance, low elastic modulus, low wear resistance at elevated temperature, and poor corrosion resistance [1]. Reinforcement of appropriate ceramic particles with magnesium alloy can prevail over such shortcomings. Nevertheless, the adverse reaction and the interfacial incoherence between the reinforcement and metal matrix alloy results in brittleness. Owing to such constraint the utilization of magnesium composites in marine application is inadequate. In this regard, an endeavor has been made to develop a nano-scale ceramic reinforced magnesium matrix composites in sequence of improving the mechanical properties for such application. Among the various Mg alloys, the die cast AZ91D magnesium alloys are primarily utilized for fabricating magnesium matrix nano-composites, due to excellent castabilty, [2]. From the literature it has been observed that the inclusion of nano-reinforcements like oxides (Al_2O_3, TiO_2, Y_2O_3, ZnO, and ZrO_2), carbides (SiC, B_4C, and TiC), nitrides (BN, AlN, and TiN), borides (TiB_2, SiB_6, and ZrB_2), CNT and graphene, is used to get better yield strength and ductility of magnesium [3]. It has been inferred that the end properties of magnesium nano-composites (MgNCs) is strongly influenced by processing methods. The selection of processing technique is very crucial, as the result of microstructure features are highly influenced by processing parameters. The liquid metallurgy technique is most economical method of synthesizing MgNCs. The stir casting method is most suitable method for synthesizing such nano-composites due to low processing cost and high production rate. This method is also used to improve wettability between the matrix and reinforcement to achieve optimum mechanical properties of MgNCs [4]. In this present study nano-sized glass particles with various weight percentage of 4,8 and 12% reinforced with AZ91D magnesium alloy is developed using stir casting method and the mechanical properties of as-cast composites has to be characterized.

2 Experiments and Methods

Stir casting method has been used for synthesis MgNCs. AZ91D magnesium alloy is used as matrix material and its chemical composition is shown in Table 1. Nano-glass particles of 30 μm of density 0.60 g/cc have been utilized as reinforcement to synthesis MgNCs. The chemical composition of nano-glass particles

Alloy designation	Alloying additives
AZ91D	9.0% Al, 0.7% Zn, 0.13% Mn

Table 1 Chemical composition of AZ91D magnesium alloy

are SiO_2, B_2O_3, CaO, MgO, and Na^+ salts of particle. Necessitate amount of magnesium alloy is melted in a crucible at 685 °C in a resistance heating furnace. The alloy is melted and is stirred with high shear impeller under CO_2 and SF_6 atmosphere. Mixture of SF_6 gas with CO_2 is commonly used as carrier gas to prevent oxidation and excess burning of magnesium during casting process, since magnesium is highly pyrophoric in nature. The flow rate of gas mixtures has been maintained at 40 L/min. The desired mass fraction of nano-glass particles are preheated at 400 °C and incorporated manually into the melt magnesium matrix. The stirring speed and stirring time is maintained at 600 rpm and 15 min to ensure the homogeneous distribution of reinforcement in matrix alloy. A stainless-steel die is coated with sulfur powder to prevent oxidation. It is placed at the bottom of the spruce which reduces the interaction between the molten nano-composite and the atmosphere during gravity pouring. The molten nano-composite is poured into the die cavity and it is allowed to cool in the die itself. The sample size of MgNCs for tensile test specimen is maintained as ASTM standard B557 M. The gauge length of 200 mm with 8 mm of width, 1.5 mm of thickness and overall specimen length of 450 mm. The sample size of compression test sample is 31.37 mm of specimen diameter with the length of 152.4 mm.

3   Results and Discussion

3.1   Microstructure of Magnesium Nano-Composites

The microstructure of MgNCs has been characterized by homogenous distribution of nano-sized glass particles in magnesium matrix alloy. Figures 1, 2 and 3 shows microstructure of magnesium matrix reinforced through nano-glass particles. It has been observed that the nano-glass reinforcement particles occur at the grain boundaries of the magnesium alloy matrix solid solution. The marginal deformed

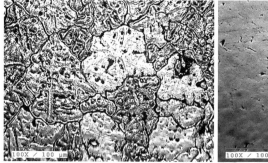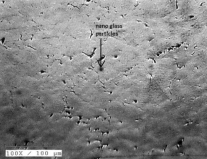

Fig. 1 AZ91D—4% of nano glass particle reinforced magnesium composites. **a** Deformed shape of nano glass particles

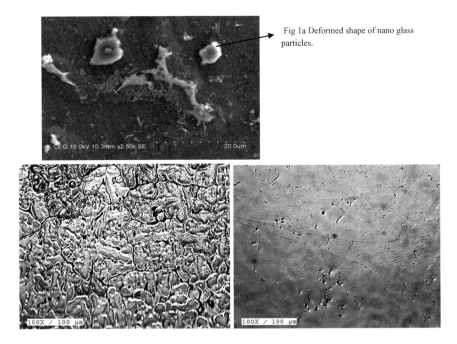

Fig 1a Deformed shape of nano glass particles.

Fig. 2 AZ91D—8% of nano glass particle reinforced magnesium composites

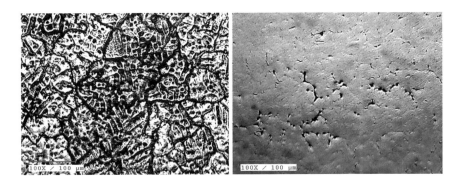

Fig. 3 AZ91D—12% of nano glass particle reinforced magnesium composites

nano-glass particles may be probably due to heat input that has been observed in the SEM image Fig. 1a. However majority of nano-particles have retained the shape of sphere. The higher concentration of the reinforcement particles in the melt seemed to have lowered the rate of cooling resulting in the reduction in the dendrite pattern of solidification. Hence, more granular metal matrix grains and distribution of nano-sized reinforcement has been inferred. The metal matrix shows the precipitated grains of $Mg_{17}Al_{12}$ phases in magnesium solid solution due to rapid solidification. Hence, these $Mg_{17}Al_{12}$ act as a strengthening mechanism for MgNCs.

3.2 Influence of Nano-Sized Glass Reinforcement on Tensile Strength

The tensile property of the magnesium nano-composites is determined as per ASTM standard B557 M. It has been observed that ultimate tensile strength of MgNCs with dissimilar mass percentage decreases compared to AZ91D magnesium alloy as shown in Table 1. The decreases in strength initiate weak adhesion between the matrix and reinforcement particles. Owing to that, it necessitates to relocate the load from the matrix to ceramic reinforcement particles and thus yielding in MgNCs behavior contains defects. It has been observed that the tensile strength of MgNCs has not been enhanced due to partial break down of nano-glass particles clusters observed in magnesium nano-composites. Owing to the lack of adequate slip activity the HCP crystal structure of magnesium alloy deformability is limited, even though the particle size of reinforcements added in the matrix alloy is in nano-length scale, nevertheless the premature failure has not observed in the MgNCs as cited in the literature [5] (Table 2).

3.3 Influence of Nano-Sized Glass Reinforcement on Compressive Strength

Compressive testing of the MgNCs has been evaluated at room temperature as per ASTM E9 standards. It has been observed that the compressive strength of MgNCs is superior to that of as-cast AZ91D magnesium alloy. The compressive strength of MgNCs has been increased owing to the increase in the percentage of mass fraction reinforcement in the matrix alloy, the presence of Mg_2Si interface between matrix and reinforcement and a typical characteristic is a linear region, plateau region, and densification region materialize in the load displacement curve which is similar to that of metallic foams composites. It is renowned that the stress value remained constant and there is no sudden drop subsequent to yielding. The compressive strength of MgNCs increases owing to the nano-scaled particle size, uniform distribution of reinforcement in the matrix alloy and cushioning effect of nano-composites. The stress strain curve illustrates that the deformation of MgNCs is characterized as linear region and densification region as cited in the literature [6–8]. In linear region the compressive strength of nano-composites deformed

Table 2 Tensile properties of magnesium nano composites with dissimilar mass percentage

Materials used	Ultimate tensile strength	% of elongation
AZ91D alloy	230 MPa	7
AZ91D—4% of nano glass particle	220 MPa	4.28
AZ91D—8% of nano glass particle	216 MPa	2.79
AZ91D—12% of nano glass particle	206 MPa	2.31

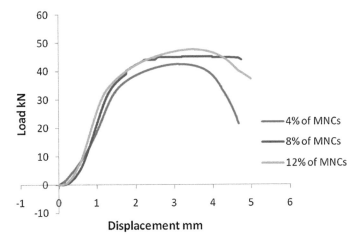

Fig. 4 Compressive properties of magnesium nano composites

linearly and then plastically followed by a drop in stress entered into the plateau region. In plateau region the compressive stress of MgNCs is quite flat and stable and increases with small slope as shown in Fig. 4. The densification strain mainly depends upon the presence of porosity entailed in the nano-composites. During compression testing the compressing material fills up the matrix porosity owing to that the nano-glass particle tends to fracture and the matrix porosity embedded within them is disclosed which can also be occupied by compressing material. This causes higher energy absorption in nano composites [9, 10]. The average compressive strength of MgNCs for various mass fraction reinforcement particles has been observed as 376.86 MPa, 396.534Mpa and 423.093 MPa respectively.

4 Conclusions

The nano-sized glass particles reinforced magnesium composites has been fabricated using stir casting method. From the experimental study the microstructure, tensile, and compressive stress has been determined. Based on that the following conclusions have been made:

(a) The nano-sized glass particles are homogeneously distributed in the magnesium matrix alloy without the sign of residual pore.
(b) The compressive strength of nano-composites has been increased due to cushioning effect of composites.
(c) The tensile strength of magnesium nano-composites decreased owing to weak adhesion among the ceramic particles between the matrix and reinforcement particulates.

References

1. Kulekei MK (2008) Magnesium and its alloys applications in automotive industry. Int J Adv Manuf Technology 39:851–865
2. Sankaranarayanan S, Gupta M (2015) Review on mechanical properties of magnesium (nano) composites developed using energy efficient microwaves. Powder Metallur 58:183–192
3. Wong WLE, Gupta M (2016) High performance lightweight magnesium nano composites for engineering and biomedical applications. Nano World J 2:78–83
4. Hasim J, Looney L, Hashmi MSJ (1999) Metal matrix composites produced by stir casting method. J Mater Process Technol 92–93:1–7
5. Tun KS, Wong WLE, Nguyen QB, Gupta M (2013) Tensile and compressive responses of ceramic and metallic nanoparticle reinforced Mg composites. Materials 6:1826–1839
6. Luong DD, Strbik OM, Hammond VH, Gupta N, Cho K (2013) Development of high performance light weight aluminium alloy/SiC hollow sphere syntactic foams and compressive characterization at quasi-static and high strain rates. J Alloy Compound 550:412–422
7. Tao XF, ZhaoYY (2009) Compressive behavior of Al matrix syntactic foam toughened with Al particles. Scripta Materilia 61:461–464
8. Shen MJ, Wang XJ, Li CD, Zhang MF, Hu XS, Zheng MY, Wu K (2013) Effect of bimodal size SiC particulates on microstructure and mechanical properties of AZ31B magnesium matrix composites. Mater Design 52:1011–1017
9. Lan J, Yang Y, Li X (2004) Microstructure and microhardness of SiC nano particles reinforced magnesium composites fabricated by ultrasonic method. Mater Sci Engg A. 386:284–290
10. Anbuchezhiyan G, Mohan B, Sathianarayanan D, Muthuramalingam T (2017) Synthesis and characterization of hollow glass microspheres reinforced magnesium alloy matrix syntactic foam. J Alloy Compound 719:125–132

Microstructure and Abrasive Wear Behavior of Copper–Boron Carbide Nanocomposites

L. Poovazhagan, H. Jeffrey Thomas and M. Selvaraj

Abstract The effects of B_4C nanoparticle reinforcements with an average size of 50 nm on the abrasive wear performance and microstructure of Cu nanocomposites were examined. Various combinations of Cu–B_4C nanocomposites were synthesized by solid state powder metallurgy technique. Abrasive wear experiments were carried out by sticking the abrasive sheets of 400 grit silicon carbide on the disc of pin-on disc-wear tester. Pins of unreinforced Cu and Cu nanocomposites were pressed and rotated against silicon carbide abrasive sheets under dry sliding conditions at different loading conditions. Addition of nano-B_4C in Cu matrix significantly enhances the hardness of the nanocomposites which in turn increases the wear resistance of nanocomposites. The nanocomposites with 1.5 wt% of B_4C possess the superior microhardness and maximum wear resistance property. Morphology of worn out surfaces were explored by scanning electron microscopy.

Keywords Abrasive wear · Nanocomposites · Powder metallurgy
Microstructure · Microhardness

1 Introduction

Copper and its alloys found numerous applications in electrical contacts, welding electrodes, and electronic packaging units attributed to their superior electrical and thermal conductivity, fine corrosion resistance properties, and high melting point. However, the relatively low wear resistance of Cu and its alloys hinder their further use in many new industrial applications. Recent developments in metal matrices reinforced with particle reinforcements (metal matrix composites—MMCs) have provided a new way to produce high-wear resistant materials [1–3]. Addition of ceramic reinforcement particles in metal or alloy matrices actually increases the

L. Poovazhagan (✉) · H. J. Thomas · M. Selvaraj
Department of Mechanical Engineer, SSN College of Engineering, Chennai, India
e-mail: poovazhaganl@ssn.edu.in

© Springer Nature Singapore Pte Ltd. 2019
A. K. Lakshminarayanan et al. (eds.), *Advances in Materials and Metallurgy*,
Lecture Notes in Mechanical Engineering,
https://doi.org/10.1007/978-981-13-1780-4_6

both strength (room temperature and elevated temperature) wear resistance properties of base matrix [4–6].

Cu composites have been actively explored in the last three decades by various researchers. Till now, majority of researchers contributed on the production and characterization techniques. To some reasonable extent, the work was carried out to improve the wear resistance of Cu matrix. Y. Tang et al. examined the wear resistance behavior of Cu–C fiber composites and found that addition in C reduces the wear of the fiber composites [7]. Several works have been carried out on Cu-based composites by adding different fillers like SiC, TiC, Al_2O_3, and TiO_2, and investigated their mechanical and tribological properties [8–11]. Soner et al. [12] fabricated enhanced wear resistance Cu–TiC powder metallurgy composites by adding TiC fillers in Cu matrix. Akhtar et al. [13] produced Cu–TiC powder metallurgy (P/M) composites and evaluated the tribological and mechanical properties. P. Jha et al. characterized the TiC reinforced Cu–Ni composites prepared by P/M route [14].

This paper describes the abrasive wear performance of Cu nanocomposites strengthened by B_4C nanoparticles. The abrasive wear performance of monolithic Cu and Cu nanocomposites were evaluated at different applied loads and nano-B_4C contents. The morphology of worn surfaces was analyzed using scanning electron microscopy.

2 Experimental Works

Copper powder with varying weight percentages of nano-boron carbide were fabricated via powder metallurgy route. Compositions including pure copper (100% Cu), 99% Cu + 1% B_4C, 98.5% Cu + 1.5% B_4C were prepared (Table 1). These powders were mixed together in a ball mill running at 400 revolutions per minute for about 45 min. Compaction was done using hydraulic press with 250 bar pressure. The green compact is then sintered at 850 °C for 3 h and 30 min. The specimens required for abrasive wear tests were cut by Wire Electrical Discharge Machining (WEDM) after oxide removal from each specimen. Abrasive wear experiments were carried out by pasting the abrasive sheets of 400 grit silicon carbide on the disc of pin-on-disc wear tester. Specimens for abrasive wear test were cut from unreinforced Cu and Cu nanocomposites of size 8 mm diameter and 8 mm length. Prepared specimens were pressed and rotated against silicon carbide abrasive sheets under dry sliding conditions at room temperatures.

Table 1 Different combinations of Cu and nano-B_4C

Sample	Copper (wt%)	Nano-boron carbide (wt%)
1	100	0
2	99	1
3	98.5	1.5

3 Results and Discussion

3.1 Vickers's Micro Hardness Test

The results of the Vickers microhardness measurements carried out on the mono-lithic and reinforced cupper nanocomposites are reported in Table 2. A load 0.5 kg with a dwell time of 10 s was applied during hardness testing. Increasing the nano-B_4C content in Cu matrix significantly increases the microhardness of the fabricated nanocomposites. Out of the three materials tested, 1.5 wt% B_4C reinforced Cu nanocomposites showed maximum hardness value of 83, which is 84% more than the microhardness of pure Cu.

In general, the improvement in hardness could be attributed to the following factors; In Cu/B_4C nanocomposites higher dislocation densities were observed than with unreinforced Cu. This is because of the huge difference in the Young's modulus mismatch and coefficient of thermal expansion (CTE) mismatch between Cu and B_4C [15, 16]. Normally, the CTE of Cu is higher than the CTE of B_4C fillers. When the Cu-composite is cooled from an elevated temperature to ambient temperature, misfit strain took place due to the dissimilar thermal contraction at the interface of matrix and filler. Residual stresses are built-up which promotes high dislocation densities in the vicinity of the B_4C fillers or at the Cu/B_4C interface. The residual stress is isotropic for equiaxed elements. This leads to the formation of density concentrated dislocations in the matrix, and hence, to a higher hardness. Escalating dislocation density leads to stronger resistance to dislocation motion, since adjacent dislocations often hinder one another.

Dislocation motions are often restricted by the addition of fillers. Therefore, the stress required for the dislocations to travel in the Cu-composite is greater than that is needed in pure Cu. This mechanism normally termed as Orowan strengthening (Fig. 1). The moving dislocation forms a loop like structure around the fillers. During deformations, dislocations are accumulated; thus, internal stresses are generated. Apart from fillers, dislocations need to trounce these stress fields from dislocation pileups [17].

3.2 Abrasive Wear Test Using Pin-On Disc Tester

Abrasive wear tests for pure copper and different weight percentages of Cu–B_4C nanocomposites were carried out on pin-on disc machine (Fig. 2). Cylindrical work

Table 2 Hardness values of samples

Sample	Field-1	Field-2	Field-3	Average
Pure Cu	45.1	43.8	46.5	45
Cu + 1% B_4C	74.6	76.6	76.8	76
Cu + 1.5% B_4C	89.1	79.4	80.9	83

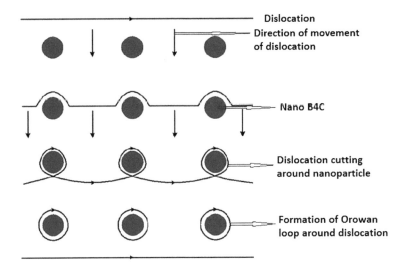

Fig. 1 Schematic representation of formation of Orowan loop

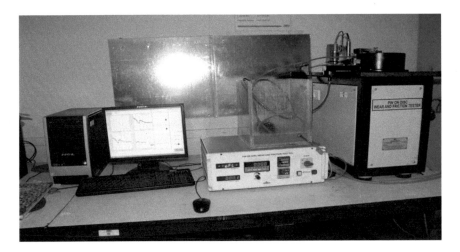

Fig. 2 Computer-assisted pin-on-disc wear tester

pieces (Ø8 mm × 8 mm) which have flat contact region with disc were used during abrasive wear tests. The abrasive material used in this study was 400 grit silicon carbide sheets which are pasted on the disc of wear testing apparatus. The pin was held against the counter face of a rotating disc running at 400 rpm speed. The test duration was fixed as a constant at 10 min per experiment. The loads were set as 1 kg (10 N), 2 kg (20 N), and 3 kg (30 N) for the successive experiments. New abrasive sheets were changed after every experiment to ensure that fresh abrasives are in contact with the test pins. Lower wear rates of Cu composites attributed to the

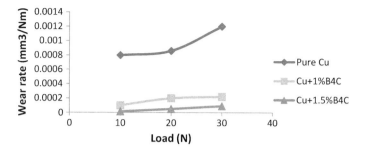

Fig. 3 Load versus wear rate

enhanced hardness and strength of composites due to the incorporation B_4C fillers. The exhibition of strong bond between Cu matrix and B_4C filler is foremost reasons that lessen the wear of the composite.

Increase in applied load tends to increase the wear rate of both pure Cu and Cu–B_4C nanocomposites (Fig. 3). Out of the three materials tested in this work, Cu nanocomposites with 1.5 wt% B_4C fillers shows lowest wear rate. The main cause for the raise in wear rate with increased loading condition is the frictional force applied on the surface of the material during abrasive wear. This shows that the percentage of reinforcement added to a composite determines the properties like wear resistance in a directly proportional manner. The choice to prepare the composite through powder metallurgy route has also effected in a good way by increasing the strength of the composite and making sure that the bond between Cu and B_4C is adequately maintained. Increase in applied load during abrasive wear tests tends to increases the temperature levels.

The relation between nano-B_4C wt% and wear rate is depicted in Fig. 4. Increase in nano-B_4C content in Cu matrix reduces the wear rate. Cu nanocomposites with 1.5 wt% B_4C shows the minimum wear rate. This is in accordance with the literature and general concept that increase in hardness decreases wear and hence, decreases wear rate.

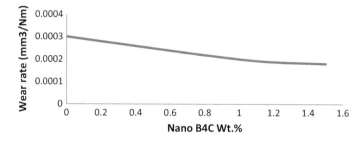

Fig. 4 Nano-B_4C wt% versus wear rate

3.3 Scanning Electron Microscopy (SEM) Characterization

For microstructural investigations, trial samples were prepared with the standard metallographic procedures. Microstructure of powder metallurgy samples were examined using a Carl Zeiss MA15/EVO 18 SEM machine. A typical microstructure of pure Cu and 1 wt% B_4C reinforced Cu-nanocomposite is shown in Fig. 5. Fairly uniform distribution of B_4C nanoparticles in Cu matrix was seen in Fig. 5b.

Abrasive wear-track patterns of nanocomposites and the parent material at three different loads were studied by taking SEM micrographs. SEM images were taken for Pure Cu at 10 N and 30 N load and for Cu + 1.5% B_4C at 10 and 30 N. This enables to understand the clear picture of abrasive wear behavior of fabricated materials. The different levels of magnification chosen were 250x, 500x, 1000x, 2500x and 5000x.

Worn out surface SEM images of pure Cu at 10 N with different magnifications is portrayed in Fig. 6a, b. It is obvious from the images; the absence of any reinforcement has resulted in severe abrasive wear caused by the free movement of abrasives on the surface of the material when it comes in contact with it during the wear test. The wear groove and pattern is unidirectional since the movement of the rotating disc was in clockwise direction. Pure copper predominantly show huge weight loss all through the abrasive sliding. Linear increase of volume loss exhibited in pure copper specimen with increasing sliding distance.

At 30 N load for the pure Cu sample, worn out surface groove is even more prevalent (Fig. 7). This is because of the extra frictional force caused due to the increase in load from 10–30 N. Thereby, proving the fact that increase in load is directly proportional to wear. The projection in the 2.5x image (Fig. 7b) denotes that some parts of the pin were depreciated to an even greater depth than others because of irregularities in micro level.

Worn out surface morphology (SEM images) of 98.5% Cu + 1.5% B_4C at 10 N load with 2500x and 5000x are indicated in Fig. 8a, b. Wear track and pattern are irregular as seen from the images. As compared to the pure Cu wear pattern, the grooves are thinner (less wear) due to the presence of hard B_4C nanoparticles in Cu matrix. Incorporation of B_4C nanoparticles increase the hardness and reduce the wear rate.

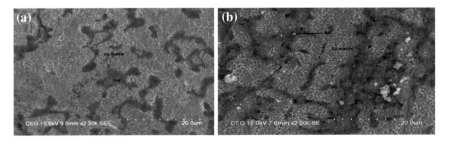

Fig. 5 SEM images **a** pure Cu **b** Cu-1 wt% B_4C

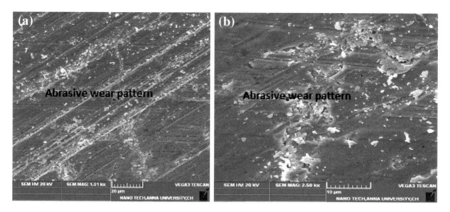

Fig. 6 Worn out surface SEM images of pure Cu at 10 N load at **a** 1000x **b** 2500x magnification

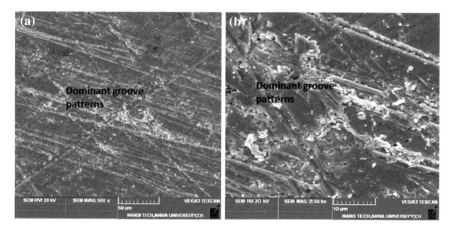

Fig. 7 Worn out surface SEM images of pure Cu at 30 N load at **a** 1000x **b** 2500x magnification

Thinner worn out groove patterns (Fig. 9) are observed in Cu + 1.5% B$_4$C at 30 N load. In fact, at 500x and 5000x resolution, there were not many differences in terms of wear pattern or amount of wear observed. This can be majorly attributed to the presence of 1.5% of B$_4$C which provides the necessary hardness to the material to deter the wear rate. Proper mixing and blending of boron carbide powder with copper powder is also a major factor in increasing the wear resistance.

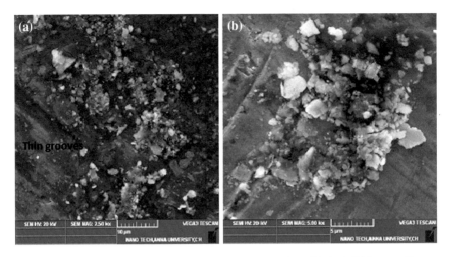

Fig. 8 Worn out surface SEM images of Cu + 1.5% B$_4$C at 10 N **a** 2500x **b** 5000x magnification

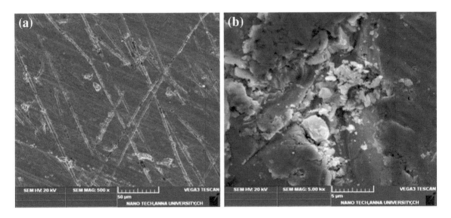

Fig. 9 Worn out surface SEM images of Cu + 1.5% B$_4$C at 30 N load **a** 500x **b** 5000x magnification

4 Conclusions

Powder metallurgical route was successfully employed to fabricate copper matrix nanocomposites reinforced with nano-B$_4$C fillers (0, 1.0 and 1.5% by weight). Scanning electron microscopic images validates the homogeneous mixing of nano-B$_4$C fillers in copper matrix. Vicker's microhardness results showed the significant improvement in microhardness of nanocomposites when compared to monolithic cupper. The abrasive wear test carried out at room temperature proved that the wear rate diminished with an addition of nano-B$_4$C fillers in copper matrix under the application of normal load. The Cu/1.5 wt% nano-B_4C possessed superior wear resistance properties. As the normal load increases, the wear rate of both monolithic copper and copper nanocomposites also increased. The worn out surface SEM images showed profound long grooves for pure copper and thin grooves for Cu nanocomposites.

References

1. Geiger AL, Walker JA (1991) J Min Met Mater Soc 43:8
2. Girot FA, Quenisset JM, Naslin R (1989) Comp Sci Tech 30:155
3. Zhan Y, Zhang G (2004) Friction and wear behavior of copper matrix composites reinforced with SiC and graphite particles. Tribol Lett 17(1)
4. Martinez MA, Martin A, Lorca JL (1993) Scr Metall 28:207
5. Lloyd DJ (1994) Particle reinforced aluminium and magnesium matrix composites. Int Mater Rev 39(1):1–23
6. Zhang ZF, Zhang LC, Mai YW (1995) Particle effects on friction and wear of aluminium matrix composites. J Mater Sci 30(23):5999–6004
7. Tang Y, Liu H, Zhao H, Liu L, Yating W (2008) Friction and wear properties of copper matrix composites reinforced with short carbon fibers. Mater Des 29:257–261
8. Sapate SG, Uttarwar A, Rathod RC, Paretkar RK (2009) Analyzing dry sliding wear behaviour of copper matrix composites reinforced with pre-coated SiCp particles. Mater Des 30:376–386
9. Sabbaghiana M, Shamanian M, Akramifard HR, Esmailzadeh M (2014) Effect of friction stir processing on the microstructure and mechanical properties of Cu–TiC composite. Ceram Int 40:12969–12976
10. Wang X, Liang S, Yang P, Fan Z (2010) Effect of milling time on electrical breakdown behavior of Al_2O_3/Cu composite. J Mater Eng Perform 19:906–911
11. Ramesh CS, Ahmed RN, Mujeebu MA, Abdullah MZ (2009) Fabrication and study on tribological characteristics of cast copper–TiO_2–boric acid hybrid composites. Mater Des 30:1632–1637
12. Buytoz S, Dagdelen F, Islak S, Kok M, Kir D, Ercan E (2014) Effect of the TiC content on microstructure andthermal properties of Cu–TiC composites prepared by powder metallurgy. J Therm Anal Calorim 117:1277–1283
13. Akhtar F, Askari SJF, Shah KA, Du X, Guo S (2009) Microstructure, mechanical properties, electrical conductivity and wear behavior of high volume TiC reinforced Cu-matrix composites. Mater Charact 60:327–336
14. Jha P, Gautam RK, Tyagi R, Kumar D (2016) Sliding wear behavior of TiC-reinforced Cu-4 wt% Ni matrix composites. JMEPEG 25:4210–4218
15. Arsenault RJ, Shi N (1986) Dislocation generation due to differences between the coefficients of thermal expansion. Mater Sci Eng, A 81:175–187
16. Arsenault RJ (1984) The strengthening of Al alloy 6061 by fiber and platelet silicon carbide. Mater Sci Eng, A 64:171–181
17. Moon J, Kim S, Jang J, Lee J, Lee C (2008) Orowan strengthening effect on the nanoindentation hardness of the ferrite matrix in microalloyed steels. Mater Sci Eng A, 487 (1–2):552–557

On Mechanical and Thermal Properties of Concretes with Rubber as Partial Replacement to Well-Graded Conventional Aggregates

Srinath Rajagopalan, P. Sreehari, B. Mahalingam and K. Mohammed Haneefa

Abstract The disposal of used automobiles tyres is a major environmental concern of the day. Annually, there are 1.2 billion tyres seeking disposal issues globally. Concrete as a major construction material can potentially adopt processed rubber waste—derived from tyres as a partial replacement for aggregates. The present study is focused on mechanical and thermal properties of concretes with rubber as partial replacement to well-graded conventional aggregates. The study includes replacement of 10, 20, 30 and 100% of coarse aggregate with waste tyre aggregates and was compared with conventional M30 grade concrete. Concretes were subjected a thermal cyclic heating of 50 °C for a period of 7 days. The results showed that the weight of the rubber concrete is reduced by 12% compared to the conventional concrete and the strength of the concrete reduces with the increase in rubber content. At ambient temperatures, the rubber concrete was found thermally stable. The compressive strengths were found increasing after cyclic heating of 50 °C for a period of 7 days. This effect may be due the effect of Portland pozzolana cement which has 30% of fly ash in it. From this study, it can be concluded that rubber waste concretes can be a potential candidate for structural and non-load bearing structures at ambient temperatures.

Keywords Compressive strength · Concrete · Rubber tyre waste
Cyclic heating · Fly ash

S. Rajagopalan (✉) · P. Sreehari · B. Mahalingam
Department of Civil Engineering, SSN College of Engineering, Chennai, India
e-mail: srinathr@ssn.edu.in

P. Sreehari
e-mail: sreeharip@ssn.edu.in

B. Mahalingam
e-mail: mahalingamb@ssn.edu.in

K. M. Haneefa
Department of Civil Engineering, IIT Madras, Chennai, India
e-mail: mhkolakkadan@gmail.com

© Springer Nature Singapore Pte Ltd. 2019   57
A. K. Lakshminarayanan et al. (eds.), *Advances in Materials and Metallurgy*,
Lecture Notes in Mechanical Engineering,
https://doi.org/10.1007/978-981-13-1780-4_7

1 Introduction

The disposal of used automobiles tyres is a major environmental concern of the day. Annually, there are 1.2 billion tyres seeking disposal issues globally. Concrete as a major construction material can potentially adopt processed rubber waste—derived from tyres as a partial replacement for aggregates. The present study is focused on mechanical and thermal properties of concretes with rubber as partial replacement to well-graded conventional aggregates.

2 Background

Modern pneumatic tyres essentially consist of synthetic or natural rubber, fabric and wire. Additionally, carbon black and chemical admixtures are added to improve durability and efficiency. The wide range of chemicals used in rubbers warrants special attention during their disposal. Leaching of rubber to soil or ground water may result in contamination and hence pronounces major environmental concern. Styrene butadiene copolymer is a most commonly used variety of rubber tyre.

2.1 Rubber Used in Concrete

The study of [1] on rubber concrete, with 38, 25 and 19 mm size chips concluded a 85% reduction in compressive strength and 50% in tensile splitting strength compared to normal concrete. However, rubber concrete absorbs a large amount of plastic energy under tensile and compressive loads. Schimizze et al. [2] developed rubberized concrete mixes and recommended to use them for stabilized base layers in flexible pavements. El Gammal et al. [3] and Kotresh and Belachew [4] studied rubber in chopped and crumbed forms. They observed significant reduction in strength and density compared to normal concrete. However, ductility of the concrete was increased. The observations were similar for fine and coarse aggregate replacements. According to their observation, the recommended optimum replacement was 10–25%. Torgal et al. [5] recommended concrete with rubber aggregate for earthquake-resistant structure and for acoustic purposes such as noise reduction. Similar observations were reported by Topcu [6], Khatib and Bayomy [7], and Zheng et al. [8].

2.2 Other Disposal Methods

Barla et al. [9] proposed rubber as a supplement fuel for paper and pulp industry and cement kiln. Since the energy obtained from the study was reasonable, the method was found sustainable and efficient. In another study by Collins et al. [10], scrap rubber tyres were used for artificial coral reef construction and found effective. In addition to that, other uses were feedstock for making carbon black [11], guard rail posts [12], noise barriers [13] and asphalt pavement mixtures [14].

3 Materials Used and Experimental Design

Well-graded granite coarse aggregate and river sand fine aggregates were use in the study. Figure 1 provides the gradation of coarse aggregate used. Fly ash-based Portland pozzolana cement (PPC) was used as binder. The design mix was designed for M30 grade concrete and used for comparison. Slump test as per IS 7320 [15], flow table test as per IS 5512 [23] and compaction factor as per IS 5515 [16] were used to asses the fresh properties. Table 1 provides the composition of the control mix used in the study. Discarded bus tyres were obtained and thoroughly washed to remove any dirt. The tyres were manually cut to approximately 20 mm strips and chopped to irregular pieces using plate cutter (Fig. 2). The rubber aggregate were sieved in accordance with IS 383 [17]. The rubber aggregates were mixed to obtain the same gradation as mineral aggregates. The study includes volume replacement of 10, 20, 30 and 100% of coarse aggregate with waste tyre aggregates and compared with refence mix. The replacements in each size fractions of coarse aggregate

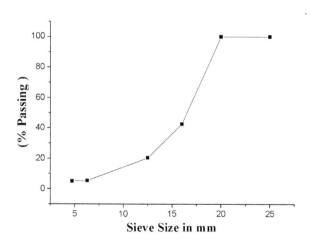

Fig. 1 Gradation of coarse aggregate

Table 1 Mixes and compositions

Mix	Cement kg/m^3	Coarse aggregate kg/m^3	Fine aggregate kg/m^3	w/c ratio
Control	360	1040	960	0.50

Fig. 2 Sieved rubber
aggregates

were done corresponding proportions used for the control mix. 7 and 28 days
strengths were observed. Concretes were subjected a thermal cyclic heating of
50 °C for a period of 7 days.

4 Results and Discussion

4.1 Fresh Properties

Table 2 gives fresh properties of concrete mixes used in the present study. Figure 3
pictorially depicts the effects of rubber replacement on fresh properties of concrete.
Effect on slump (Fig. 3a) and compaction factor (Fig. 3b) are marginal. However,
the flow table tests indicated that addition of rubbers improves the flow (Fig. 3c).
The percentage of increments in flows were 64.7, 66.7, 72.0, 72.0 and 84.0%,
respectively, for the mixes of control, 10, 20, 30 and 100% replacement of coarse
aggregate with rubber. Replacement of rubber does not alter the fresh properties of
concrete, and even some cases it improves the workability.

Table 2 Fresh properties of concrete used

Mix	Slump (mm)	Compaction factor	Slump flow (mm)	Average flow (%)
Control	75	0.87	39.0, 41.5, 43.0	64.7
10% rubber concrete	72	0.85	41.0, 40.0, 44.0	66.7
20% rubber concrete	65	0.87	41.0, 44.0, 44.0	72.0
30% rubber concrete	68	0.90	44.0,41.0,44.0	72.0
100% rubber concrete	70	0.86	47.0, 46.0, 45.0	84.0

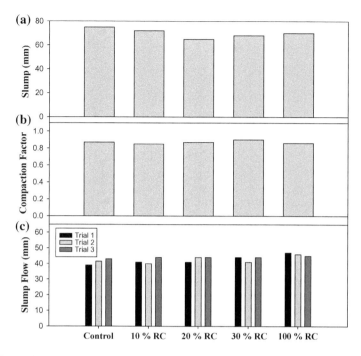

Fig. 3 Properties of fresh concretes used in the study

4.2 Hardened Properties

Figure 4 represents 7 days and 28 days compressive strengths of concretes investigated in the study. Upon replacing a 10% coarse aggregate by rubber chips; the reduction in compressive strength was marginal. The compressive strength was changed from 30.33 to 30.0 MPa for reference mix and 10% replacement, respectively. However, beyond 10% replacement; the reduction in strength was drastic. Concrete with 20% replacement of rubber chips exhibited a compressive strength of 11.33 MPa. Similar trends were observed for 30 and 100% replacements. The trends of 28 days compressive strengths were very much similar to the 7 days compressive strength. The compressive strengths were found to be 38.0, 33.3, 14.0, 8.7 and 4.67 MPa for reference mix, 10, 20, 30 and 100% replacements, respectively. The reductions in compressive strength for 10, 20, 30 and 100% replacements with respect to control were found to be 12.3, 63.2, 77.2 and 87.7%, respectively. The reductions in compressive strengths may be resulted from poor bonding between the binder phase and the aggregates. Moreover, the flexible nature of the rubber might have resulted in initiation or activation of interfacial transition zone (ITZ) in the concrete with more replacements of rubber with coarse aggregate and subsequent drastic reduction compressive strengths. However, the concrete with 10% coarse aggregate replaced by rubber chips (with a compressive

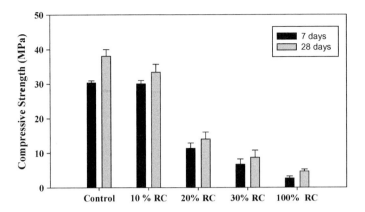

Fig. 4 Compressive strength after 7 days and 28 days curing

strength 33.3 MPa) has enough potential to be used for residential or infrastructural purposes. Even though ductility of concrete with rubber chips exhibits promising results (as discussed in the literature review), more studies may be required to ensure the durability aspects of these concretes before they can be used for structural purposes. Other concretes with low compressive strengths can be used for non-structural purposes such as partition wall and acoustic components.

4.3 Thermal Performance Studies

Rubber is a polymeric material which may be susceptible to temperature variations. Thermal performance studies may be required for rubber concrete for validating their use for ambient temperature applications. With respect to this objective, rubber concrete specimens were subjected to a maximum of 50 °C for 7 days. The mode of heat treatment was cyclic; which resembles day time heating and subsequent cooling in the night. Heat treatment was performed in an electric oven as illustrated in Fig. 5.

Figure 6 elucidates thermal performance of rubber concrete with considerable increment in compressive strengths upon thermal treatment. There were no signs of cracks formation in the specimens (Fig. 7). The increment in strengths were marginal for reference mix and 10% replacement (7 and 6%, respectively). Meanwhile, the increments in strengths were significant for 20 and 30% replacements (30.9 and 30.6%). The one reason for increment in strength is the use of Portland pozzolana cement (PPC) for making concrete. Fly ash-based concrete exhibits increment in strength upon an exposure to 20–200 °C [18]. Pozzolanic activity leads to more formation of calcium silicate hydrate (CSH) by consuming calcium hydroxide. Moreover, formation of a tobermorite gel is found in moderately heated fly ash concretes, which is three times stronger than normal CSH [19–22]. Additionally the

Fig. 5 Concrete cubes in hot air oven for cyclic heating @ 50 °C for 7 days

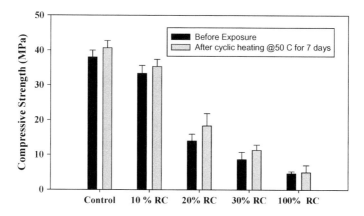

Fig. 6 Thermal performance of rubber concretes

enhanced bonding between cement matrix and rubber upon thermal treatment may be another reason for the increment in strength. The concrete with 100% replacements of coarse aggregate with rubber exhibited similar trend with increment in strength corresponding to reference mix and 10% replacement. On a concluding note with the results of cyclic thermal study, rubber replaced concrete is thermally stable in ambient temperatures, and can be used for civil and infrastructural purposes. More studies based on fracture mechanics and microstructure may be required to improve this material further.

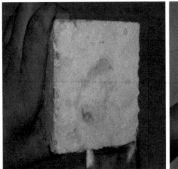

Fig. 7 Concrete cubes before and after cyclic heating @ 50 °C for 7 days

5 Conclusions

From the current study flowing conclusions are drawn,

(a) Rubber incorporated concretes exhibited similar or improved fresh properties compared to normal concrete
(b) Replacement of rubber in concrete reduces the compressive strength. However, the reduction in strength was marginal for 10% replacement of coarse aggregate. The reductions in strengths were significant for 20 and 30% replacement. 100% replacement resulted in drastic reduction in compressive strength.
(c) Rubber replaced concrete exhibited thermal stability at ambient temperatures. Moreover, increment in strengths observed may be attributed from advantage of pozzolana cement and improved bond between rubber aggregate cement matrix.
(d) Rubber replaced concrete may be a potential candidate for sustainable housing and infrastructural development. Additionally, the product can be proposed for partition walls or acoustic components.

Acknowledgements The authors acknowledge the contributions of Mr. K. Hutesh Reddy, Mr. K. Madhan Gopal, and Mr. Tammineni Sai Krishna Undergraduate students in Department of Civil Engineering, SSN College of Engineering for this project. The authors are grateful to SSN Trust for their financial support through student internal funding from student project scheme.

References

1. Eldin NN, Senouci AB (1993) Rubber-tire practices as concrete aggregate. J Mater Civ Eng 5 (4):478–496
2. Schimizze RR, Nelson JK, Amirkhanian SN, Murden JA (1994) Use of waste rubber in light-duty concrete pavements. In: Proceedings of the third material engineering conference, infrastructure: new materials and methods of repair. San Diego, CA, pp 367–374

3. El-Gammal A, Abdel-Gawad AK, El-Sherbini Y, Shalaby A (2010) Compressive strength of concrete utilizing waste tire rubber. J Emerg Trends Eng Appl Sci 1(1):96–99
4. Kotresh KM, Belachew MG (2014) Study on waste tyre rubber as concrete aggregates. Int J Sci Eng Technol 3(4):433–436
5. Torgal FP, Shasavandi A, Jalali S (2011) Tyre rubber wastes based concrete: a review. In: Wastes: solutions, treatments and opportunities. 1st international conference. Guimarães, Portugal
6. Topcu JB (1995) The properties of rubberized concretes. Cem Concr Res 25(2):304–310
7. Khatib ZK, Bayomy FM (1999) Rubberized portland cement concrete. J Mater Civ Eng 11 (3):206–213
8. Zheng L, Huo XS, Yuan Y (2008) Strength, modulus of elasticity, and brittleness index of rubberized concrete. J Mater Civ Eng 20:692–699
9. Barla MA, Eleazer WE, Whittle DJ (1993) Potential to use waste tires as supplemental fuel in pulp and paper mill boilers, cement kilns and in road pavement. Waste Manage Res 11 (6):463–480
10. Collins KJ, Jensen AC, Mallinson JJ, Mudge SM, Russel A, Smith IP (2001) Scrap tyres for marine construction: environmental impact. In: Dhir RK, Limbachiyya MC, Paine KA (eds) Recycling and reuse of used tyres. Thomas Telford, London, pp 149–162
11. Wójtowicz MA, Bassilakis R, Serio MA (2004) Carbon black derived from waste tire pyrolysis oil. Advanced Fuel Research, Inc
12. Seckinger NR, Abu-Odeh A, Bligh RP, Roschke PN (2005) Performance of guardrail systems encased in pavement mow strips. J Transp Eng 131(11):851–860
13. Lakušić S, Bjegović D, Haladin I, Baričević A, Serdar M (2011) RUCONBAR—innovative noise protection solution made of recycled waste tyre. Mech Transp Commun 3:X-76–X-82
14. Xiao F, Amirkhanian SN, Shen J, Putman B (2009) Influences of crumb rubber size and type on reclaimed asphalt pavement (RAP) mixtures. Constr Build Mater 23(2):1028–1034
15. IS 7320 (1974) Specification for concrete slump test apparatus. Bureau of Indian Standards, New Delhi
16. IS 5512 (1983) Specification for flow table for use in tests-of hydraulic cements and pozzolanic materials. Bureau of Indian Standards, New Delhi
17. IS 383 (2016) Coarse and fine aggregate for concrete specification(third revision). Bureau of Indian Standards, New Delhi
18. Poon CS, Azhar S, Anson M, Wong YL (2001) Comparison of the strength and durability performance of normal- and high-strength pozzolanic concretes at elevated temperatures. Cem Concr Res 31:1291–1300
19. Haneefa KM, Santhanam M, Parida FC (2013) Review of concrete performance at elevated temperature and hot sodium exposure applications in nuclear industry. Nucl Eng Des 258:76–88
20. Haneefa KM, Santhanam M, Parida FC (2013) Thermal performance of limestone mortars for use in sodium cooled fast breeder reactors. Indian Concr J 87(12):25–41
21. Hossain KMA (2006) High strength blended cement concrete incorporating volcanic ash: Performance at high temperatures. Cem Concr Compos 28:535–545
22. Nasser KW, Marzouk HM (1979) Properties of mass concrete containing fly ash at high temperatures. ACI J 76(4):537–551
23. IS 5515 (1983) Specification compacting factor apparatus. Bureau of Indian Standards, New Delhi

Effect of Crumb Rubber on Inorganic and High Compressible Clay

S. V. Sivapriya and N. Charumathy

Abstract The generation of crumb rubber is abundant in our country and safer disposal has become a challenging job. Hence, a study has been attempted by mixing the crumb rubber in two different types of soil (CI and CH) in different percentages and their properties were studied. They were mixed with the soil sample in 10, 15, 20, 30, 40 and 50%. The standard Proctor compaction, CBR, UCC and permeability tests were conducted. From the experiments conducted, it is observed that maximum dry density decreases with increase in optimum moisture content. The CBR and UCC value increases with increase in percentage of crumb rubber till 15% for both the type of soils and the coefficient of permeability increases with increase in percentage of crumb rubber.

Keywords Crumb rubber · Standard compaction · CBR · Permeability
UCC · Reuse

1 Introduction

Development of transportation is a main criteria of a developing nations with increasing the number of lanes and proposal of new connecting roads in roadways, developing new port, harbours and airports. The development in roadway leads to more usage of multi-tyred vehicles causing environmental pollution. From previous literatures, over 10 billion tyres are leftover worldwide every year [1]. The main challenge is to dispose the wearied and tared tyre as it will not decompose or decay with time. Hence, dumping of tyres become non-viable, the study of recycle and reuse of used tyre has become emerging study. According to Waste Tyres

S. V. Sivapriya (✉)
Department of Civil Engineering, SSN College of Engineering, Chennai, India
e-mail: sivapriyasv@ssn.edu.in

N. Charumathy
Chennai Metro Water, Chennai, India
e-mail: charumathy.n@rediffmail.com

© Springer Nature Singapore Pte Ltd. 2019
A. K. Lakshminarayanan et al. (eds.), *Advances in Materials and Metallurgy*,
Lecture Notes in Mechanical Engineering,
https://doi.org/10.1007/978-981-13-1780-4_8

Management Rules draft guidelines [2], proposed to generate mechanisms for collection and disposal of used tyres. Earlier, it was used as ground cover for playground equipment, surface material for running tracks and athletic fields or dumped in landfill; which will not be possible in near future.

In geotechnical aspect, Baykal et al. [3] studied the combination of rubber and kaolinite in development of large pores or cracks using water gasoline as permeates. The hydraulic conductivity tests show the clay-tyre mix in water is lower compared to gasoline as permeate. By adding crumb rubber by 10–20% maximum as admixture significantly reduces the surface hardness [4–6]. Usage of shredded tyres and rubber with sand was used as a lightweight backfill and such fills are lighter than conventional fills. The strength of these backfill materials was as equal to traditional material [7]. If an aqueous polymer was used as soil stabilizer in unsaturated clay, it is highly economical due to its potential in time and cost saver due to its low curing period [8].

In highway aspect, Bosscher et al. [9] suggest the tyre chip need not be in a specific specification in construction activities; however, it was made easily when particle size is less than 75 mm (maximum) in a highway embankment. As per AASHTO [10], suggested to have modulus of subgrade reaction as 16 MN/m^3 for pure tyre chip in rigid pavement design. In asphalt mix, crumb rubber was used in additives and it increased the strength and quality. The asphalt mix when mixed with crumb rubber needed less asphalt content with increase in permeability and less durable [11].

In material aspect, with increase in combination of crumbed tyre from 5, 10, 15 and 20% by volume it reduces the unit weight of concrete by partially replacing sand [12]. However, the addition of crumb rubber to the mix had limited effect towards workability and compressive strength, the mixture showed required strength for lightweight concrete [13]. The concrete with waste rubber materials in structures was more prone to more earthquakes and in railway sleeper [14]. Rubber waste in concrete by 15% maintains high resistance to acid attack [15]. As, in general, an acceptable percentage of 25% by volume was suggested as fine aggregates with crumb rubber and concrete [16].

Besides huge study of crumb rubber, the existing million amounts of crumb rubber generated cannot be recycled in one area. Hence, more study is suggested to have better reuse of recycled material.

2   Experimental Work

For laboratory study, typical cohesive soil was taken from Potheri (S1) and Vadapalani (S2), India to conduct the experiments. To facilitate recycle process, rubber tyre is crumbed (CR) consisting of 40–50% rubber, 25–40% carbon black and 10–15% low molecular weight was used as replacement in soil by 10, 15, 20, 30, 40 and 50% weight.

Table 1 Index properties of the soil

Soil	S1	S2
Liquid limit (w_L), %	63	39
Plastic limit (w_P), %	31	16
Shrinkage limit (w_S), %	12	8

Initially to categorize the soil, index property tests such as liquid (w_L), plastic (w_P) and shrinkage limit (w_S) test were conducted as per standards (IS 2720 parts 5 and 6) [17, 18] and their corresponding properties are listed in Table 1. The laboratory results show S1 as highly compressible clay (CH) and S2 as inorganic clay according to IS 1498 [19].

Laboratory tests such as standard Proctor compaction, unconfined compressive tests, California bearing ratio and permeability tests were conducted as per standards (IS 2720 part 7, 10, 16 and 17, respectively) [20–23] to study behaviour of crumb rubber in cohesive soil.

3 Results and Discussion

3.1 Influences in Compaction

A quantity of 3 kg air-dried sample passing 4.75 mm sieve was used to compact in a 994 cc volume mould in three layers; each layer is compacted with a height of fall of 310 mm with a energy of 25 blows. Figure 1 shows the relation of maximum dry unit weight (γ_d) with different mix of CR with soil and then it is compared with virgin soil. The maximum dry density and optimum moisture content of the soil sample S1 is 16 kN/m^3 and 18.6%, respectively: similarly for S2 it is 19.2 kN/m^3 and 15.5%, respectively. It clearly shows that the maximum dry unit weight reduces with increase in percentage of CR as it starts giving a cushion effect to the soil. This observation is similar to that Signes et al. [24].

Fig. 1 Influence of CR in compaction

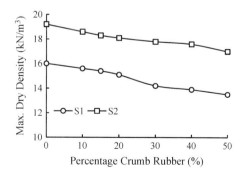

3.2   Influences in Shear Strength

The Unconfined compressive test was conducted for both the samples to find the compressive strength of the clay with different mix. Till 15% mix of CR with soil, it is scattered away from each other and after that CR overlap with each other and form a larger size particle which compresses when loaded; in turn gives lesser shear strength. The UCC value for 15% of S1 is 260 kN/m^2 and for S2 is 263 kN/m^2 (Fig. 2), this shows the influence of CR is better in Inorganic clay. As with increase in compressibility of soil and CR percentage, the effect will not be significant.

3.3   Influences in Penetration Ratio

The California bearing ratio (CBR) test was conducted to check the feasibility of reusing the CR as a stabilizing agent in subgrade under closed moisture and density conditions. The soil was compacted in OMC and maximum dry density obtained from compaction test with different ratios. Value corresponding to 2.5 mm penetration is taken as CBR value: until 15%, there is increase in CBR value (Fig. 3). However, due to the reduction in compression, the influence of CR reduces in further addition.

3.4   Influences in Permeability

One of the main properties of the soil 'permeability' is defined as the flow of water between the voids in the soil that can be reduced with the addition of CR material to the soil. The size of the particle will vital major role in reducing the permeability, in turn leads to seepage. With increase in percentage of CR until 20% (Fig. 4), there is greater increase in permeability, then after the permeability behaviour is unaltered majorly.

Fig. 2 Influence of CR in compressive strength

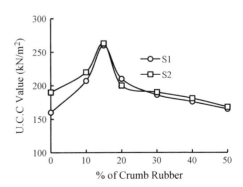

Fig. 3 Influence of CR in penetration

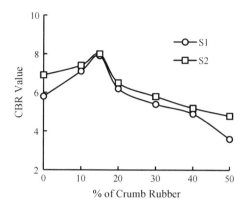

Fig. 4 Coefficient of permeability with variation in percentage of crumb rubber

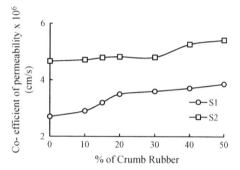

4 Conclusions

From the experimental study, the following observations are made:

1. From the Standard Proctor Compaction test, it is observed that the maximum dry density reduce with the increase in percentage of Crumb rubber for both the soils S1 and S2.
2. The CBR value increased with increasing percentage of crumb rubber up to 15% for both the soils S1 and S2. Sample S1 mixed with crumb rubber of 20% gave higher CBR value when compared with the soil S1 alone. However, for soil S2, this trend is observed in 15% crumb rubber added with soil.
3. The UCC values increased with increasing percentage of crumb rubber up to 15% for both the soils S1 and S2. Sample S1 mixed with crumb rubber of 50% gave higher UCC value when compared with the soil S1 alone. However, for soil S2 this trend is observed in 20% crumb rubber added with soil.
4. The coefficient of permeability increases with increasing percentage of crumb rubber mixed with the soils S1 and S2.

The interaction of crumb rubber is good in inorganic clay when compared to highly compressible clay. Hence, crumb rubber can be used as replacement in backfill in an inorganic type of soil.

References

1. Alamo-NoleLuis A, Perales-Perez O, Roman-Velazquez FR (2011) Sorption study of toluene and xylene in aqueous solutions by recycled tires crumb rubber. J Hazard Mater 185:107–111
2. U.S. Environmental protection agency (2017). https://www.epa.gov/
3. Baykal I, Yesiller N, Koprulu K (1992) Rubber-clay liners against petroleum based contaminants, environmental geotechnology. In: Proceedings mediterranean conference on environmental geotechnoogy, pp 477–481
4. Groenevelt PH, Grunthal PE (1998) Utilisation of crumb rubber as a soil amendment for sports turf. Soils Tillage Res 47(2):169–172
5. Zhao S, He T, Duo L (2011) Effects of crumb rubber waste as a soil conditioner on the nematode assemblage in a turfgrass soil. Appl Soil Ecol 49:94–98
6. Tapas Das, Baleshwar S (2013) Benefits and impacts of scrap tyre use in geotechnical engineering. J Environ Res Dev 7(3):1262–1271
7. Lee JH, Salgado R, Bernal A, Lovell CW (1999) Shredded tires and rubber-sand as lightweight back fill. J Geotech Geoenvironmental Eng 125(2):132–141
8. Naeini SA, Naderinia B, Izadi E (2012) Unconfined compressive strength of clayey soils stabilized with waterborne polymer. KSCE J Civil Eng 16(6):943–949
9. Bosscher PJ, Edil TB, Kuraoka S (1997) Design of highway embankments using tire chips. J Geotech Geoenvironmental Eng 123(4):295–304
10. AASHTO (1993) Guide for design of pavement structures. American Association of State Highway and Transportation Officials, Washington, D.C
11. Wulandari PS, Tjandra D (2017) Use of crumb rubber as an additive in asphalt concrete mixture. Sustain Civil Eng Struct Constr Mater, SCESCM, Procedia Eng 171:1384–1389
12. Balaha MM, Badawy AAM, Hashish M (2007) Effect of using ground tire rubber as fine aggregate on the behaviour of concrete mixes. Indian J Eng Mater Sci 14:427–435
13. Batayneh MK, Marie I (2008) Asi IPromoting: the use of crumb rubber concrete in developing countries. Waste Manag 28(11):2171–2176
14. Zheng L, Huo S, Yuan Y (2008) Experimental investigation on dynamic properties of rubberized concrete. Constr Build Mater 22:939–947
15. Azevedo F, Pacheco-Torgal F, Jesus J, Barroso de Aguiar L, Camões LAF (2012) Properties and durability of HPC with tyre rubber wastes, vol 34, pp 186–191
16. Camille A, Salem IG (2013) Utilization of recycled crumb rubber as fine aggregates in concrete mix design. Constr Build Mater 42:48–52
17. IS (1985) 2720—part 5: Indian standard methods of tests for soils—Determination of liquid and plastic limit. Bureau of Indian Standards, New Delhi (Reaffirmed 2006)
18. IS (1987) 2720—part 6: Indian standard methods of tests for soils—determination of shrinkage factors. Bureau of Indian Standards, New Delhi (Reaffirmed 1995)
19. IS (1970) 1498: Indian standard classification and identification of soils for general engineering purposes. Bureau of Indian Standards, New Delhi (Reaffirmed 2002)
20. IS (1980) 2720—part 7: Indian standard methods of tests for soils—determination of water content and dry density relation using light compaction. Bureau of Indian Standards, New Delhi (Reaffirmed 2011)
21. IS (1991) 2720—part 10: Indian standard methods of tests for soils—determination of unconfined compressive strength. Bureau of Indian Standards New Delhi (Reaffirmed 2006)

22. IS (1987) 2720—part 16: Indian standard methods of tests for soils—determination of California bearing ratio. Bureau of Indian Standards, New Delhi (Reaffirmed 2002)
23. IS (1986) 2720—part 17: Indian standard methods of tests for soils—determination of permeability. Bureau of Indian Standards, New Delhi (Reaffirmed 2002)
24. Signes CH, Garzo´n-Roca J, Martı´nez Ferna´ndez P, Garrido de la Torre ME, Insa Franco R (2016) Swelling potential reduction of Spanish argillaceous marlstone facies tap soil through the addition of crumb rubber particles from scrap tyres. Appl Clay Sci 132–133:768–773

Hot Deformation Studies of Al–Cu–Mg Powder Metallurgy Alloy Composite

Sai Mahesh Yadav Kaku and Asit Kumar Khanra

Abstract High-performance requirements of materials in automobile industry gave the scope of alloying of monolithic metal powders which plays a significant role in the enhancement of properties. In the present study, monolithic powder matrix aluminum is alloyed with respective proportions of copper and magnesium by mechanical alloying. With 4 wt% of Cu as constant, different wt% (0.25, 0.5, 0.75, and 1) of Mg are added to form various powder alloy preforms. The alloy preforms are triaxially compacted in a tool steel die and sintered at 550 °C for 1 hour in continuous flow high pure argon gas atmosphere. The sintered alloy preforms are subjected to uniaxial deformation at two different temperatures, (400 and 500 °C). The hardness and the relative density increases with the extent of deformation. Potentiodynamic polarization is performed on deformed preforms to evaluate the effect of deformation. Corrosion rates and hardness of the alloy preforms were correlated with the relative densities. The extent of deformation and compositional variations on all the properties were analyzed.

Keywords Al–Cu–Mg · Sintering · Relative density · Deformation and Potentiodynamic polarization

1 Introduction

During the past three decades extensive research has been done on alloying which resulted in the evolution of different set of alloys. Among all the alloys, Al alloys acquired a prominent place in automobile, aviation, and other structural industries. This is because of the versatile properties of aluminum like lightweight, ductile, malleable metal, and nonmagnetic nature. Aluminum is also known for its corrosion resistant nature and its low density. Copper being the most ductile material is

S. M. Y. Kaku (✉) · A. K. Khanra
Department of Metallurgical and Materials Engineering, NIT Warangal,
Hanamkonda, India
e-mail: saimaheshkaku@gmail.com

© Springer Nature Singapore Pte Ltd. 2019
A. K. Lakshminarayanan et al. (eds.), *Advances in Materials and Metallurgy*,
Lecture Notes in Mechanical Engineering,
https://doi.org/10.1007/978-981-13-1780-4_9

alloyed into aluminum to offer good ductility nature. While magnesium offers strength and toughness to the material are the main alloying elements. The densification behavior of Al–Cu is studied with different amounts of copper content in the composite. The maximum densification is attained in Al–4% Cu which is around 94% relative density approximately [1].

Al 2024 aluminum alloy was made through PM route which is subsequently extruded and annealed over a range of temperature 300–500 °C. To study the effect of strain rate on the morphological changes due to heat treatment, torsion tests were performed [2]. Deformation tests at elevated temperatures are performed on PM2324 and AA2014 at 350–500 °C. Highest densification was achieved in PM2324 alloy by hot deformation which has subsequently increased the mechanical properties than its counterpart AA2014 alloy [3]. The effect of heat treatment and the duration of heat treatment on the deformation of Al–4.4% Cu–1.4% Mg over a temperature of 105–200 °C are reported [4]. The nonuniform strains and temperature during the deformation of the Al–Cu clad composite layers were correlated to the results between temperatures and the grain morphology [5]. The effect of formation of secondary phase particles on the cold working and plastic deformation in the Al–Cu–Li–Mg–Ag alloys has been reported. The effect of precipitation hardening on strain tare and slip localization is analyzed [6, 7]. The flow behavior of Al–Cu–Mg–Ag alloy during hot compressive deformation test was studied and its effect on morphological changes was studied. The subgrain size increased and the dislocation density decreased as an effect of morphological changes. The transfer of the mechanism from dynamic recovery to dynamic recrystallization is predicted [8].

In the present study, the Al–Cu–Mg alloy samples are made through powder metallurgy route (PM). The effect of alloying copper (Cu) and magnesium (Mg) with aluminum (Al) is studied. The alloy powder preforms of all the compositions are subjected to compression test in incremental steps up to failure. The effect of load during compressive deformation test is correlated with the height strain at the respective steps. To evaluate the effect of temperature on deformation process and strain produced during deformation. Compressive deformation test is done at 25, 400, and 500 °C. The hardness of the deformed preforms is measured at each interval of deformation process and is correlated with the height strain. This is to study the effect of extent of strain on the hardness of the alloy preform.

2   Experimental Work

The Aluminum (Al), Copper (Cu), and Magnesium (Mg) are the precursors used for the preparation of alloy preforms. Cu at 4 wt% fixed in the matrix Mg is added in 0.25 wt% incremental steps to produce Al–4 Cu, Al–4 Cu–0.25 mg, Al–4 Cu–0.5 Mg, Al–4 Cu–075 Mg, and Al–4 Cu–1 Mg alloy power preforms. The homogenous alloy mixture is compacted in manual pellet press followed by sintering in tubular furnace at 550 °C for 1 h in continuous argon flow atmosphere. The micro Vickers hardness tester (SCHIMADZU) is used to determine the Vickers hardness of sintered and deformed performs. A load of 500 g and 15 s of hold time are used for each microhardness test. In the present study, an average of six readings for each sample is reported. The sintered powder alloy preforms of all the different compositions are used as raw materials for the deformation tests. In the process of deformation test, six samples of each composition are deformed to different levels up to failure as shown in Fig. 1. The appearance of a visible crack on the bulge of the deformed preform is considered as failure.

3 Results and Discussion

The SEM micrographs of precursors Al, Cu, and Mg used for preparation of Al–Cu–Mg powder preforms are shown in Fig. 2. The Al powder particles used as matrix in composites prepared are 40 μm and irregular shape. While the size of Cu and Mg particles are around 30–40 μm and of irregular shape. Figure 3 shows the relationship between the load applied during deformation and the strain produced in the preform as a result of deformation. The extent of strain increased with increase in the temperature during deformation which is observed to be common in all

Fig. 1 A sample set of deformed preforms

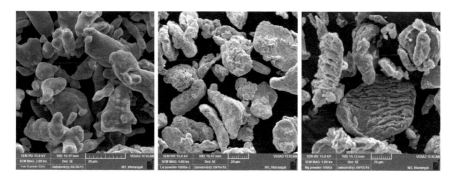

Fig. 2 SEM micrographs of Al, Cu, and Mg powders

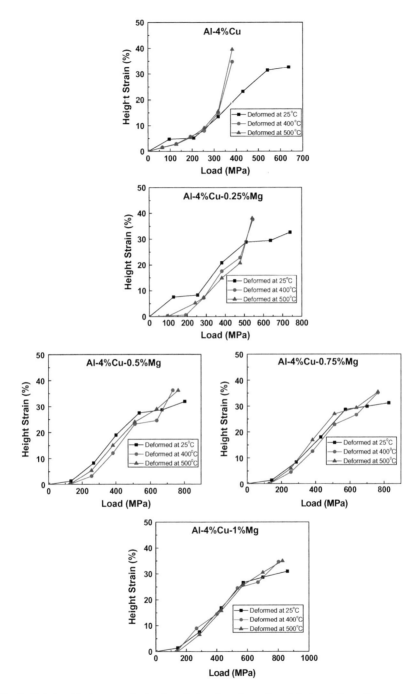

Fig. 3 Relationship between the load applied and strain produced during deformation

Fig. 4 Comparison between load and strain of deformation at three different temperatures

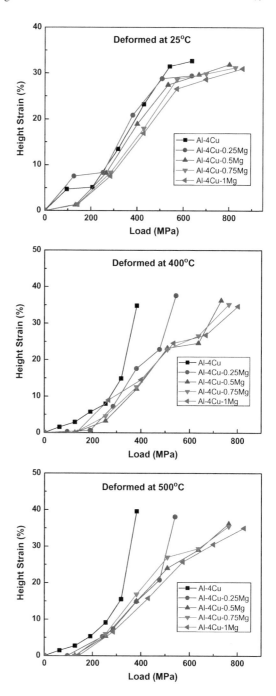

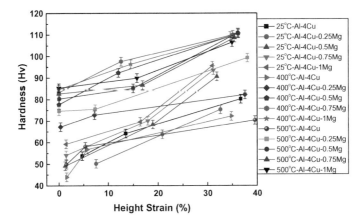

Fig. 5 Effect of strain on the hardness of deformed alloy preforms

compositions. Deformation at high temperatures requires low stresses than deformation at room temperature and subzero temperatures. When compared between different temperatures of deformation process, the extent of deformation decreased with increase in Mg content in the alloy matrix as shown in Fig. 3. This is because of the increase in toughness induced into the matrix with Mg alloying addition at the cost of ductility. During deformation at 25 °C, all the compositions deformed approximately to equal extent of strain at similar load levels. This trend of deformation changed at higher temperatures of deformation at 400 and 500 °C (Fig. 4). Al–4 Cu deformed at lower loads and exhibited high extent of strains. With increase in the temperature the loads required to deform decreased.

In Fig. 5, the hardness of the composite increased with the extent of strain. Hardness increased with the amount of Mg in the matrix. This is due to the inherent toughness nature induced by the Mg added to the matrix. Hardness also increased with the extent of strain. This increase in hardness as a function of strain is due to the high densification achieved with the high extent of strain. As extent of strain increased with the temperature used during deformation process. Alloy preforms deformed at higher temperatures are measured to have high hardness than the other alloy preforms deformed at low temperatures. With both alloying addition of Mg and deformation process hardness of the composites are enhanced 2.5 times approximately when compared to bare sintered alloy preforms.

4 Conclusions

From the above results, the following are the summary of conclusions:

- The load applied to deform decreased with the increase in temperature during deformation.

- The load required to deform increased with increase in Mg content in the alloy preform.
- The extent of strain decreased with the increase in the Mg in the alloy preform.
- Hardness of the composite increased with the extent of strain and increased with the Mg in the matrix.

Acknowledgements The authors are grateful to Lab Technicians of MMED, NIT Warangal for their continuous support during the execution of experimentation.

References

1. Wolla DW, Davidson MJ, Khanra AK (2014) Studies on the formability of powder metallurgical aluminum–copper composite. Mater Des 59:151–159
2. Bardi F, Cabibbo M, vangelista EE, Spigarelliand S, Vukc˘evic M (2003) An analysis of hot deformation of an Al–Cu–Mg alloy produced by powder metallurgy. Mater Sci Eng A339:43–52
3. Mann RED, Hexemer RL Jr, Donaldsona IW, Bishop DP (2011) Hot deformation of an Al–Cu–Mg powder metallurgy alloy. MaterSci Eng A 528:5476–5483
4. Wang H, Li C, Li J, Wei X, Mei R (2013) Effect of deformation and aging on properties of Al–4.1% Cu–1.4% Mg aluminum alloy. ISRN Mater Sci Volume 2013, Article ID902970
5. Kocicha R, Kocicha L, Davis CF, Lowe TC, Szurman I, Macháčková A (2016) Deformation behavior of multilayered Al–Cu clad composite during cold-swaging. Mater Des 90:379–388
6. Deschamps A, Decreusa B, De Geuser F, Dorin T, Weyland M (2013) The influence of precipitation on plastic deformation of Al–Cu–Li alloys. Act Mater 61:4010–4021
7. Ringer SP, Muddle BC, Polmear IJ (1995) Effects of cold work on precipitation in Al–Cu–Mg–(Ag) and AI–Cu–Li–(Mg–Ag) Alloys. Meta Mater Transvol 26A:1659–1660
8. Xiao Y, Qing LP, Yun BH, Wen BL, Wen J, Zhi MY (2009) Flow behavior and microstructural evolution of Al–Cu–Mg–Ag alloy during hot compression deformation. Mater Sci Eng A 500:150–154

Evaluation of Vacuum Arc Melted-Powder Metallurgy Al–ZrB$_2$ Composite

Sai Mahesh Yadav Kaku

Abstract There are numerous ways to produce Metal Matrix Composites (MMC). In the present study, a new approach has been designed with the combination of vacuum arc melting and powder metallurgy to produce non-equilibrium MMCs. The Al–2%ZrB$_2$ homogenous composite mixture was compacted in uniaxial direction and used as raw material for vacuum arc melting. The green compacts are subjected to high temperature within a short span of time. The melts were assured for homogenous composition by melting from different faces. Hardness of the composite was measured and which appeared to be higher than its powder metallurgy counterpart. The XRD patterns were analyzed for the phases present in the composite after melting. The composite has been tested to evaluate the corrosion rate through potentiodynamic polarization study.

Keywords MMC · Green compact · XRD · Potentiodynamic polarization

1 Introduction

The continuous advancement in the automobile, aviation, defense, and marine industry has prompted the need for light weight, fuel efficiency, and other specific property requirements. This requirement led to the development of advanced materials such as metal matrix composites (MMCs). Among different MMCs, aluminum has the great potential to fulfill all set of requirements. Aluminum metal matrix composites gained scope for applications in aerospace, defense, automotive industries, and various other fields. Aluminum MMCs, because of its light weight and other specific properties have a great potential in automotive engineering components. Axle tubes, reinforcements, blade, gear box casing, turbine, fan and compressor blades are some significant applications in military and civil aviation

S. M. Y. Kaku (✉)
Department of Metallurgical and Materials Engineering, NIT Warangal,
Warangal, India
e-mail: saimaheshkaku@gmail.com

© Springer Nature Singapore Pte Ltd. 2019
A. K. Lakshminarayanan et al. (eds.), *Advances in Materials and Metallurgy*,
Lecture Notes in Mechanical Engineering,
https://doi.org/10.1007/978-981-13-1780-4_10

for aluminum MMCs. MMCs and alloys are extensively used over conventional metals and alloys as they offer higher specific properties (properties/unit weight) of strength, stiffness, higher specific modulus, thermal stability, tribological properties and various other mechanical properties which enhances the product performance [1, 2].

Powder metallurgy is a route to produce non equilibrium composition of alloys and composites. This process has various advantages than the conventional composite production techniques. The pores and the density are the reasons that hinder the performance of the PM components [3]. The equiatomic Ti–50.0 at.%Ni alloy was prepared by vacuum arc remelting and subjected to heat treatment cycle and the same has been tested for evaluation of mechanical properties and morphological changes [4]. NiTi shape memory alloys are prepared through vacuum induction melting. Graphite dies are used as crucibles in melting. The effect of carbon pickup from the crucible during melting was studied and its effect on the properties has been reported [5]. Aluminum and its alloys ZrB_2 sintered in different techniques spark plasma, microwave and pressure less sintering both by in situ and powder metallurgy route. SPS composites appeared to achieve more densification [6–8]. Alumina–ZrB_2 prepared by in conventional and by SHS technique. Composite made by SHS resulted in maximum densification [9, 10]. Densification increased with the extent of deformation. Corrosion rate of deformed Al–ZrB_2 powder metallurgy composite decreased with densification. Hardness of the Al–ZrB_2 increased with the extent of strain hardening [11–13]. Hassani et al. [14] reported the effect of pores and the size of pores on the formability and densification behavior of the Al–SiC preforms. Appa Rao et al. [15] reported the deformation and workability behavior of pure copper and reported the effect of aspect ratio on the formability. Effect of aspect ratio on the densification and formability behavior Cu–TiB_2 is reported by Gadakary et al. lower the aspect ratio better is the densification [16].

The family of borides have significant properties like high chemical stability, high hardness and good thermo-electric properties. Among the family of boride powders titanium diboride (TiB_2) and zirconium diboride (ZrB_2) are significant ceramics. They exhibit superior properties over other ceramics like high melting point, extremely high hardness and wear resistance, low specific gravity, magnetic, electrical properties, with high mechanical properties and chemical inertness at elevated temperatures. The composites containing ZrB_2 reinforcements exhibit great dimensional stability which has great significance in case of mating components in automobile industry.

In the present study the Al–ZrB_2 composite was produced through combination of powder metallurgy and vacuum arc melting, a unique approach. The composite was metallographically prepared to see microstructure. The micro Vickers hardness of the composite was measured. To evaluate the life time of the composite in marine applications, the corrosion behavior of the composite was evaluated for prolonged immersion times in 3.5% Nacl aqueous solution.

2 Experimental Work

The powder precursors are mixed in proper weight proportions to draw Al–2ZrB$_2$ homogenous mixture for sufficient time. The homogenous mixture is then compacted in a 20 mm stainless steel die to make green compacts. The compaction of raw materials was carried out in a 25 ton manual pellet press (KIMAYA ENTERPRISES). The green compacts produced by compaction were used as raw material to produce melts of the composites by vacuum arc melting. Once the green compacts were placed in the depression of the copper hearth (die) the chamber was evacuated with the help of high performing vacuum pump (Fig. 1). A little amount of Ar gas was purged into the chamber to avoid any reaction of melt within the chamber. Melting is carried out around 1600 °C and repeated for quiet number of times to ensure homogenous composition in the final melt (Fig. 2). The power source used for generation of arc between electrode and metal is basic tungsten inert gas setup (TIG—600 A). The surfaces of the composite was ground and then polished to mirror surface finish for polarization studies. The surface of composite was exposed to 3.5% NaCl aqueous solution and constant surface area was maintained during the whole immersion time. Among the three electrodes, composite is taken as the working electrode, platinum acts as counter electrode and calomel acts as the reference electrode as shown in Fig. 3. Open circuit potential (OCP) was performed for 30 min to stabilize the flow of electrons followed by Tafel polarization between −0.5 and +0.5 V with a scan rate of 0.166 mV/s as a standard. I_{Corr} and E_{Corr} are extracted from the tafel extrapolation.

3 Results and Discussion

Figure 4 shows the SEM micrographs of precursors Al and ZrB$_2$ which are used in the preparation of composite. The size of Al powder particles were around 40 μm with irregular shape. The ZrB$_2$ powder was produced by the self-propagating high temperature synthesis. The size of the ZrB$_2$ particles produced was of sub-micron size and exist in agglomerate form. The composite obtained by the vacuum arc melt

Fig. 1 Experimental setup of vacuum arc melting and copper die

Fig. 2 Vacuum arc melted Al–ZrB$_2$ composite

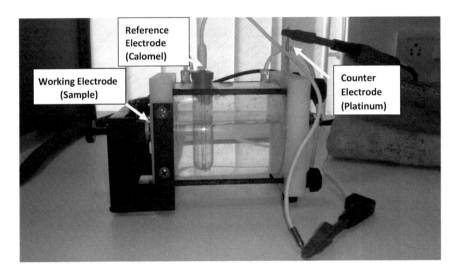

Fig. 3 Experimental setup for polarization study

was ground for flat surfaces and polished for mirror surface finish. The composite was tested for micro hardness by micro Vickers hardness tester (SCHIMADZU). A load of 500 g for 15 s of dwell time was used for micro hardness test. The composite was measured to have 57.5 Hv (Table 1).

The SEM micrograph of the vacuum arc melted (VAM) Al–ZrB$_2$ before and after corrosion study were shown in Fig. 5. From the SEM micrographs it has been observed that the sub-micron ZrB$_2$ particles were homogenously distributed in the

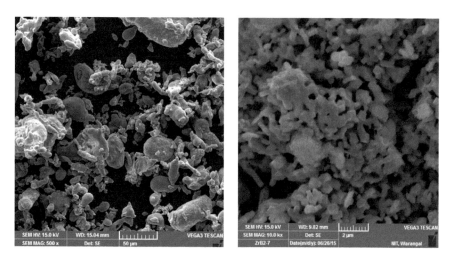

Fig. 4 SEM micrographs of Al and ZrB$_2$ powders

Table 1 EDX of VAM Al–ZrB$_2$ for 120 h of immersion in 3.5% NaCl electrolyte

S. No.	Element	Weight percentage
1	O	44.38
2	Al	55.34
3	Zr	1.18

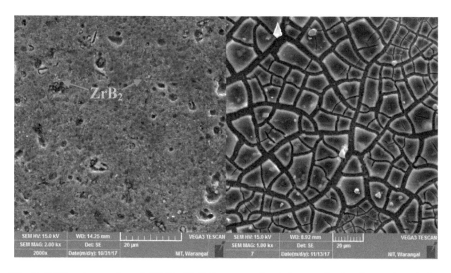

Fig. 5 SEM micrographs of VAM Al–ZrB$_2$ composite before and after immersion during corrosion study

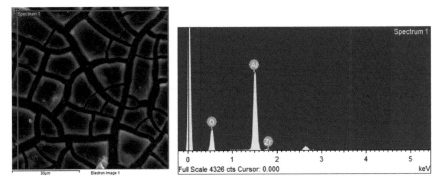

Fig. 6 EDX of VAM Al–ZrB$_2$ composite after 120 h of immersion in electrolyte

Al matrix. The high pressure of the arc produced during vacuum arc melting process left some blow holes in the bulk of the composite. The size of the blow holes range between 0.5 and 2 µm. Due to the prolonged immersion of 120 h in potentiodynamic polarization study a layer of corrosion products were formed. Alumina layer was formed over the surface of the composite during immersion in the electrolyte solution. This acts as the protective layer and inhibits the further corrosion of the composite. The EDX results confirms the formation of Al$_2$O$_3$ layer on the surface exposed to electrolyte (Fig. 6). Continuous crack like morphology of alumina layer was observed in the SEM micrographs of VAM composite after 120 h of immersion. This was due to the contraction forces of corrosion product surface layer withdrawn from the electrolytic solution. The width of the continuous cracks range between 1 and 4 µm.

In Fig. 7 Tafel extrapolation curves of Al–ZrB$_2$ composite at different immersion conditions were compared. From the Tafel extrapolation curve the I_{corr} and E_{corr} are obtained by intercept method. The shifting of extrapolation curve towards anodic side up to 24 h of immersion time. With further increase in the immersion time the curve shifted towards cathodic side. This shift is due to the formation of continuous corrosion product layer. The formation of Alumina layer which is the corrosion product acted as a protective layer which inhibited further corrosion.

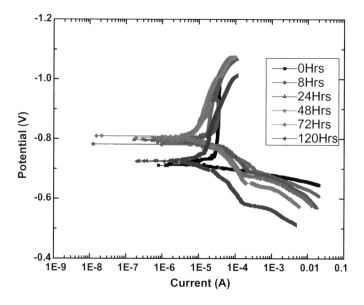

Fig. 7 Tafel extrapolation curves of Al–ZrB$_2$ composite with different immersion times

4 Conclusions

- Micro-structural analysis show ZrB$_2$ particles are homogenously distributed in the composite made through powder metallurgy-vacuum arc melting route.
- The reinforcement ZrB$_2$ particles exist both as individual and in lumps in the VAM Al–ZrB$_2$ composite.
- Hardness of the composite increased by 100% with the addition of ZrB$_2$ particles to the composite.
- A protective corrosion layer of Al$_2$O$_3$ is formed during immersion. This layer inhibits the further corrosion.

Acknowledgements The author is grateful to Prof. M. K. Mohan, HOD, MME, NIT Warangal for his continuous support during the experimentation.

References

1. Ahamed R, Asokan P, Aravindan S (2009) EDM of hybrid Al–SiCp–B$_4$Cp and Al–SiCp–Glassp MMCs. J Adv Manuf Technol 44:520–528
2. Ramulu M, Paul G, Patel J (2001) EDM surface effects on the fatigue strength of a 15vol% SiC$_P$/Al metal matrix composite material. Compos Struct 54:79–86

3. Chandramouli R, Kandavel TK, Shanmugasundaram D, Ashok Kumar T (2007) Deformation, densification, and corrosion studies of sintered powder metallurgy plain carbon steel preforms. Mater Des 28:2260–2264
4. Foroozmehr A, Kermanpur A, Ashrafizadeh F, Kabiri Y (2012) Effects of thermo-mechanical parameters on microstructure and mechanical properties of Ti–50 at.%Ni shape memory alloy produced by VAR method. Mater Sci Eng A 535:164–169
5. Frenzel J, Zhang Z, Neuking K, Eggeler G (2004) High quality vacuum induction melting of small quantities of NiTi shape memory alloys in graphite crucibles. J Alloy Compd 385:214–223
6. Ghasali E, Pakseresht A, Safari-Kooshali F, Agheli M, Ebadzadeh T (2015) Investigation on microstructure and mechanical behavior of Al–ZrB$_2$ composite prepared by microwave and spark plasma sintering. Mater Sci Eng A 627:27–30
7. Zhao D, Ilu X, Liu Y, Bian X (2005) In-situ preparation of Al matrix composites reinforced by TiB$_2$ particles and sub-micron ZrB$_2$. J Mater Sci 40:4365–4368
8. Foagagnolo JB, Robert MH, Ruiz-Navas EM, Torralba JM (2004) 6061 Al reinforced with zirconium diboride particles processed by conventional powder metallurgy and mechanical alloying. J Mater Sci 39:127–132
9. Mishra SK, Das SK, Pathak LC (2006) Sintering behavior of self-propagating high temperature synthesised ZrB$_2$–Al$_2$O$_3$ composite powder. Mater Sci Eng A 426:229–234
10. Mishra SK, Das SK, Sherbacov V (2007) Fabrication of Al$_2$O$_3$–ZrB$_2$ in situ composite by SHS dynamic compaction: a novel approach. Compos Sci Technol 67:2447–2453
11. Kaku SMY, Khanra AK, Davidson MJ (2017) Effect of deformation on densification and corrosion behavior of Al–ZrB$_2$ composite. Metall Mater Eng 23(17):47–63
12. Kaku SMY, Khanra AK, Davidson MJ (In press) Strain hardening behavior and its effect on properties of ZrB$_2$ reinforced Al composite prepared by powder metallurgy technique. J Inst Eng (India)-D
13. Kaku SMY, Khanra AK, Davidson MJ (In press) Micro-structural analysis and densification behavior of Al–ZrB$_2$ powder metallurgy composite during upsetting. Transac Indian Institute Metals
14. Hassani A, Bagherpour E, Qods F (2014) Influence of pores on workability of porous Al/SiC composites fabricated through powder metallurgy + mechanical alloying. J Alloy Compd 591:132–142
15. Appa Rao J, Babu Rao J, Kamaluddin S, Sarcar MMM, Bhargava NRMR (2009) Studies on the workability limits of pure copper using machine vision system and its finite element analysis. Mater Des 30:2143–2151
16. Gadakary S, Veerababu R, Khanra AK (2016) Workability studies on Cu–TiB$_2$ powder preforms during cold upsetting. Mater Technol 50(3):373–380

Interlaminar Fracture Behaviour of Hybrid Laminates Stacked with Carbon/Kevlar Fibre as Outer Layers and Glass Fibre as Core

A. Arockia Julias and Vela Murali

Abstract The mode I interlaminar fracture toughness of hybrid laminate was investigated by conducting double cantilever beam test as per ASTM D5528. Unidirectional carbon or kevlar fibre was stacked as outer layers and glass fibre layers in quasi isotropic orientation was used as core. Layer configuration of the 12 ply symmetric laminate was taken as $[H/H/G_0/G_{135}/G_{90}/G_{45}]_s$. H/H was varied as G_0/G_0, C_0/G_0, C_0/C_0, K_0/G_0 and K_0/K_0. These hybrid laminates were prepared by hand lay-up technique and post cured by compression at high temperature and pressure. The strain energy release rate (G_{IC}) calculated using modified beam theory method was found to be better for the glass/carbon-epoxy hybrid laminates under mode I loading. Fractographic analysis of the delaminated surfaces shows fibre imprints and hackles. In glass/kevlar-epoxy hybrid laminate cusp was observed that could be formed due to shear loads acting on that laminate.

Keywords Mode I · Double cantilever beam · Delamination · Hybrid
Fibre · SEM

1 Introduction

Application of composite materials in aerospace, marine and automobile industry are mostly in the form of laminates. These laminates are fabricated by binding multiple layers together. Delamination may occur in layered composites due to poor interlaminar fracture toughness. Double Cantilever Beam (DCB) test is a viable

A. Arockia Julias (✉)
Department of Mechanical Engineering, B S A Crescent Institute
of Science & Technology, Vandalur, Chennai, India
e-mail: juliasarok@gmail.com

V. Murali
Department of Mechanical Engineering, CEG Campus, Anna University,
Chennai, India
e-mail: velamurali@gmail.com

© Springer Nature Singapore Pte Ltd. 2019   91
A. K. Lakshminarayanan et al. (eds.), *Advances in Materials and Metallurgy*,
Lecture Notes in Mechanical Engineering,
https://doi.org/10.1007/978-981-13-1780-4_11

method to measure the interlaminar mode I fracture resistance. In DCB test the opening displacement between two DCB arms, reaction force at the pulling point, and the delamination crack length can be obtained [1]. The critical energy release rate calculated from these data indicates the crack resistant force required to create a new surface [2]. For hybrid laminates containing carbon and glass fibre at 1:1 volume fraction, failure may occur in three stages, namely brittle cracking, crack growth in waves and large displacement during crack propagation [3, 4]. Hwang and Shen [5] measured the interlaminar fracture toughness of a glass/carbon fibre reinforced interply hybrid composite and found that the crack growth is dominated by mode I failure. The interlaminar fracture toughness was found to be higher for the interply hybrid composites in comparison with that of the carbon fibre composites. Kim et al. [6] found that the interlaminar fracture toughness value was higher for the carbon-epoxy laminates in comparison with that of the kevlar-epoxy and carbon/kevlar-epoxy hybrid laminates. In glass/kevlar-epoxy hybrid composites the fracture toughness is found to be better along the fibre direction compared to across the fibre direction [7]. The fracture characteristics of Auxetic kevlar-epoxy laminated composites was recently studied using DCB test by Sen Yang [8]. The Auxetic kevlar composites shows significant improvement in fracture toughness compared to regular woven kevlar composites.

The fibre orientation of adjacent layers play a major role in determining the delamination behavior of laminated composites. In carbon-epoxy specimens containing mid-layers (crack plane) with $0°//\theta°$ fibre orientation, extensive fibre bridging occurs as a result of transverse cracking. This increases with increase in θ and reached the maximum for $0°//90°$ interface [9]. The critical strain energy release rate (G_{IC}) of carbon-epoxy laminate was found to be better for the specimens with $0°//90°$. In glass-epoxy laminate, specimens stacked with $0°$ fibres at the mid-plane offered less resistance to crack propagation, while the specimens with $90°$ oriented fibres at the mid-plane offered maximum resistance to crack propagation [10]. In most of the unidirectional FRP composites the failure was characterised by stable crack propagation with noticeable fibre bridging behind the crack tip [11]. Even in multidirectional carbon fibre reinforced polymer laminates, the fracture toughness depends on both the amount of fibre bridging and the interface angles [12]. However, delamination migrating from one lamina to an adjacent lamina by deviating its direction makes it difficult to find out the true fracture toughness value [13]. Fractographic study using Scanning Electron Microscopic (SEM) on multi-directional fibre reinforced carbon-epoxy laminates reveal that mode I fracture surface has a large amount of fibre pull-out, but mode II fracture surfaces has cusp formation [14]. Analysis on the new surfaces generated due to mode I fracture shows resin rich regions in the case of glass fibre laminate and brittle fracture in carbon fibre laminate [15]. In unidirectional glass-epoxy laminate broken fibre pulled-out from the resin and scarps that form opposing faces are the indications of mode I fracture failure [16]. Most of the studies on hybrid composites concentrate on the fracture behaviour between the carbon and glass fibres by placing them on either side of the crack plane. In this work the fibre type and orientation of the crack plane is maintained constant. The effect of hybridization is studied by varying the outer layer with glass, carbon and kevlar fibre.

2 Materials and Methods

Fibre reinforced composite laminates are fabricated at desired thickness by stacking a number of thin layers of fibre and consolidating them into a solid using a resin. In this study, laminates were stacked with glass fibre layer in quasi-isotropic arrangement as core and unidirectional carbon or kevlar fibre as outer layers. Layer configuration of the 12-ply symmetric laminate was taken as $[H/H/G_0/G_{135}/G_{90}/G_{45}]_s$. H/H was changed as G_0/G_0, C_0/G_0, C_0/C_0, K_0/G_0 and K_0/K_0 to form five different laminates namely GGQ, CGQ, CCQ, KGQ and KKQ respectively. The reinforcement materials used are unidirectional E-glass mat of 400 gsm, T300 carbon mat of 300 gsm, and kevlar 49 mat of 200 gsm. Epoxy resin (Araldite LY556) was used as matrix material. The resin was mixed with the curing agent (hardener) HY951 at 10:1 weight ratio to initiate the chemical reaction. The laminates were prepared by hand lay-up technique in an open mold. Fibre layers were stacked one over the other by applying the resin-hardener mixture between the layers. The delamination (pre-crack) in the mid-plane required for the DCB test was created by inserting a 100 μm thick Teflon film on one side of the laminate in the longitudinal direction for the required length [3]. The laminates were partially cured under atmospheric condition for 4 h followed by curing in a compression molding machine for 10 min at 70 °C under a constant pressure of 8 bars. The laminates were then allowed to cool in atmospheric condition for 24 h. Higher curing temperature was maintained to initiate and sustain the chemical reaction, while higher pressure removes the entrapped air and consolidate the individual plies into a bonded laminate [17]. The composition of the laminates are given in Table 1.

The DCB test was conducted as per ASTM standard D5528 on five types of laminates namely GGQ, CGQ, CCQ, KGQ and KKQ. The laminates were trimmed as per guidelines specified in the standard and specimens were prepared. The length of the specimen was taken as 160 mm including the delamination length of 70 mm

Table 1 Composition of laminates

Nomenclature	Density (g/cm³)	Thickness (mm)	Fibre volume fraction %			Hybrid ratio
			Glass	Carbon	Kevlar	
GGQ	1.82	3.8	52	0	–	0
CGQ	1.71	4	42	10	–	23.8
CCQ	1.61	4.2	34	20	–	58.8
KGQ	1.56	4	45	–	8	17.7
KKQ	1.32	4.1	37	–	16	43.2

(a)   (b)   (c)

Fig. 1 DCB test specimen **a** glass-epoxy, **b** glass/carbon-epoxy, **c** glass/kevlar-epoxy

from one end. Width of the specimen was taken as 25 mm. The thickness varied from 3.8 to 4.2 mm depending on the type of fibre used. Length of the loading hinges used to transfer the load from UTM to the specimen was maintained as 20 mm. This reduces the pre-crack length at loading point to 50 mm. The loading hinges made up of aluminum is riveted to the specimen carefully after drilling two holes of 2 mm diameter in the specimen. The sides of the specimens were marked at an interval of 5 mm from the crack tip using a fine tip permanent marker.

The test was conducted in an Instron UTM at a constant cross-head speed of 5 mm/min. The glass-epoxy specimen, glass/carbon-epoxy hybrid specimen and glass/kevlar-epoxy hybrid specimen being tested is shown in Fig. 1a–c respectively. The crack extension was monitored through a 12.1 megapixel optical camera with 20 × optical zoom. The load and displacement values corresponding to each 5 mm crack extension were manually recorded by visual inspection from the pre-crack length (50 mm) to a maximum length of 90 mm. This was repeated for all the five samples tested in each type and the average value was taken.

Modified Beam Theory (MBT) method given in Eq. (1) was preferred over other methods to find the mode I strain energy release rate (G_{IC}).

$$G_{IC} = \frac{3P\delta}{2b(a + |\Delta|)} \tag{1}$$

In this Equation, P is the applied load in N, δ is the load point displacement in mm, b is the width of the specimen in mm, a is the delamination length in mm and Δ is the correction factor in mm. The correction factor Δ was found by plotting the cube root of compliance against the crack length and measuring the deviation in the x intercept. Compliance C is the ratio of load point deflection (δ) and applied load (P).

3   Results and Discussion

In this study on hybrid laminates the type of fibre in the outer layers alone was varied without changing the fibre orientation. Hence in all the specimens, glass fibre layer was present on both side of the crack plane (pre-crack) with fibre orientation of the interface as 45°//45°. Thus the bending resistance of the outer layers will play a major role in deciding the interlaminar fracture toughness. The load taken by the specimen and the corresponding displacement are recorded for every 5 mm increment in crack length. The test was conducted on five samples and the average load vs displacement is plotted in Fig. 2. From the graph it is observed that the glass/carbon-epoxy hybrid specimen CGQ take more load, whereas displacement is more in the glass/kevlar-epoxy hybrid specimen. The carbon fibre stacked as outer layer resist bending and hence the deflection in the arms of the DCB specimen is less compared to that of the glass/kevlar hybrid laminates. If the deflection is very high compared to the crack propagation, then shearing load may increase leading to mode II fracture.

In DCB test, the crack propagate between the layers at the mid-plane and separates them, to create two new surfaces. In some layered composites the crack might not propagate strictly along the intermediate layer where the pre-crack was laid [4]. If the crack extend in the transverse direction by breaking one or two layers and start propagating in another plane then it is called crack jumping. This occurs due to strong interface bonding between the layers along the crack plane and variation in fibre orientation between the adjacent layers. In this study crack jumping from mid-plane to adjacent layers as shown in Fig. 3a, b was observed for all the specimen irrespective of the fibre type. This could be due to the 45°//45° interface at the mid-plane and change in fibre orientation of the adjacent layers. As a result of crack jumping the thickness of the specimen increases in one side and reduces on the other side leading to unsymmetrical bending.

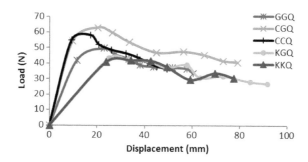

Fig. 2 Load-displacement curve obtained from the DCB test

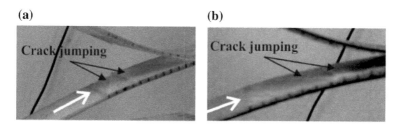

Fig. 3 Crack jumping during DCB test **a** glass-epoxy, **b** glass/carbon-epoxy hybrid

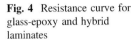

Fig. 4 Resistance curve for glass-epoxy and hybrid laminates

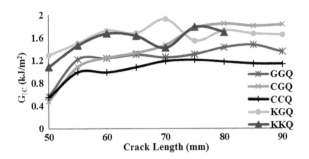

The mode I strain energy release rate G_{IC} calculated using modified beam theory method is plotted in Fig. 4 (resistance curve). The G_{IC} value at crack initiation is found to be 50% more for both the glass/kevlar-epoxy hybrid specimens KGQ (1.29 kJ/m^2) and KKQ (1 kJ/m^2) in comparison with that of the other specimens. This could be due to the lower bending resistance offered by the kevlar fibre stacked as outer layer. However, it should be noted that during the test, deflection of the DCB arms in the loading direction was found to be very high compared to the crack extension in the transverse direction. This could lead to increase in shear load which is a component of mode II fracture. The resistance curve is also found to be nearly similar for the specimens KGQ and KKQ even though, there is 8% difference in kevlar fibre volume-fraction between the two specimens. The resistance curve became nearly flat after reaching a crack length of 60 mm for all the specimens except the glass/carbon-epoxy hybrid specimen CGQ. In this particular specimen, the G_{IC} value increases with increase in crack length. It is found to be 0.49 kJ/m^2 for crack initiation at 50 mm and 1.82 kJ/m^2 when the crack reaches 90 mm. The G_{IC} at 90 mm for this specimen CGQ is even higher than that of the glass/kevlar hybrid specimens.

3.1 Fractographic Observations of DCB Specimen

Macro-level information like crack jumping and fibre bridging can be obtained through visual inspection, but micro-level analysis using SEM is required to understand the mechanism of failure in detail. The fracture surfaces of DCB specimen was examined using Hitachi S-3400 scanning electron microscope. Images were captured at the pre-crack region, where initial crack opening occurred and the crack jumping region. The direction of crack propagation was represented by an arrow. The SEM images of the specimen GGQ are shown in Fig. 5a, b. In general mode I fracture surfaces are very smooth as observed in the micrographs. Fibre imprints are found in both the images, which could be due to the fibre peel-off that occurs during crack opening. Fibre imprints of cross fibres placed in the

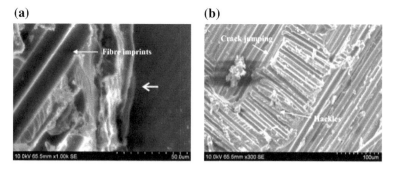

Fig. 5 SEM images of glass-epoxy laminate GGQ

adjacent layers is also visible in Fig. 5b. Hackles are formed due to the resistance offered by the matrix during crack propagation [16]. The presence of hackles oriented perpendicular to the fibre direction indicate brittle nature of the matrix with good bonding strength.

The SEM observations of glass/carbon-epoxy hybrid specimens CGQ and CCQ are shown in Figs. 6a, b and 7a, b respectively. In both the specimens, initial crack propagation region given in Figs. 6a and 7a shows smooth fibre surface covered by bare fibres with little deformation in the surrounding matrix. Due to higher bending resistance offered by the carbon fibre stacked as outer layer, crack propagation occurs without much fibre imprints or pull-out. The other regions are not flat as crack jumping occurs from one layer to other. The fracture surface has been complicated due to the presence of multi-plane delamination, ply splitting and fibre fracture. This shows little resemblance to the fracture surfaces generally observed in unidirectional laminates. The change in fibre orientation of adjacent layers could be one of the reason for this phenomenon. During crack jumping, matrix failure with striations (river markings) are observed in the specimen CGQ as shown in Fig. 6b. These resin rich regions can introduce a blunting effect for the crack growth. Fibre fracture and cracks extending into the underlying plies are observed in the specimen

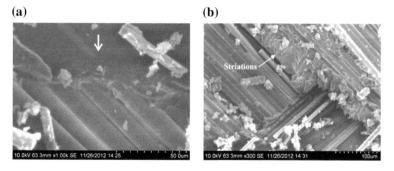

Fig. 6 SEM images of glass/carbon-epoxy hybrid laminate CGQ

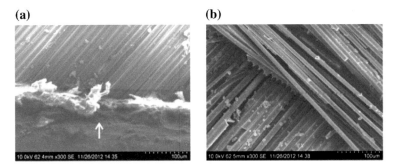

Fig. 7 SEM images of glass/carbon-epoxy hybrid laminate CCQ

CCQ as shown in Fig. 7b. The bunch of fibre pull-out from the matrix and fibre fracture at different places indicate a possible occurence of fibre bridging during crack propagation.

In the glass/kevlar-epoxy hybrid specimen KGQ cusp is observed at the pre-crack region during crack propagation as shown in Fig. 8a. The cusp can resist the crack propagation and increase the fracture toughness [14]. This indicates that the local condition at the crack tip may not be pure tension because cusp occurs due to shear stresses. Few micro voids visible near the pre crack region in Fig. 8a could be formed during fabrication of laminates. Advanced fabrication techniques can be used to avoid these defects and maintain uniform properties through out the laminates. The region where crack jumping occured with matrix breakage and fibre imprints on the top layer is shown in Fig. 8b. In the specimen KKQ, fibre pull-out is observed as shown in Fig. 9a. It appears like the matrix is torn away in this region causing the fibre to break and pull-out. The evidence for deep crack jumping over many layers in the transverse direction leading to breakage of fibre as a bundle is shown in Fig. 9b. Bare fibre surfaces with deep fibre pull-out from the matrix observed could be due to the higher deflection of the cantilever beam as indicated in the load-displacement curve. This is due to the poor bending resistance of the kevlar

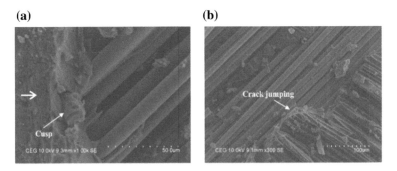

Fig. 8 SEM images of glass/kevlar-epoxy hybrid laminate KGQ

(a)   **(b)**

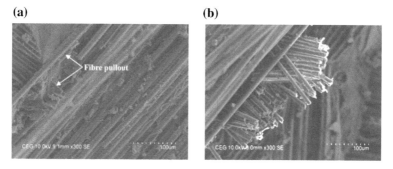

Fig. 9 SEM images of glass/kevlar-epoxy hybrid laminate KKQ

fiber stacked as outer layer. If the deflection increases beyond a limit the crack propagation rate may reduce and the mode I fracture condition will become invalid due to the increase in shear component.

4 Conclusions

- The mode I strain energy release rate G_{IC} is found to be 50% higher for the glass/kevlar-epoxy hybrid laminate (1 kJ/m^2) in comparison with the glass-epoxy laminates (0.5 kJ/m^2). This is due to the higher strain-to-failure exhibited by the kevlar fibre stacked as outer layers.
- Crack jumping from mid-plane to adjacent layers is observed for all the specimens irrespective of the fibre type. This could be due to the 45°//45° interface at the mid-plane and change in fibre orientation of the adjacent layers.
- The glass/kevlar-epoxy hybrid laminates show better fracture resistance but the cusp formation observed in the fracture surface is an outcome of shear load that indicate mode II fracture.
- The interlaminar fracture toughness of glass/carbon-epoxy hybrid laminate CGQ is found to be better as indicated by the SEM images and the increasing resistance curve.

Acknowledgements The authors are grateful to B S Abdur Rahman Crescent Institute of Science and Technology and CEG campus of Anna University for their support to carryout this work.

References

1. Kravchenko SG, Kravchenko OG, Carlsson LA, Byron Pipes R (2015) Influence of through-thickness reinforcement aspect ratio on mode I delamination fracture resistance. Compos Struct 125:13–22
2. Lim JI, Rhee KY, Kim IIJ, Jung DII (2014) Effect of stacking sequence on the flexural and fracture properties of carbon/basalt/epoxy hybrid composites. Carbon Lett. 15:125–128
3. Zhu XY, Li ZX, Jin YX (1993) Laminar fracture behaviour of (carbon/glass) hybrid fibre reinforced laminates-I. Laminar fracture process. Eng Fract Mech 44:545–552
4. Zhu XY, Li ZX, Jin YX (1993) Laminar fracture behaviour of (carbon/glass) hybrid fibre reinforced laminates-II. Laminar fracture criterion. Eng Fract Mech 44:553–560
5. Hwang SF, Shen BC (1999) Opening-mode interlaminar fracture toughness of interply hybrid composite materials. Compos Sci Technol 59:1861–1869
6. Kim SC, Kim JS, Yoon HJ (2011) Experimental and numerical investigations of mode I delamination behaviours of woven fabric composites with carbon, kevlar and their hybrid fibres. Inter J Prec Eng Manf 12:321–329
7. Maheswaran J, Velmurugan T, Mohammed Mohaideen M (2013) An experimental and numerical study of fracture toughness of kevlar—glass epoxy hybrid composite. In: Proceedings of the international conference on energy efficient technologies for sustainability. India, pp 936–941
8. Yanga S, Chalivendraa VB, Kim YK (2017) Fracture and impact characterization of novel auxetic kevlar®/epoxy laminated composites. Compos Struct 168:120–129
9. de Pereira AB, de Morais AB (2004) Mode I interlaminar fracture of carbon/epoxy multidirectional laminates. Compos Sci Technol 64:2261–2270
10. Shetty MR, Vijay Kumar KR, Sudhir S, Raghu P, Madhuranath AD, Rao RMVGK (2000) Effect of fibre orientation on mode-I interlaminar fracture toughness of glass epoxy composites. J Reinf Plast Compos 19:606–620
11. Compston P, Jar PYB (1998) Comparison of interlaminar fracture toughness in unidirectional and woven roving marine composites. Appl Compos Mater 5:189–206
12. Sebaey TA, Blanco N, Costa J, Lopes CS (2012) Characterization of crack propagation in mode I delamination of multidirectional CFRP laminates. Compos Sci Technol 72:1251–1256
13. Greenhalgh ES, Rogers C, Robinson P (2009) Fractographic observations on delamination growth and the subsequent migration through the laminate. Compos Sci Technol 69:2345–2351
14. Gilchrist MD, Svensson N (1995) A fractographic analysis of delamination within multidirectional carbon epoxy laminates. Compos Sci Technol 55:195–207
15. Zenasni R, Bachir AS, Arguelles A, Castrillo MA, Vina J (2007) Fracture characterization of woven fabric reinforced thermoplastic composites. J Eng Mater Technol 129:190–193
16. Marat-Mendes R, de Freitas M (2013) Fractographic analysis of delamination in glass epoxy composite. J Compos Mater 47:1437–1448
17. Arockia JA, Ram Kumar N, Vela M (2015) Evaluation of lamina properties and fractographic studies on glass/epoxy, carbon/epoxy and kevlar/epoxy composites. Appl Mech Mater 813:46–50

Microstructural Analysis and Simulation Studies of Semi-solid Extruded Al–Cu–Mg Powder Metallurgy Alloys

Katti Bharath, Asit Kumar Khanra and M. J. Davidson

Abstract Semi-solid extrusion of Al–Cu–Mg Powder Metallurgy (P/M) alloys had simulated under three different temperatures and extrusion angles in the present investigation. Al, Cu, and Mg powders were taken in different ratios in order to produce strong and light weight P/M alloys. Billets were prepared with an aspect ratio of one (φ 15 $\times$ 15 mm) to get good deformation results. Al–4Cu–0.5Mg alloy composition was optimized to do semi-solid extrusion after considering density, hardness, and strength as best parameters to optimize. Alloys were sintered at 550 °C and prepared samples with Initial Relative Density (IRD) of 90% for densification and deformation studies. The working temperature range for semi-solid extrusion test was derived from TG/DTA analysis. Extrusion tests were performed on a hydraulic press under different deformation temperatures (550, 575 and 600 °C) and different solid fractions (0.93, 0.76, and 0.56) respectively. All the extrusion tests were performed with a low extrusion ratio of 1.44, die approach angles of 30°, 45°, and 60° and strain rate of 0.1 s^{-1}. High density (>95%) and high hardness (>1000 MPa) extruded Al alloys were produced with good microstructures. Microstructural analyses were done for all Al alloys and found uniform distribution of grains at different temperatures. Dynamic recrystallization of grains was found with increasing liquid fraction during extrusion experiments. For an accurate prediction of microstructure evolution the strain rate, strain and temperature have to be considered and these can be calculated by FEM simulation. Simulation studies had been performed at three selected temperatures using Deform-2D software. Simulation and experimental results have been shown good agreement between them.

Keywords Powder metallurgy · Semi-solid extrusion · Simulation
Deformation · Densification · Grain size distribution

K. Bharath (✉) · A. K. Khanra
Department of Metallurgical and Materials Engineering, NIT Warangal, Warangal, India
e-mail: katti.bharath@gmail.com

M. J. Davidson
Department of Mechanical Engineering, NIT Warangal, Warangal, India

© Springer Nature Singapore Pte Ltd. 2019
A. K. Lakshminarayanan et al. (eds.), *Advances in Materials and Metallurgy*,
Lecture Notes in Mechanical Engineering,
https://doi.org/10.1007/978-981-13-1780-4_12

1   Introduction

Aluminium alloys are light in weight and most appropriate materials for the construction of heavy transport vehicles to increase the fuel efficiency and reduce emission gases. Al alloys are largely exploited for construction of aircraft, motorcar, and ship bodies due to the much higher workability and material properties. Most of the Al alloys are usually provided as profiled products or flat sheets which are produced by metal forming process. Semi-solid extrusion process has become superior shape forming process to produce Al products compared with conventional forming processes like forging, rolling, and casting. Semi-solid extrusion process is a new technology of metal forming process which combines both casting and forging advantages [1]. Non-ferrous metals such as Al, Cu, and Mg are mostly used to produce different products. Semi-Solid extrusion process is used to produce products between its liquidus and solidus temperature ranges. This process reduces macrosegregation, load, forming forces during shaping process and porosity and increase the strength and hardness [2].

The mechanical properties of products produced by semi-solid extrusion are strongly affected by change in microstructure of final products with their final shape, grain size and recrystallization conditions. Since the mechanical properties of material are directly dependent on microstructural change, it is important to understand the grain size evolution during extrusion and the effect of process parameters on such evolution. Several researchers have analysed the relation between the process conditions and microstructural evolution since years. The effect of process parameters on the grain size evolution and on the subgrains have shown by Sweet et al. [3]. Because of high solubility and strengthening effect, Cu has become prime alloying element for aluminium. According to the Al–Cu phase diagram, the strength of aluminium alloys increases with increasing Cu content up to 6%. Mg is an another alloying element used in Al–Cu combination to reduce surface oxides and to accelerate age hardening at room temperature [4]. Grain growth occurs due to grain boundary relocation to reduce the stored energy of boundaries by growing the grain size and reducing the boundary size [5]. No dynamic recrystallization have seen in aluminium alloy 6060 due to low dislocation densities [6]. Billets extruded at less than 350 °C basically gives recrystallized structure after the solution treatment while billets extruded at 450 °C appear to retain much of its original grain structure [4].

The coarse grains produced after the extrusion and heat treatment process cause reduction in mechanical properties. The recrystallization grain size and volume fraction recrystallization affect the aerospace industry parts due to the occurence of corrosion, damage tolerance, and fatigue crack propagation. Many researchers, who worked on extrusion process usually concentrated on controlling the product shape and quality or tool design/die design, only a few have focused on microstructural evolution during extrusion process [7, 8]. Two different types of deformation modes can be seen during any extrusion process, i.e., axisymmetric deformation inside the extrudate and shear deformation close to the surface, i.e., contact between material and working tool. Extruded product explains the changes occured in the cross-section by its microstructure and crystallography [9].

In the present investigation, the authors conducted semi-solid extrusion on Al–Cu–Mg powder metallurgy alloy billets and investigated the effect of extrusion temperature and extrusion angle on the microstructure of the extrudates. And the experimental data has correlated with the simulation analysis.

2 Experimental Work

In the present study, aluminium, cupper, and magnesium pure powders were used each with mesh size of ~ 325, ~ 325, and ~ 36 respectively. In the first step, Al powder was blended with addition of 4% Cu and 0.5% Mg in an mortar mixture for 60 min in order to get homogeneous mixture of powders. The blended Al–4% Cu–0.5% Mg alloy powder was taken into 15 mm diameter die which is made of high strength tool steel and compacted using 25 tonne manual hydraulic press with controlled pressure of 520 MPa. Zinc stearate powder dispersed in alcohol was used as a lubricant while compacting to avoid friction between die wall and alloy powder.

Cylindrical green compacts were sintered at 550 ± 10 °C for a period of one hour of soaking time and then allowed to cool in furnace itself. Relative density of sintered samples was analysed and measured by Archimedes principle with an accuracy of $\pm 1\%$. Semi-solid extrusion experiments were performed on sintered samples having 90% Initial Relative Density (IRD) to evolute the densification and deformation studies. Extrusion experiments were done on 50 Tonne capacity hot press. Semi-solid extrusion tests were performed between solidus and liquidus temperatures obtained from TG/DTA analysis. The extrusion tests of sintered samples done using dies with low extrusion ratio (1.44) and with three approach angles those are 30°, 45° and 60° respectively at three optimized temperatures such as 550, 575 and 600 °C respectively. The samples before and after extrusion were characterized by different methods those included density measurement by Archimedes principle, microstructural analyses by optical microscopy and Scanning Electron Microscopy (SEM) and hardness test.

In this study, two-dimensional static implicit finite element method, i.e., DEFORM-2D was used as a simulation tool. An Arrhenius type constitutive relation involving different solid fractions and temperatures has developed. This relation has been used to simulate semi-solid extrusion process in DEFORM-2D.

3 Results and Discussion

In general, densification and deformation conditions play a vital role in changing morphology and mechanical properties of powder metallurgy materials. All the samples were prepared with 90% IRD and with an aspect ratio of one

(φ 15 × 15 mm) by applying trial and error method. Al–4Cu–0.5Mg green compacted samples were sintered at 550 °C temperature and produced 90% sintered
density samples with the density of 2.48 g/cc (theoretical density: 2.76 g/cc).
Micro-hardness of sintered samples measured using Vickers micro-hardness testing
machine by applying 500 g load. The micro-hardness value of sintered samples was
observed as 647 ± 20 MPa. Sintered samples were then gone foe TG/DTA (Fig. 1)
analysis. Solidus and liquidus temperatures have to be identified to accomplish the
semi-solid state extrusion at various temperatures. Solidus and liquidus temperatures of sintered Al–4Cu–0.5Mg alloy were identified as 542.7 and 662.8 °C
respectively. Deformation temperature and solid fraction play vital role in
semi-solid extrusion process. Scheil equation was used to identify the solid fractions of an alloy at any given temperature. Three working temperatures were
selected in between the solidus and liquidus temperature range, those are 550, 575
and 600 °C respectively. The solid fractions at these temperatures were computed
as 0.96, 0.79 and 0.58 respectively from the below equation [10].

$$f_s = 1 - (T_s - T/T_s - T_l)^{(1/1-k)},$$

where k = Partition coefficient = 0.17
  T_l = Liquidus temperature
  T_s = Solidus temperature

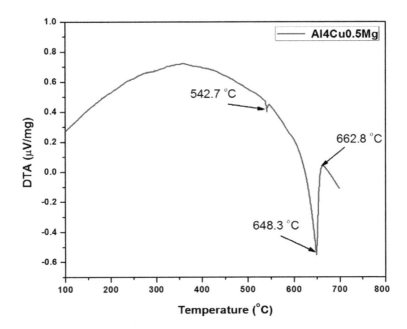

Fig. 1 TG/DTA analysis of Al–4Cu–0.5Mg

The samples prepared with 90% IRD were then subjected to extrusion process and extruded at three optimized temperatures and solidus fractions with dies having low extrusion ratio (1.44) and die approach angle of 30°, 45°, and 60°, respectively at low strain rate of 0.1 s^{-1} (Fig. 2). The propagation of cracks on the surface and edge of the sample have resulted in the samples extruded at high temperature and liquid fraction which can be seen in Fig. 2c, f, i. Liquid segregation might be the main cause to propagate cracks on the surface of the samples extruded above 550 °C.

The microstructural analysis of semi-solid extruded samples at three different temperatures and three different die approach angles are shown in Figs. 3 and 4. During the extrusion process, solid and liquid fractions of an alloy play an important role in changing the mechanical and metallurgical properties of any alloy.

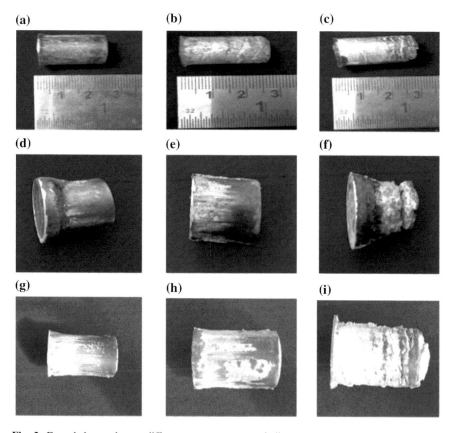

Fig. 2 Extruded samples at different temperatures and die approach angles, **a** 550 °C, 30°; **b** 575 °C, 30°; **c** 600 °C, 30°; **d** 550 °C, 45°; **e** 575 °C, 45°; **f** 600 °C, 45°; **g** 550 °C, 60°; **h** 575 °C, 60°; **i** 600 °C, 60°

(a)   (b)   (c)

(d)   (e)   (f)

(g)   (h)   (i)

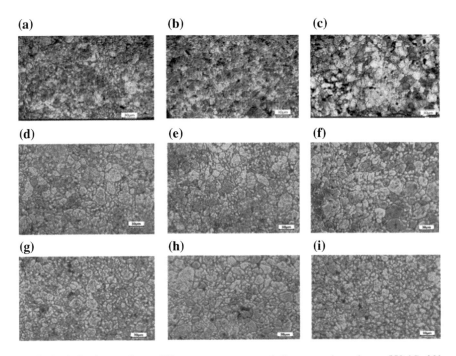

Fig. 3 Optical micrographs at different temperatures and die approach angles, **a** 550 °C, 30°; **b** 575 °C, 30°; **c** 600 °C, 30°; **d** 550 °C, 45°; **e** 575 °C, 45°; **f** 600 °C, 45°; **g** 550 °C, 60°; **h** 575 °C, 60°; **i** 600 °C, 60°

The liquid fraction of a material increases with the increasing extrusion temperature. The samples extruded at higher liquid fraction resulted the formation of uniform and globular grains. Coarser and nonuniform grains could be seen in samples extruded at 550 °C because of lower liquid fraction. The solid particles glide over the liquid when the liquid fraction is high in materials during the extrusion process and it provides uniform grain structure and grain boundaries. Grain size distribution has been changed with the changing extrusion temperature, the size of the grains has enhanced with liquid fraction increment. The size of the grains are less in the materials extruded at 550 °C compared to the materials extruded at 575 and 600 °C because of higher deformation at lower temperatures. Precipitate formation can be observed in case of SEM analysis. This precipitate has confirmed as Al_2Cu phase by XRD and EDS analyses which are not included in this paper. The precipitate formation has decreased with increasing extrusion temperature because of increment of liquid fraction with temperature. The strength and hardness of a material enhance with the formation of precipitate along the grain boundary which can be seen in SEM microstructures. The samples extruded at 550 °C have shown high precipitate formation compared to other two temperatures (Fig. 4a, d, g).

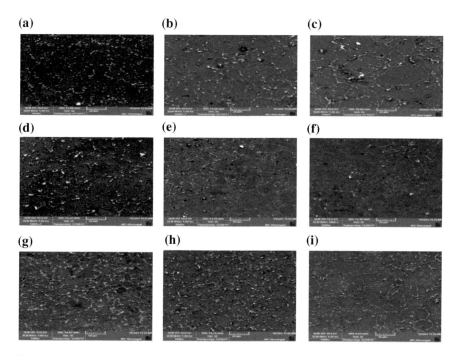

Fig. 4 SEM micrographs at different temperatures and die approach angles, **a** 550 °C, 30°; **b** 575 °C, 30°; **c** 600 °C, 30°; **d** 550 °C, 45°; **e** 575 °C, 45°; **f** 600 °C, 45°; **g** 550 °C, 60°; **h** 575 °C, 60°; **i** 600 °C, 60°

The density of extruded samples was measured by Archimedes principle. The samples extruded at 550 °C have shown higher densification in case of 30°, 40° and 60° approach angles which can be seen in Fig. 5. The liquid fraction of the samples extruded at 550 °C was less compared to the samples extruded at 575 and 600 °C. At 550 °C, the density values are high because of higher deformation with high solid fraction. The same trend has followed by micro-hardness profile shown in Fig. 6. Micro-hardness values of samples extruded at 550 °C are high as compared to the samples extruded at 575 and 600 °C.

The quality of a material can be analysed by extensive finite element studies at different solid and liquid fractions. FEM based simulations have conducted on Al–Cu–Mg alloys extruded in semi-solid state. The process parameters used for FEM studies are as follows:

    I. Die temperature (550, 575 and 600 °C)
    II. Strain rate (0.1 s^{-1})
    III. Extrusion ratio (1.44)
    IV. Die approach angle (30°, 45° and 60°)

Figure 7 shows the FEM model of semi-solid extrusion process. DEFORM-2D simulation software has been used as a tool to simulate the semi-solid extruded Al

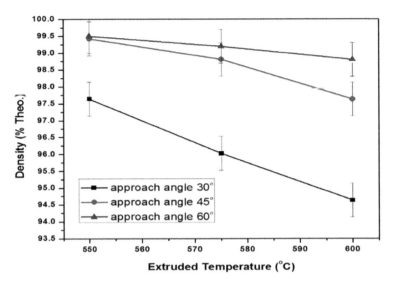

Fig. 5 Density graph of extruded samples at different temperatures and approach angles

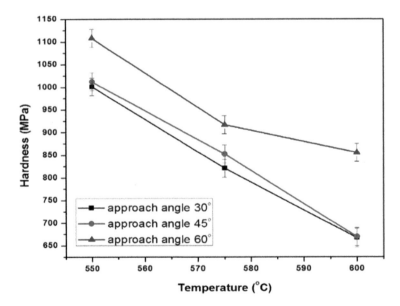

Fig. 6 Hardness graph of extruded samples at different temperatures and approach angles

alloy samples. The workpieces with cylindrical shape (φ 15 × 15 mm) were used for simulation studies. Axisymmetric analysis was used for the FEM simulation to decrease the simulation time. An Arrhenius type strain rate relation has used in all simulation studies i.e., $\acute{\varepsilon} = A[\sinh(\alpha\sigma)]^{n}\exp[-Q/RT]$ [10]. The workpiece was set

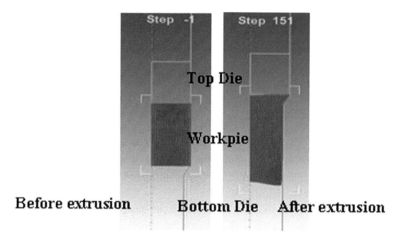

Fig. 7 Finite element simulation model (DEFORM-2D)

as porous for all three temperatures and die approach angles. The bottom and top die were set as rigid. Constant shear friction was considered between die and workpiece and coefficient of friction $m = 0.4$ was assumed between top die to workpiece and workpiece to bottom die in all simulations.

The experimental results have verified with the simulation results. Figure 8 shows the density of simulated samples at 550, 575 and 600 °C respectively with

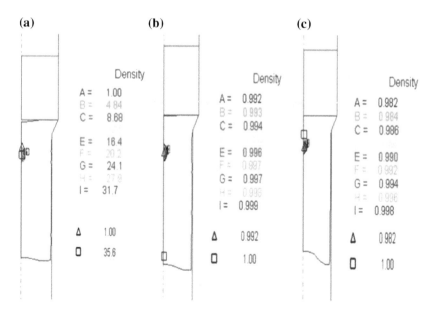

Fig. 8 Density of simulated samples extruded at **a** 550 °C, **b** 575 °C, **c** 600 °C with 30° approach angle

30° approach angle. The average density of simulated samples show the density values same as got in experimentation. The density of simulated samples extruded at 550 °C have shown high densification results compared to other simulated samples extruded at 575 and 600 °C respectively.

The distribution of strain is uniform in case of extrusion at 600 °C due to high liquid fraction compare to the other two temperatures (Fig. 9). The effective stress of a material depends on the solid and liquid fractions and it is high at higher solid fraction. At 550 °C, the effective stress in material is high and showed 38.5 MPa and at 600 °C, it is observed as 25 MPa which can be seen in Fig. 10. The lower stress distribution at 600 °C was observed because of the existence of liquid surrounded between the solid particles. The solid particles glide over the liquid medium and resulted as liquid flow.

The die approach angle makes easy to move the heated metal along the die wall. Frictional conditions will vary with varying die approach angle at the interface of the die wall and workpiece. Inhomogeneity in the metal flow and microstructures increases with the increasing die approach angle (Fig. 3). However, globular microstructure was found even at the 600 °C extrusion temperature because of gliding nature of solid particles in the liquid medium. High deformation has taken place in case of high solid fraction (i.e., semi-solid extrusion at 550 °C). Dynamic recrystallization has observed in materials extruded at 600 °C in which nucleation and grain growth have occured during deformation and it is facilitated by high angle

(a)   **(b)**   **(c)**

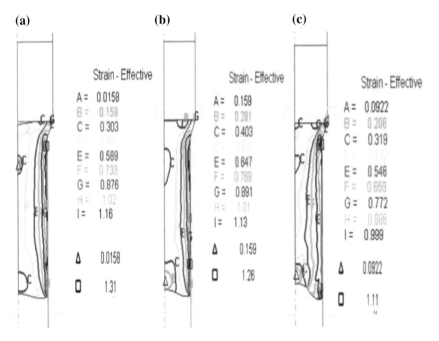

Fig. 9 Effective strain of simulated samples extruded at **a** 550 °C, **b** 575 °C, **c** 600 °C with 30° approach angle

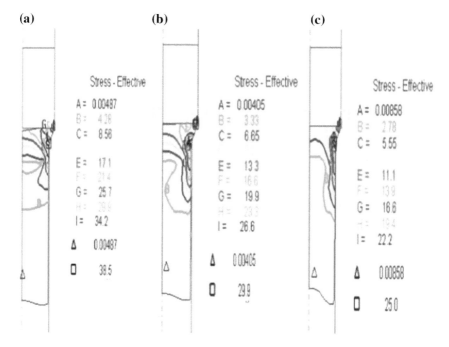

Fig. 10 Effective stress of simulated samples extruded at **a** 550 °C, **b** 575 °C, **c** 600 °C with 30° approach angle

grain boundaries. Dynamic recrystallization was occurred during deformation process so that the fine recrystallized grains are able to observe along the grain boundaries. The samples extruded at different conditions were cooled in atmosphere to provide complete recrystallization process. Grain boundaries became straight and clear, that might be due to transformation of the low-angle grain boundaries into high angle grain boundaries through the dislocation absorption in case of extrusion at 600 °C temperature. The high angle grain boundaries and migration of dislocations might be the reason for the growth of recrystallized grains.

Density and hardness (Figs. 5 and 6) of extruded materials have increased with increasing die approach angle because of high deformation. The same phenomenon has observed in simulation results which are shown in Fig. 11. The effective strain is observed to be more for the material extruded with 60° and less with die having 30° approach angle. However, the metal extruded with 30° approach angle was pressed stronger than the metal pressed with 60° approach angle. This indicates that, along the narrow die approach angle, higher solid fraction provide metal with higher strength. The variation in effective strain with die approach angle have shown in Fig. 12. The distribution of strain is high in case of samples extruded with 30° die approach angle compare to other two angles. The effective stress of a material has increased with increasing die approach angle at a given temperature

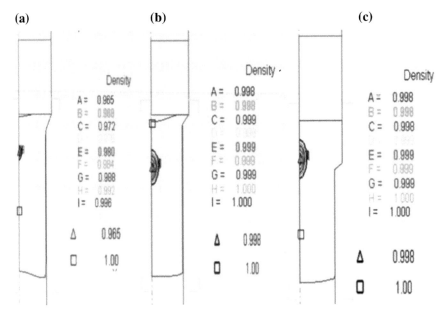

Fig. 11 Density of simulated samples extruded at 550 °C with **a** 30°, **b** 45°, **c** 60° approach angle

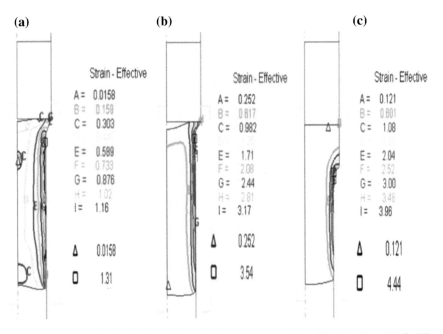

Fig. 12 Effective strain distribution of simulated samples extruded at 550 °C with **a** 30°, **b** 45°, **c** 60° approach angle

(a) **(b)** **(c)**

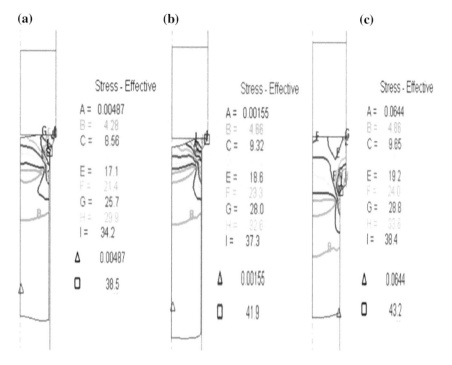

Fig. 13 Effective stress distribution of simulated samples extruded at 550 °C with **a** 30°, **b** 45°, **c** 60° approach angle

which can be seen in Fig. 13. Effective stress of a material extruded with 30° approach angle has shown as 25 and 43.2 MPa with 60° approach angle.

4 Conclusions

In the present investigation, experimental and FEM simulation studies have been performed on semi-solid extruded Al–Cu–Mg powder metallurgy alloys. Higher density and hardness values were observed in experimental and simulated alloys at high solid fraction, i.e., at 550 °C. Effective stress and strain distributions were high in case of high die approach angle at a given temperature. However, uniform distribution of properties could be produced with the combination of narrow die approach angle and high solid fraction. The dynamic recrystallization was observed in case of extrusion at 600 °C, i.e., with high liquid fraction.

Acknowledgements The authors are grateful to all the lab technicians for their help in carrying out various experiments.

References

1. Kaneko S, Murakami K, Sakai T (2009) Effect of the extrusion conditions on microstructure evolution of the extruded Al–Mg–Si–Cu alloy rods. Mat Sci Eng A 500:8–15
2. Salleh MS, Omar MZ, Syarif J, Mohammed MN (2012) An overview of semisolid processing of aluminium alloys. ISRN Mat Sci 2013·9
3. Sweet ED, Caraher SK, Danilova NV, Zhang X (2004) Proceedings of the eighth international aluminium extrusion technology seminar. Aluminum Association, Orlando, FL, USA, pp 115–126
4. Sheppard T (2006) Prediction of structure during shaped extrusion and subsequent static recrystallization during the solution soaking operation. J Mat Proc Tech 177:26–35
5. Parvizian F, Kayser T, Klusemann B, Svendsen B (2010) Modelling and simulation of dynamic microstructure evolution of aluminium alloys during thermomechanically coupled extrusion process. Int J Mater Form 3(suppl 1):363–366
6. McQueen HJ, Evangelista E, Bowles J, Crawford G (1984) Hot deformation and dynamic recrystallization of Al–5 Mg–0.8 Mn alloy. Metal Sci 18(8):395–402
7. Laue K, Stegner H (1981) Extrusion, American society for metals. Materials Park, 217
8. Storen S, Thomas Moe P, Totten GE, Mackenzie DS (eds) (2003) Handbook of aluminum, vol. 1 physical metallurgy and processes. New York, p 419
9. Van Geerteruyden WH, Brownie HM, Misiolek WZ, Wang PT (2004) Evolution of surface recrystallization during indirect extrusion of 6xxx aluminum alloys. Met Mat Trans A 36:1049–1056
10. Shashikanth CH, Davidson MJ (2014) Simulation studies on deformation behaviour of AA2017 alloy in semi-solid state using FEA. Mat High Temp 31:274–281

Mechanical and Tribological Properties of A356/Al$_2$O$_3$/MoS$_2$ Hybrid Composites Synthesized Through Combined Stir and Squeeze Casting

K. Sekar, M. Manohar and K. Jayakumar

Abstract Investigation of mechanical and tribological properties of A356 alloy reinforced with Al$_2$O$_3$ nano (average size 30 nm) and MoS$_2$ micro-particles (10 μm) are presented in this study. The percentage of Al$_2$O$_3$ was constant (1 wt%) and MoS$_2$ varied from 0.5 to 2 wt% with the interval of 0.5%. The ceramic particles were added when stirring at 300 rpm and 800 °C with squeeze casting pressure of 43 MPa in stir and squeeze casting machine. Characterization of A356/Al$_2$O$_3$/MoS$_2$ hybrid composites were conducted and Tribology studies on the samples were also investigated at 10, 30 and 50 N load, sliding speed of 1.3 m/s and sliding distance of 1100 m in dry condition. Microstructure analysis of the composite showed that the Al$_2$O$_3$ (1 wt%) and MoS$_2$ (0.5 and 1%) particles were dispersed uniformly in the A356 alloy matrix. Partial agglomeration was observed in the synthesized metal-matrix composite (MMC) with higher MoS$_2$ (2%) and Al$_2$O$_3$ (1%). The MMC containing 0.5 and 1 wt% of MoS$_2$ and Al$_2$O$_3$ (1 wt%) exhibited the higher bending strength, lesser wear, and coefficient of friction.

Keywords A356 alloy/Al$_2$O$_3$/MoS$_2$ hybrid MMCs · Stir and squeeze casting Microstructure analysis · Hardness · Bending strength and wear studies

K. Sekar
Department of Mechanical Engineering, National Institute of Technology,
Calicut, Kerala 673601, India
e-mail: sekar@nitc.ac.in

M. Manohar
MME, Vikram Sarabhai Space Centre, ISRO, Thiruvananthapuram 695022,
Kerala, India

K. Jayakumar (✉)
Department of Mechanical Engineering, SSN College of Engineering,
Kalavakkam, Chennai 603110, India
e-mail: kjayakumar@ssn.edu.in

© Springer Nature Singapore Pte Ltd. 2019
A. K. Lakshminarayanan et al. (eds.), *Advances in Materials and Metallurgy*,
Lecture Notes in Mechanical Engineering,
https://doi.org/10.1007/978-981-13-1780-4_13

1 Introduction

Among different aluminum alloys, A356 alloy is widely used in different industries like aircraft, aerospace, automobiles, electronics, and various other fields because of its low density, higher electrical and thermal conductivity and corrosion resistance. The present need for lightweight materials with high strength and stiffness have attracted much interest in the development of processes for developing aluminum-based metal-matrix composites (Al MMCs). Hence, an effort is made to synthesize A356 alloy reinforced with nano Al_2O_3 and micro MoS_2 particulate composites through squeeze casting with stirring is attempted.

Limited studies were carried out in synthesis of hybrid aluminum-based composites using combined stir and squeeze casting processes. K. Sekar et al. [1] conducted experimental studies on mechanical and tribological properties of A356 alloy-Al_2O_3 MMCs produced using combined stir and squeeze casting process. They concluded that among Al_2O_3 nano-particle based composites, 1wt% addition of Al_2O_3 gave better properties such as higher hardness, compressive strength, and lesser wear loss.

Miskovic et al. [2] did experimental study to analyze the structural and mechanical properties of an A356 alloy composite with Al_2O_3 particle additions and reported improvement in mechanical properties of the MMC in comparison to the base alloy. Suresh et al. [3] investigated on LM25 alloy reinforced with different wt% of micro- and nano-sized Al_2O_3 particles using stir casting. The composites synthesized were characterized for mechanical properties and distribution of Al_2O_3 particles. Results showed that the MMC with nano particles exhibited better hardness and strength compared to micro particles reinforced composites.

Synthesis of A356/nano-Al_2O_3 composites were proposed by Mazahery et al. [4] and they claimed that presence of nano-Al_2O_3 particle leads to improvement in hardness, ultimate tensile strength (UTS) and ductility. The properties of nano Al_2O_3 particulate-reinforced aluminum based composites was studied by Yung-Chang Kang and Lap-Ip Chan [5] and reported that the hardness and tensile behaviors of aluminum MMCs reinforced with Al_2O_3 particle have been found to be increased with increase of reinforcement.

Vinoth et al. [6] presented a study on mechanical and wear characteristics of stir-cast Al-Si10 Mg and Al-Si10 Mg/MoS_2 composites. They concluded that the UTS, hardness and elongation percentage were decreased with the addition of self lubricating MoS_2 particles to Al-Si10 Mg alloy. Also the wear rate showed an enormous decrease for MoS_2 addition due to the presence of a MoS_2 layer, which forms a film on the wear surface. Al-Qutub et al. [7] carried experimental studies on wear and friction of Al-Al_2O_3 composites at various sliding speeds. The results illustrated that higher load and concentration of Al_2O_3 particles lead to higher wear rates. Winer [8] reported that the low friction and easy cleavage of MoS_2 is intrinsic to the material crystal structure and condensable vapor in the review of MoS_2 as a lubricant.

From the literature, difficulties faced during synthesis of particulate MMCs by liquid phase processes are identified as uniform mixing of reinforcement in the matrix with no sinking, floating and wettability of particle reinforcement in the matrix with minimal porosity and higher density. To overcome these challenges while adding nano and micro particles on molten matrix metal, stirring with squeeze casting procedure is implemented in the present study. Moreover, past literature discussed the mechanical properties of MMCs in terms of microstructure and mechanical properties. Therefore, the aim of the present study is to analyze the microstructure, mechanical and tribological properties of the A356 alloy reinforced with Al$_2$O$_3$ nano and MoS$_2$ micro particles hybrid MMC.

2 Experimentation

2.1 Materials and Manufacturing Process

The matrix alloy used for this study is A356 alloy and Al$_2$O$_3$ particles of 30 nm size and MoS$_2$ particles size of 10 μm were added as the reinforcement on the matrix. The A356 alloy billets were melted in electric furnace at 800 °C. After complete melting of base alloy, the stirrer was brought down into the molten metal and stirred at a constant speed of 300 rpm. Preheated nano Al$_2$O$_3$ (900 °C) and MoS$_2$ (250 °C) powders were gradually added to melt to improve the wettability in the molten base metal. The stirring action was carried for 10 min to improve uniform dispersion of two different reinforcements. After melting, bottom pouring valve of the furnace was operated to pour the molten metal into the die steel mold. Then, squeeze punch was concurrently activated to squeeze the molten composite materials and is shown in Fig. 1 with component number 6. After squeezing the molten metal, the punch and die setup were cooled slowly to enhance the mechanical properties of the cast composites by the cooling effect. The final blank has diameter of 46 mm and length 260 mm after stir and squeeze casting and is shown in Fig. 1b. Figure 1a shows the newly designed combined stir and squeeze casting machine which was used in the current work.

2.2 Specimens Preparation

Different samples from the processed composites were prepared for microstructural analysis, hardness, bending strength and wear test. As per ASTM E3 standard, the fine-polished samples were etched using Keller's reagent. Microstructure analysis was carried out using scanning electron microscopy (SEM). Hardness test was conducted in Rockwell machine and Bending tests were conducted on the samples using 200 kN capacity universal testing machine (UTM). Bending test sample sizes

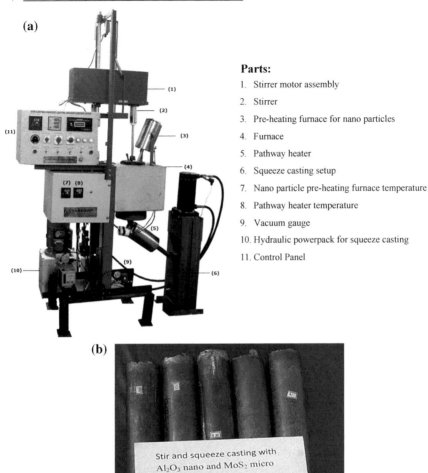

Fig. 1 a Stir and squeeze casting machine, **b** hybrid composites blanks

of 10 mm in diameter and 100 mm in length were prepared as per ASTM B769-11 standards. Wear tests were conducted on the composite specimen using a pin-on-disc tribometer and the sample size was 8 mm in diameter and 27 mm in length as per ASTM G-99 standard. After wear test, worn out surfaces were examined by SEM to understand the wear mechanisms of the composites.

3   Results and Discussions

3.1   *Microstructure Analysis of A356 Alloy/Al2O3/MoS2 Composites*

From the optical micrographs shown in Fig. 2a–e, it is observed that the fine grain structure for the stir and squeeze cast specimen due to the use of squeeze casting pressure and higher heat transfer coefficient. By adding reinforcement particles, the grain size reduces because the particles act as nucleation sites. It was also found that at some clustering of MoS$_2$ particulates as shown in Fig. 2e.

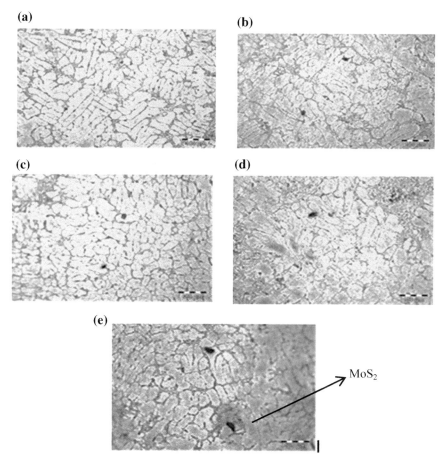

Fig. 2 Optical micrographs of A356 alloy and MMCs at 200X: **a** A356 alloy, **b** MMC-1% Al$_2$O$_3$ & 0.5% MoS$_2$, **c** MMC-1% Al$_2$O$_3$ & 1% MoS$_2$, **d** MMC-1% Al$_2$O$_3$ & 1.5% MoS$_2$, **e** MMC-1% Al$_2$O$_3$ & 2% MoS$_2$

By effective squeezing process, breaking up of the agglomerates and redistribution of the reinforced particles were occurred. This occurred when the amount of liquid inside the slurry was large enough and the particles slide over each other and breaking up occurs. At higher % of reinforcements, the pressure applied was not enough to break up all the MoS_2 clusters, therefore clusters were retained in it after squeezing. The light colored region in the microstructure is the Aluminum matrix and the gray region is the silicon (Si). The large size of Si phase will degrade the strength of the castings made due to its brittle nature. Due to the addition of magnesium, $MgAl_2O_4$ has been formed in the composite and was located in the interface of the matrix and reinforcement. Good wettability and interfacial bonding were achieved by pre-heating of Al_2O_3 and MoS_2 particulates before adding to matrix.

In order to see the reinforcement's distribution in microscopic levels, SEM analysis was carried out and is shown in Fig. 3. From the images (Fig. 3b–e), it is

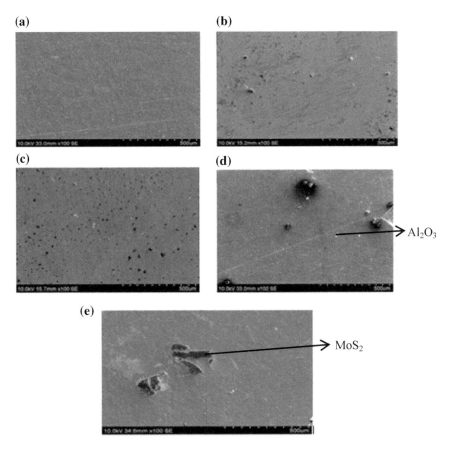

Fig. 3 SEM images: **a** A356 alloy, **b** MMC-1% Al_2O_3 & 0.5% MoS_2, **c** MMC-1% Al_2O_3 & 1% MoS_2, **d** MMC-1% Al_2O_3 & 1.5% MoS_2, **e** MMC-1% Al_2O_3 & 2% MoS_2

observed that, white colored Al_2O_3 particles were well dispersed in all composites. Black colored particles are MoS_2 and clustering effect of MoS_2 is more in composite with higher % of MoS_2 which is shown in Fig. 3e.

3.2 Hardness

From Fig. 4, as Al_2O_3 being a hard phase, hardness of the composite is increased with its addition. An increase in hardness of the order of 10% was noticed with 1% addition of Al_2O_3 and 0.5% of MoS_2 and this may be due to uniform distribution of reinforcement and resistance to deformation.

By adding 1% of MoS_2, hardness was found to be decreased and with further increase in wt% addition of MoS_2, the tendency was observed as marginal from 57 HRB to 53.4 HRB. It is due to the soft MoS_2 reinforcement causing easing the movement of grains along the slip planes rendering the composites material easily deformable under load and also clustering effect of ceramic particles within the matrix [9].

3.3 Bending Strength

Figure 5 shows the variation of bending strength for the different prepared samples. By adding 0.5 and 1% of MoS_2 particles to aluminum alloy, the bending strength is increased by 8% compared to A356 alloy. Bending strength is increased because of the strong bond due to cohesion between reinforcement and matrix to carry the load applied to the material. Beyond 1% MoS_2 bending strength showed a decrease in

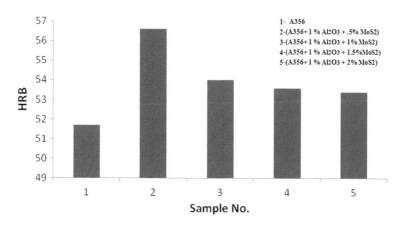

Fig. 4 Variation of hardness for different samples

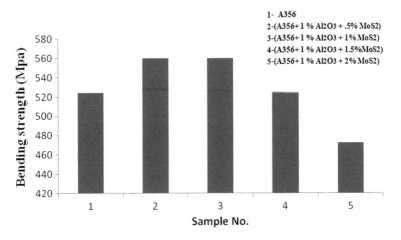

Fig. 5 Variation of bending strength for MMCs

trends. This may be due to decrease in wettability with increase in particle addition which eventually leads to low interfacial and bending strength [10].

3.4 Wear Test

Nowadays, Al based MMCs are used in brake liners and drums due its wear resistance and low coefficient of thermal expansion. Therefore, dry wear test analysis was carried out on the prepared composites to check its wear resistance. The sliding wear behavior of the A356 alloy and composites was carried out with different loads (10, 30 and 50 N) at a constant sliding speed (1.3 m/s) and sliding distance (1100 m). From Figs. 6 and 7, the wear and coefficient of friction (COF) are more at a higher load in Al alloy and composites.

Hybrid composite with 2% MoS_2 shows an enormous decrease in the wear by 78% compared to A356 alloy. Wear loss was reduced by 37, 46, 67.8, and 78% with 0.5, 1, 1.5, and 2% MoS_2, respectively compared to A356 alloy. The wear loss and coefficient is decreased, due to addition of MoS_2 and reason for this is due to the formation of MoS_2-lubricated tribolayer between the contact of prepared specimens and wear test disk materials [11].

3.5 Worn Surface Morphology

In order to understand the wear mechanism of the processes composites, worn surface morphology of the base alloy and composites were examined by SEM and is shown in

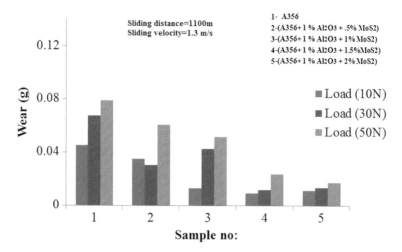

Fig. 6 Variation of wear for different samples

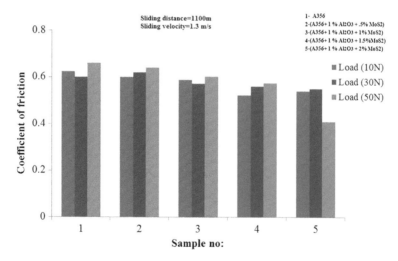

Fig. 7 Variation of coefficient of friction MMCs

Fig. 8a–e. Among Fig. 8a–e, one can see the appearance of the brittle fractures in Fig. 8e due to sliding, the nucleation and propagation of cracks within MoS_2 and Al_2O_3 particles and matrix alloy occurred causing the brittle fracture. It can be observed that increasing of applied load resulted in the increased proportion of the contact surface which was not protected by mechanically mixed layer and increased extent of surface damage (Fig. 8b, c). SEM analysis of worn surfaces of pin shows fine wear debris, grooves and oxide layers on worn out surface confirming abrasion, cracks, and delamination as the dominant wear mechanism involved [11].

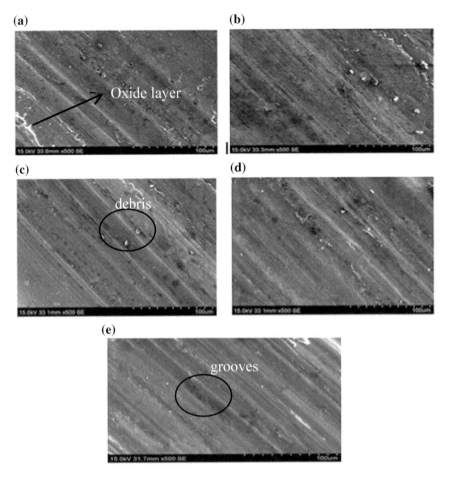

Fig. 8 Worn surface images at 50 N applied load: **a** A356 alloy, **b** MMC-1% Al_2O_3 & 0.5% MoS_2, **c** MMC-1% Al_2O_3 & 1% MoS_2, **d** MMC-1% Al_2O_3 & 1.5% MoS_2, **e** MMC-1% Al_2O_3 & 2% MoS_2

4   Conclusions

Based on the results obtained from the present work, the following conclusions are drawn:

1. Aluminum-based hybrid composites have been successfully fabricated using combined stir and squeeze casting technique. From the microstructure analysis, the uniform distribution and better wettability of reinforcements on the matrix were observed.
2. Reinforcement improved the hardness and bending strength to an extent of wt% addition of reinforcement when compared with base alloy.

3. The incorporation of MoS$_2$ particles in the base metal as a second reinforcement reduced the wear loss of the hybrid MMCs compared to composite with Al$_2$O$_3$ alone. Considerable improvement in wear resistance of about 78% was found with 2% addition of MoS$_2$ particles. Coefficient of friction of hybrid composites was decreased with increase in % of reinforcements due to formation of MoS$_2$ tribolayer at sliding interface.

References

1. Sekar K, Allesu K, Joseph MA (2015) Mechanical and wear properties of Al–Al$_2$O$_3$ metal matrix composites fabricated by the combined effect of stir and squeeze casting method. Trans Indian Inst Met 68:115–121
2. Miskovic Z, Bobic I, Tripkovic S, Rac A, Vencl A (2006) The structure and mechanical properties of an Aluminium A356 alloy base composite with Al$_2$O$_3$ particle additions. Tribol Ind 28:3–9
3. Suresh SM, Mishra D, Srinivasan A, Arunachalam RM, Sasikumar R (2011) Production and characterization of micro and nano Al$_2$O$_3$p reinforced LM25 aluminium alloy composites. ARPN J Eng Appl Sci 6:94–98
4. Mazahery A, Abdizadeh H, Baharvandi HR (2009) Development of high-performance A356/ nano-Al$_2$O$_3$ composites. Mater Sci Eng A 518:61–64
5. Kang YC, Chan SLI (2004) Tensile properties of nanometric Al$_2$O$_3$ particulate-reinforced aluminum matrix composites. Mater Chem Phys 85:438–443
6. Vinoth KS, Subramanian R, Dharmalingam S, Anandavel B (2012) Mechanical and tribological characteristics of stir-cast Al-Si10 Mg and self-lubricating Al-Si10 Mg/ MoS$_2$ composites. Mater Technol 46:497–501
7. Al-Qutub AM, Ibrahim M, Allam MA, Samad M (2008) Wear and friction of Al-Al$_2$O$_3$ composites at various sliding speeds. J Mater Sci 43:5797–5803
8. Winer WO (1967) Molybdenum disulphide as a lubricant: a review of the fundamental knowledge. Wear 10:422–452
9. Pitchayyapillai G, Seenikannan P, Raja K, Chandrasekaran K (2016) Al6061 hybrid metal matrix composite reinforced with alumina and molybdenum disulphide. Adv Mater Sci Eng 2016:1–9
10. Demir A, Altinkok N (2004) Effect of gas pressure infiltration on microstructure and bending strength of porous Al$_2$O$_3$/SiC-reinforced aluminium matrix composites. Compos Sci Technol 64:2067–2074
11. Monikandan VV, Joseph MA, Rajendrakumar PK (2016) Dry sliding wear studies of aluminum matrix hybrid composites. Resour Efficient Technol 2:S12–S24

Synthesis and Characterization of TiB$_2$–SiC Ceramic Composite Produced Through Spark Plasma Sintering

P. G. HariKrishnan and K. Jayakumar

Abstract The ceramic matrix is quite rigid and strong, but its fracture toughness has to be increased in order to fully realize its potential possibilities. This difficulty can be resolved by developing ceramic matrix composites (CMCs). CMCs have been processed to realize quasi-ductile fracture behavior and advantages of monolithic ceramics at high temperature. From different CMCs, TiB$_2$–SiC CMC is used in automotive brakes, cutting tools, propulsion engine exhaust, etc. In the present study, TiB$_2$–SiC CMCs with varying SiC particle reinforcement of 0, 5, 10, 15 vol.% were synthesized using powder metallurgy (P/M) consolidation method with the help of spark plasma sintering (SPS) furnace at IIT, Madras. Raw materials particles sizes are TiB$_2$ (average size of 14 µm-matrix) and SiC (average size of 1 µm-reinforcement). SPS process parameters used were sintering temperature 1450 °C, 40 MPa pressure with 10 min as hold off time. Microstructural analysis was carried out using scanning electron microscope (SEM) to observe the homogeneous distribution of reinforcement over the matrix. From the characterization studies, the CMC specimen with 15% SiC gave a good fracture toughness of 6.3 MPa√m and vickers hardness of 22.1 GPa.

Keywords TiB$_2$–SiC ceramic matrix composite · Spark plasma sintering Microstructure analysis · Hardness · Fracture toughness

1 Introduction

A ceramic material is an inorganic, non-metallic solid material, which may consist of metal, non-metal, or metalloids. The ceramic is quite rigid and strong, but its fracture toughness is very poor. The problem can be tackled by developing CMCs. Composites produced have many desirable characteristics such as its superior

P. G. HariKrishnan · K. Jayakumar (✉)
Department of Mechanical Engineering, SSN College of Engineering,
Kalavakkam, Chennai 603110, India
e-mail: kjayakumar@ssn.edu.in

© Springer Nature Singapore Pte Ltd. 2019
A. K. Lakshminarayanan et al. (eds.), *Advances in Materials and Metallurgy*,
Lecture Notes in Mechanical Engineering,
https://doi.org/10.1007/978-981-13-1780-4_14

strength compared to its individual parent components. A composite helps to improve the property of one parent material with aid of other, thus resulting in a compound that have more strengths in its properties than weakness compared to the properties of their parent material.

CMCs can be processed by several methods which includes: Hot Pressing, Inductive Heating, Indirect Resistance Heating, Field Assisted Sintering Technique, Spark Plasma Sintering (SPS), etc. Out of these methods. SPS sinters the given powders close to theoretical density at lower sintering temperature compared to conventional sintering processes. Also, the heat generation is internal, in contrast to the conventional hot pressing, where the heat is given by external heating sources. This facilitates high heating (up to 727 °C/min) and sintering in faster rate (few minutes).

SiC is a strong, hard, and chemically inert which is used in several applications. In armor applications, the high hardness of SiC, which is in the range of 20–27 GPa, is advantageous for projectile defeat. However, SiC is brittle due to its low fracture toughness (2–5 MPa$\sqrt{m}$), whereas high fracture toughness is beneficial for multi-hit capability. However, an increase in toughness is typically accompanied by a drop in hardness. TiB_2 exhibits a high hardness (25–35 GPa) and may help combat the hardness/toughness trade off when added as a reinforcing phase in SiC. TiB_2 ceramics are potential candidates for heat engine applications. However, they have a low fracture toughness value, which causes them to suffer from damage under contact stress. Hence, most important problems of SiC particle is to improve their fracture toughness. Recent survey have showed that the addition of SiC_p is successful in raising the fracture toughness of TiB_2.

Vasanthakumar and Bakshi [1] recently in 2018 used SPS for processing C/Ti CMC. They found that the vickers hardness and elastic modulus were shown to increase significantly with increase in C/Ti ratio. Karthiselva et al. [2] synthesized CMC from ZrB_2 and TiB_2 through SPS process. They achieved maximum densification of 98% at 1100 °C heating with 2 stages. A experimental study was carried on sintering of SiC/Carbon fiber using SPS [3]. Results from the study were: Density of $97 \pm 0.8\%$ and mechanical properties (bending strength of 427 ± 26 MPa, the hardness of 2992 ± 33 Vickers, and fracture toughness of 4.2 ± 0.3 MPa$\sqrt{m}$) were obtained for SiC/C samples sintered at 2200 °C. Researchers have used SPS for synthesize of multi wall CNT + Alumina (50% + 50%) combined CMCs [4]. They found optimum sintering parameter: heating rate 383 °C/min; sintering time 3 min; holding temperature 1600 °C. Use of the optimum sintering parameter produced a nanocomposite with high densification and ultimate flexural strength of 485 MPa. Low temperature sintering of TiB_2 compacts was carried out using SPS by Karthiselva et al. [5]. They concluded that high nano-hardness above 28 GPa and elastic modulus above 570 GPa were obtained from the processed CMC. Hence from the available literature, processing of CMCs from SPS process is still in developing stage and standardization of SPS process parameters for different CMCs is still in research level. Therefore, current work focused on synthesize and characterization of TiB_2 + SiC ceramic composites through SPS process for getting higher density and improving mechanical properties of TiB_2.

2 Experimental Procedure

2.1 Specimen Preparation

TiB$_2$–SiC CMC with varying SiC reinforcement of 0, 5, 10, 15 vol.% are the specimens considered in this experiment. Raw materials particles sizes are TiB$_2$ (average size of 14 μm-matrix) and SiC (average size of 1 μm-reinforcement). The composite is made using powder metallurgy powder consolidation method. The consolidation method used is SPS. First the theoretical densities and the theoretical weight are found out in order to have a benchmark or base on which the density calculation of the prepared specimen can be calculated. The weights of the composite are found out and the weights of the individual powders are found out. The powders are individually weighed and mixed. The powders are mixed in a planetary mill for 3 h. Wet milling is used for mixing TiB$_2$ and SiC. Wet milling is done to avoid any chance of powders breaking down and forming an oxide. Isopropyl alcohol was used as the medium for wet milling. Titanium carbide ceramic balls are used in the mixing of the powders. The ratio of number of balls to the weight of powder is 1:10, i.e. one ball for every 10 g of the powder. The powders are milled for 3 h at 650 rpm. The milled powders form a slurry which is transferred to a alumina crucible where it is dried either by keeping it in atmosphere under room temperature or in a drier for fast drying.

2.2 Sintering of Specimens

The required amount of powder is weighed carefully using an electronic balance. The weighed powder is poured into the high density graphite die.

The powder is added into the die and closed with the upper punch and a little squeeze. The punches and die assembly is carefully loaded into the vacuum chamber of the SPS furnace which is shown in Fig. 1.

The axis of punch and die assembly should coincide with the axis of the punches of the SPS furnace so that the heating is uniform. The heating is monitored using a thermocouple. The sintering is done at 1100 °C for 40 MPa pressure and hold off time of 10 min. SPS sintering parameters were chosen based on references 2 and 5. The process are carefully monitored, where system information such as voltage, current etc. can be monitored and process parameters such as the pressure, temperature and displacement of the punches can be monitored. Figure 2 shows the processed CMCs billets and its dimension are 25 × 10 mm. Density of the processed composites was measured using Archimedes principle and the maximum value obtained as 90%.

Fig. 1 Die and punch
assembly inside the SPS
furnace

Fig. 2 Processed CMCs

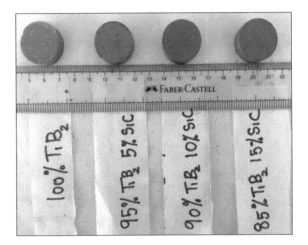

2.3   Characterization of Specimens

The testing of the specimen can be categorized into 2, mechanical testing and
structural analysis. For mechanical testing properties such as the hardness and
fracture toughness of the specimen were measured. For structural analysis SEM
imaging is used to identify the distribution of the reinforcement particles in the
matrix. Hardness testing was conducted using Vickers micro hardness testing
equipment with loading force of 1 kg. The crack the indentation produced was used
for measuring fracture toughness of the material in units of $N/m^{1.5}$ using Ansitis [6]
equation.

$$K_c = 0.016(E/H_v)^\wedge 0.5\left(P/\left(C^{1.5}\right)\right), \tag{1}$$

where K_c is the fracture toughness, E its elastic modulus, H_v the measured vickers hardness of material at that specific point, P is the applied pressure which is 1 kg, C is the crack length.

3 Results and Discussions

3.1 Microstructure Analysis

Figure 3a–d shows the SEM images of the processes specimens. SEM image gives an idea about the distribution of the reinforcements in the matrix and analyze the microscopic properties of the specimen. From the SEM image of the CMC (Fig. 3b) with 2000X, it can be seen that the TiB$_2$ got sintered completely and formed an uniform mass with little bit of SiC reinforcement here and there Fig. 3c, d clearly shows that SiC getting attached together to the plate like TiB$_2$ particles. In the images, SiC are in white and gives a fungi like structure on the TiB$_2$ matrix reducing the porosity in the specimen.

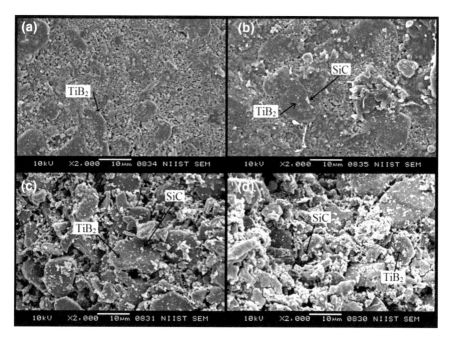

Fig. 3 SEM image of CMC at 2000X magnification, **a** 100% TiB$_2$, **b** 95% TiB$_2$ + 5% SiC, **c** 90% TiB$_2$ + 10% SiC, **d** 85% TiB$_2$ + 15% SiC

From the SEM images it can be said that the increasing the amount of SiC reinforcement in the TiB$_2$ matrix the porosity in the CMC is reduced, increasing the density of the specimen. This helped in achieving the acceptable densification of a CMC.

3.2 Hardness

Hardness Test was carried out at Sri chittira thirunal center for medical sciences, Thiruvananthapuram Kerala and it was further used for calculation of fracture toughness. Using a digital vickers hardness machine of maximum 1 kg was used to determine the hardness of the material. A diamond indenter was used for indentation and a 500X magnification microscope was used for viewing the indentation. The hardness was obtained by adjusting a red rectangular box over the diagonals of the indentation which is shown in Fig. 4a. Figure 4b shows the crack formed during measurement for sample 1 (100% TiB$_2$). Using the crack length, vickers hardens number and applied load, fracture toughness is calculated using Eq. 1.

In Fig. 5 the Vickers hardness Vs the specimen have been plotted, where the specimen 1 is being 0% SiC, 2–5% SiC, 3–10% SiC, and 4–15% SiC as reinforcement in TiB$_2$ matrix.

From the graph 5, it can be observed that the hardness of the CMC is decreased with increase in the amount of reinforcement. From the standard hardness values of 27 GPa for SiC and 33 GPa for TiB$_2$, a rule of mixtures calculation can be used to estimate the decrease in the hardness with increase in SiC$_p$ reinforcement.

Also, the fall in the hardness with increasing SiC quantity indicates a reduction in the work of indentation which may be associated to the distribution of areas with residual tensile stress [7].

 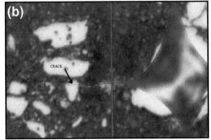

Fig. 4 **a** Indentation spot marked using red colored rectangle (500X). **b** Crack formation during hardness testing

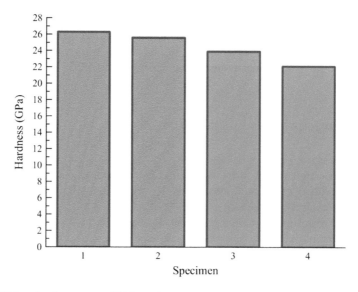

Fig. 5 Vickers hardness versus CMC specimens

3.3 Fracture Toughness

In Fig. 6, there is an increase in the fracture toughness of the specimen compared to pure TiB$_2$ with 5 MPa√m, addition of 15% SiC to the matrix gave rise to increase in fracture toughness to 6.3 MPa√m. The change/increase in number of SiC particles appeared to have affected the measured toughness due to the residual compressive stress generated in the TiB$_2$ matrix and the particles resisting to crack propagation [8].

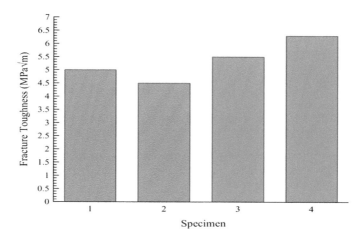

Fig. 6 Fracture toughness versus CMC specimens

In SiC–TiB_2 ceramic composites, crack deflection is the major toughening mechanism. Indeed, as a crack propagates through a TiB_2 matrix that is in compression, the radial tensile stress in the SiC_p draws the cracks, which reduced the driving force for crack transmission in the tensile regions [9].

4 Conclusions

- Study on characteristics on TiB_2 + SiC CMC is carried out with TiB_2 as the matrix and SiC as the reinforcement with varying 0, 5, 10, and 15 vol.% synthesized using SPS method.
- The specimens obtained have good densification of 83–90%. As the amount of reinforcement increased the densification also increased with 15% SiC as reinforcement giving the maximum densification, by occupying the vacant spaces in TiB_2 matrix.
- While testing for Vickers hardness in all four specimens it gave a declining graph as there is an increase in the reinforcement. And for fracture toughness it is observed that there is an increase in the fracture toughness of all four specimen as there was an increase in the SiC reinforcement vol.%. Overall, the specimen with 15% SiC gave good fracture toughness of 6.3 MPa√m and hardness of 22.1 GPa.

Acknowledgements Authors would like to thank Prof. B. S. Murty and Prof. Srinivasa R. Bakshi of IIT Madras for providing the SPS facility for carrying out the work.

References

1. Vasanthakumar K, Bakshi SR (2018) Effect of C/Ti ratio on densification, microstructure and mechanical properties of TiCx prepared by reactive spark plasma sintering. Ceram Int 44:484–494
2. Karthiselva NS, Kashyap S, Yadav D, Murty BS, Bakshi SR (2017) Densification mechanisms during reactive spark plasma sintering of Titanium diboride and Zirconium diboride. J Philos Mag 97:1588–1609
3. Ghasali E, Alizadeh M, Pakseresht AH, Ebadzadeh T (2017) Preparation of silicon carbide/carbon fiber composites through high-temperature spark plasma sintering. J Asian Ceram Soc 5:472–478
4. Fedosova NA, Kol'tsova EM, Popova NA, Zharikov EV, Lukin ES (2016) Ceramic matrix composites Reinforced with carbon nanotubes: spark plasma sintering, modeling, optimization. Refract Indus Ceram 56:636–640
5. Karthiselva NS, Murty BS, Bakshi SR (2015) Low temperature synthesis of dense TiB_2 compacts by reaction spark plasma sintering. Int J Refract Met Hard Mater 48:201–210
6. Anstis GR, Chantikul P, Lawn BR, Marshall DB (1981) A critical evaluation of indentation techniques for measuring fracture toughness: I, direct crack measurements. J Am Ceram Soc 64:533–538

7. ASTM International (2008) Standard test method for knoop indentation hardness of advanced ceramics. ASTM Standard C 1326-08E1, West Conshohocken, PA
8. Fahrenholtz WG, Neuman EW, Brown-Shaklee HJ, Hilmas GE (2010) Super hard boride-carbide particulate composites. J Am Ceram Soc 93:3580–3583
9. King DS, Fahrenholtz WG, Hilmas GE (2013) Silicon carbide–titanium diboride ceramic composites. J European Ceram Soc 33:2943–2951

Finite Element Analysis of High-Speed Machining of CFRP Material

K. Gobivel, K. S. Vijay Sekar and G. Prabhakaran

Abstract In recent time, carbon fiber reinforced plastics (CFRP) are used in important applications in aerospace, automobile, sporting equipment's, biomedical instruments, etc. High-speed machining of this material can regulate the cutting conditions to maximize production output. Investment for this process is at minimal cost which has been the generic aim of manufacturing industries all over the world. In order to accomplish this, Finite Element (FE) models have been developed for critical applications. Such numerical analysis negates the need for exhaustive experimental trials needed to estimate various parameters involving in machining. This has been the reason instrumental in coercing industries to resort to FE analysis for simulating cutting processes. The present work aims to assess and validate the deformation behavior of carbon fiber reinforced epoxy composite during high speed machining. Orthogonal turning was performed for varied cutting conditions by varying cutting speed and feed at a constant depth of cut. An FE model was constructed using ABAQUS V6.13 and the effective stress–strain response and deformation were analyzed. The simulated results for cutting force, thrust force and feed force showed good correlation with experiments.

Keywords CFRP · High-speed machining · Finite element modelling
Simulation

K. Gobivel (✉) · G. Prabhakaran
Department of Mechanical Engineering, KCG College of Technology, Chennai, India
e-mail: gobivel@gmail.com

G. Prabhakaran
e-mail: g_prabha2006@yahoo.com

K. S. Vijay Sekar
Department of Mechanical Engineering, SSN College of Engineering, Chennai, India
e-mail: vijaysekarks@ssn.edu.in

© Springer Nature Singapore Pte Ltd. 2019
A. K. Lakshminarayanan et al. (eds.), *Advances in Materials and Metallurgy*,
Lecture Notes in Mechanical Engineering,
https://doi.org/10.1007/978-981-13-1780-4_15

1 Introduction

Carbon fiber reinforced plastics (CFRP) composites have been used in the applications of structural components in passenger aircrafts, hybrid cars, sporting and biomedical instruments due to its excellent mechanical properties, corrosion and thermal resistant and durability. Because of its inhomogeneous, the problems associated with cutting composites and their interaction with the cutting tool is a complex process. Experimental studies of FRP composites have been extensively carried out in the past decades to assess cutting forces, surface integrity, tool wear, and chip morphology.

On the other hand, approaches for Finite Element Modeling (FEM) offers possible ways to do experimental trials in simulation to avoid technical problems and associated costs. Arola and Ramulu [1] were the first to investigate the numerical model for the orthogonal cutting of FRP using the Finite Element Method. The simplified two-dimensional models for the trimming process of FRP predicted the principal cutting forces that agreed well with experimental results.

Mahdi and Zhang [2] presented a paper to predict the cutting force of composites with respect to fiber angle in 2D model. Also the chip separation criterion and the properties assigned to the material were particularly addressed. Rao et al. [3–5] investigated micro and macro mechanical modelling approach to study the simulation response. Nayak et al. [6] carried out FE analysis of orthogonal machining of UD-GFRP to understand the tool geometry, orientation and subsurface damage. Santiuste et al. [7] investigated the modelling of long fiber composite material in GFRP and CFRP for aeronautical components. Hashin proposed the composite material failure criteria which include four failure modes: tensile fiber failure, compressive fiber failure, tensile matrix failure and compressive matrix failure. Calzada et al. [8] investigated the modelling of CFRP and interpret the fiber orientation based failure mechanisms in the machining process. A new approach to interfacial modelling was introduced where the material interface was modeled using continuum elements, allowing failure to take place in either tension or compression.

Ali Mkadeem et al. [9] investigated the combined approach of micro-macro modelling of machining of GFRP composites and analyzed the cutting forces. This was simulated using Tsai-Hill theory to characterize failure in plane stress conditions and isotropic behavior. Simulation of surface milling and drilling [10–12] process was carried out to analyze torque, surface damage and chip formation. Dhandekar and Shin [13] revealed that three primary approaches of modelling have been developed: (i) micro-mechanical approach, (ii) an equivalent homogeneous material (EHM) approach and (iii) combination of the two approaches for simulation of machining of composites and it have their own advantages and disadvantages.

In this study, a Finite Element (FE) model of turning in carbon fiber reinforced polymer was developed. This paper organized as follows Sect. 2 employs experimental study to validate the FEM results. Section 3 explains numerical study in which the material property, damage model and friction has been discussed. The results were analyzed in Sects. 4 and 5 presents the conclusions of this work.

2 Experimental Work

Cutting experiments were performed on a tube of carbon fiber/epoxy composite material by varying different cutting conditions. The fifteen different cutting conditions were obtained by changing the cutting speed and feed. The cutting speeds were 52, 79, 120, 240 and 346 m/min under feed rates of 0.149, 0.243, and 0.446 mm/rev at a constant depth of cut 0.5 mm. So high cutting speed and feed rate ranges up to 346 and 0.446 mm/rev respectively were taken as input for experimental trials (according to data [14]). Titanium Nitride coated carbide insert (CCMT 09T304) with 10° rake angle and 7° clearance angle was used in this work. The turning was done under dry cutting conditions and the forces like cutting force, thrust force and feed force were measured using lathe tool dynamometer.

3 Numerical Study

3.1 Finite Element Modelling

Finite Element Modelling (FEM) of the cutting process was developed using the commercial software ABAQUS/EXPLICIT V6.13. The work piece and tool were considered to be plastic and it was modeled with a 10 × 5 rectangular cross section, nominal depth of cut was 0.5 mm and meshed with (S4) four node quadrilateral elements with curved thin shell, hourglass control, reduced integration and finite membrane strain. The FE simulations were run at varying grid of mesh densities to fine tune the step interactions. The tool was modeled with rake angle, clearance angle and edge radius as used in experimental trials and TiN properties were assigned to the tool. The work material was considered as Equivalent Homogeneous Orthotropic Material (EHOM) where the individual carbon fiber and epoxy properties were replaced by combined property. To analyze the deformation pattern the cutting conditions was made as plane strain problem. Material geometry and boundary conditions were given before analysis. Boundary conditions were restrained in bottom cutting direction and perpendicular direction in left extreme of the work piece. Figure 1 shows the FE model view of the tool and work piece to generating the mesh. This simulated model was analyzed to get the required results with the post processor.

The machining parameters like cutting speed, feed, depth of cut, and tool geometry were defined in FE simulation with same conditions used as in experimental work. The interaction between the tool and the work material was node to

Fig. 1 FE model of tool and work piece in ABAQUS V6.13

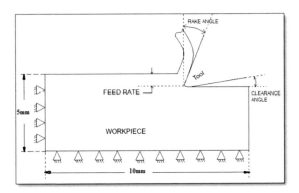

node contact. The cutting tool was moved towards the work material in negative x direction as per displacement boundary conditions. The simulation were carried at different cutting conditions and the cutting force, thrust force and feed force were analyzed. The effective stress and deformation plot were plotted.

3.2 Hashin Damage Model

Hashin proposed the damage model criteria for the fiber reinforced composites [15]. The failure criteria for the composite material include "Fiber tensile failure, Fiber compressive failure, Matrix cracking, and Matrix crushing". The equation shown below the failure damage for fiber reinforced composites:

Tensile fiber failure for $\sigma_{11} \geq 0$

$$\left(\frac{\sigma_{11}}{X_T}\right)^2 + \frac{\sigma_{12}^2 + \sigma_{13}^2}{S_{12}^2} = \begin{cases} \geq 1 & \text{failure} \\ < 1 & \text{no failure} \end{cases}$$

Compressive fiber failure for $\sigma_{11} < 0$

$$\left(\frac{\sigma_{11}}{X_C}\right)^2 = \begin{cases} \geq 1 & \text{failure} \\ < 1 & \text{no failure} \end{cases}$$

Tensile matrix failure for $\sigma_{22} + \sigma_{33} > 0$

$$\frac{(\sigma_{22} + \sigma_{33})^2}{Y_T^2} + \frac{\sigma_{23}^2 - \sigma_{22}\sigma_{33}}{S_{23}^2} + \frac{\sigma_{12}^2 + \sigma_{13}^2}{S_{12}^2} = \begin{cases} \geq 1 & \text{failure} \\ < 1 & \text{no failure} \end{cases}$$

Table 1 Macroscopic material data assigned to the workpiece (according to [16] data)	Mechanical properties	CFRP
	Tensile strength X_T (MPa)	1950
	Compressive strength X_C (MPa)	1480
	Transv. tensile strength Y_T (MPa)	48
	Transv. compressive strength Y_C (MPa)	200
	Shear strength S_L (MPa)	79
	Shear strength S_T (MPa)	79
	Fracture energy (tensile) G_{LT}	0.01
	Fracture energy (compressive) G_{LC}	0.005
	Fracture energy (transv. tensile) G_{TT}	0.0009
	Fracture energy (transv. compr.) G_{TC}	0.006
	Hashin coefficient α	1
	Viscose regularization VR (all)	1E-7

Compressive matrix failure for $\sigma_{22} + \sigma_{33} < 0$

$$\left[\left(\frac{Y_C}{2S_{23}} \right)^2 - 1 \right] \left(\frac{\sigma_{22} + \sigma_{33}}{Y_C} \right) + \frac{(\sigma_{22} + \sigma_{33})^2}{4S_{23}^2} + \frac{\sigma_{23}^2 - \sigma_{22}\sigma_{33}}{S_{23}^2} + \frac{\sigma_{12}^2 + \sigma_{13}^2}{S_{12}^2}$$
$$= \begin{cases} \geq 1 & \text{failure} \\ < 1 & \text{no failure} \end{cases},$$

where σ_{11} and σ_{22} denotes the fiber stress and transverse stress respectively. According to this Hashin damage model, when the stress of the work material exceeds unity, then the failure occurred in matrix and fiber and chip formation can takes place. Table 1 shows the mechanical properties of CFRP material to give as an input to the software which collected from literature paper.

3.3 Coulomb Friction

The coulomb friction model was applied in contact pair between the cutting tool and work material which plays an essential role in the machining process. The relative motion (slip) was occured between tool and work contact point when the shear stress of fiber material at the interface τ was more than or equal to the critical frictional stress $\mu\sigma_n$ where σ_n is the normal stress at the same point. In this work, macro-mechanical approach was followed, the coefficient of friction was taken as 0.3 for Unidirectional-CFRP composites.

4 Results and Discussions

4.1 Cutting Force

Figure 2 shows the comparison of experimental and simulated values of cutting forces for different feed rates. In Fig. 2a, at a lower feed rate of 0.149 mm/rev, the cutting force increasing beyond 100 m/min suggesting an unexpected cutting force requirement at low feed for machining of CFRP tube. The FE result shows the gradual decreases of cutting force with respect to the cutting speed. The force requirement also less as compared with the experiment result at higher speed. In Fig. 2b the cutting force generally decreases with increasing cutting speed for intermediate feed rate of 0.243 mm/rev experimentally. Also the FE results showed the same trend at increasing cutting speed; it matches good agreement with experimental results. The average deviation between experimental and FE values of cutting force was found less. In Fig. 2c, at high feed rate the cutting forces generally decreases with increasing speed. In FE results the cutting forces decrease up to 217 m/min and sudden increase at 346 m/min. The highest error of (−) 55.52% was reported (negative sign indicates higher FE value) at cutting speed 79 m/min and feed 0.446 mm/rev. The lowest error of (+) 0.74% was reported (positive sign indicates higher experimental value) at cutting speed 346 m/min and feed 0.243 mm/rev.

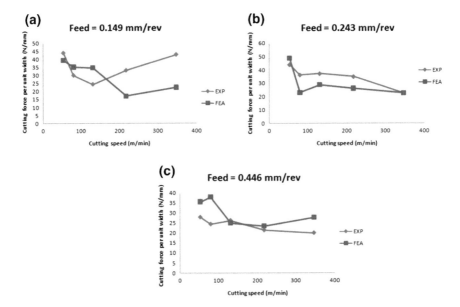

Fig. 2 Cutting force versus cutting speed at different feed rates

4.2 Thrust Force

Figure 3a showed the thrust force versus cutting speed at feed = 0.149 mm/rev. At the lower feed rate of 0.149 mm/rev, the thrust force decreases with cutting speeds gradually increasing at higher speeds. The FE predictions of the thrust force showed the same pattern in the variation of thrust force as shown by the experimental results. The maximum variation of thrust force was shown at high cutting speed with error (−) 27%. In Fig. 3b, at intermediate feed the thrust force initially increases at 79 m/min and decreases gradually at high cutting speed. The FE results were close to the experiment result beyond the 130 m/min but at the low cutting speed, the thrust force decreases. The higher error % of 53.49 was observed at a cutting speed 79 m/min. At higher feed rates Fig. 3c, the thrust forces increased initially, but decreased at higher cutting speeds suggesting the enhanced sensitivity to cutting speeds than feed rates. Higher thrust forces were seen at lower feed rates during simulation, but the impact softens with enhanced cutting speeds. The FE results were compared with the experiment at low cutting speeds, but showed an increased error of ±45% at high cutting speeds of 217 and 346 m/min and didn't match with the experimental values.

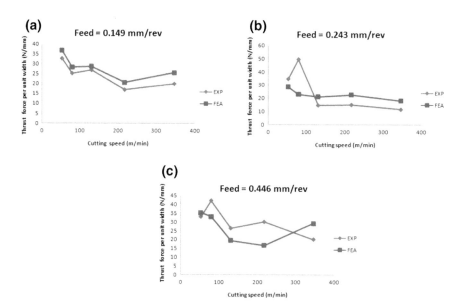

Fig. 3 Thrust force versus cutting speed at different feed rates

4.3 Feed Force

In Fig. 4a, the feed forces showed a fluctuating trend with respect to cutting speeds
unlike cutting forces which generally decreased with higher cutting speeds. At a
low feed rate of 0.149 mm/rev, the material needed higher feed forces in machining
even at a higher cutting speed of 346 m/min. The FE results also has close match
with the experimental result in low feed rate. In Fig. 4b the experiment result at an
intermediate feed rate of 0.243 mm/rev, the feed forces reduced gradually with
increasing cutting speeds. FE prediction showed the variation of low cutting speed
as compared to experiment results and at high cutting speed of 346 m/min the FE
and experiment results showed excellent correlation with an error of just 2.22%
which was very low. In Fig. 4c the experiment result showed that highest feed rate
of 0.446 mm/rev, the feed force drastically increased at intermediate cutting speeds,
but decreased at higher cutting speeds. The trend suggests that feed force
requirements decreased at high cutting speeds and feed rates. The FE results
showed good correlation at low cutting speeds at 52 and 79 m/min and also showed
a similar pattern at high cutting speed 346 m/min. But at intermediate cutting speed,
the simulated feed force showed variation with the experimental results. The
highest error (%) (+) 57.42 was observed at cutting speed 79 m/min and feed
0.243 mm/rev. The lowest variation of error (%) (+) 0.07 was observed at cutting
speed 130 m/min and feed 0.149 mm/rev.

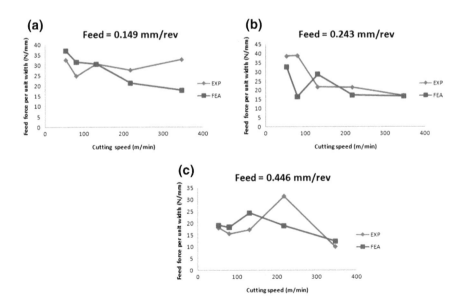

Fig. 4 Feed force versus cutting speed at different feed rates

4.4 Stress Distribution

The calculation of surface damage failure was based on Hashin criteria as described previously in this paper. Material failure was caused mainly due to the stress acted when the cutting tool just ahead on the work material. The maximum von mises stress was observed at a high feed rate and cutting speed of 217 m/min and the minimum stress at a low feed and speed of 0.149 mm/rev and 52 m/min respectively. Damage occurs, once the stress in the material reaches the yield strength and exceeds the critical limits. The damage was assumed to be isotropic such that micro cracks and voids were oriented uniformly in all directions. It seems that the cutting forces play a significant role on damage as compared to other two forces. The stress distribution provides an indication as to where the fracture could initiate. At higher feed rates the gradual increase of stress can be seen with respect to increasing cutting speed. But at the high cutting speed of 346 m/min and feed of 0.446 mm/rev, the stress was decreased because at increasing cutting speeds, the requirement of force decreases so the stress also decreases as shown in Fig. 5. However, the material has inhomogeneous in nature, prediction of an actual crack on the fiber was difficult.

4.5 Deformation Plot

The maximum deformation was observed at low feed and cutting speed. The plastic deformation in FRP composites were radically different than conventional materials due to their inhomogeneity in nature which causes poor machinability that affects the surface finish as well as accelerating tool wear. Also, the deformation was decreasing when the cutting speed increasing as shown in Fig. 6. This shows that during higher speed and feed, the material deformation was decreased. Due to the brittleness nature of the material, discontinuous chips were formed so, it was difficult to visualize the areas of low and high strains during machining.

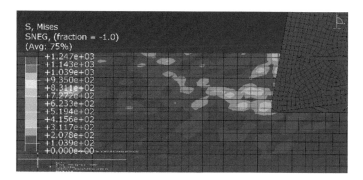

Fig. 5 Von mises stress at feed = 0.446 mm/rev, cutting speed = 346 m/min

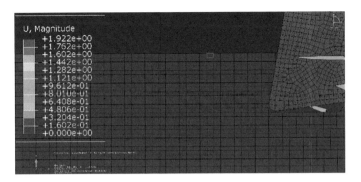

Fig. 6 Deformation at feed = 0.446 mm/rev, cutting speed = 346 m/min

5 Conclusions

A finite element modeling of CFRP material was successfully simulated at high
cutting speeds and feed rates by ABAQUS V6.13 software using an EHM
methodology and studied the cutting forces, thrust forces, and feed forces. The FE
model was built using Hashin model and able to analyze the stress distribution and
deformation rate successfully.

The range of cutting force predicted by FE model was well correlated with
experimental results. At higher feed rate this model showed a consistent negative
deviation from the experimental cutting force which means the FE results have
higher force value as compared with experimental values. The thrust force, which
has an important factor in machining composite materials generally decreases with
respect to cutting speeds, but showed higher sensitivity at intermediate feed rates
and lower cutting speeds. The FE results showing the close match at the low feed
rates with the experimental results and at higher feed rate fails to predict the thrust
force. The feed forces were generally found to decrease with cutting speeds, but
showed higher fluctuations about feed rates, than cutting forces. At intermediate
cutting speed, the feed force was not that much close to the experimental results.
The stress value was maximum at the higher feed rates and maximum stress distribution occurred on the primary shear zone (i.e.) work tool contact area as shown
in FE results. The deformation on work decreased when increasing cutting speed
and due to the broken chips, it was difficult to visualize the strain rate acted during
machining. Overall the simulation of anisotropic material can be possible at optimized cutting conditions.

References

1. Arola D, Ramulu M (1996) Orthogonal cutting of fiber reinforced composites: a finite element
   analysis. Int J Mech Sci 39:597–613

2. Mahdi M, Zhang L (2001) A finite element model for the orthogonal cutting of fiber-reinforced composite materials. J Mater Process Technol 113:373–377
3. Rao GVG, Mahajan P, Bhatnagar N (2007) Micro-mechanical modelling of machining of FRP composites—cutting force analysis. Compos Sci Technol 67:579–593
4. Rao GVG, Mahajan P, Bhatnagar N (2007) Machining of UD-GFRP composites chip formation mechanism. Compos Sci Technol 67:2271–2281
5. Rao GVG, Mahajan P, Bhatnagar N (2008) Three dimensional macro-mechanical finite element model for machining of unidirectional-fiber reinforced. Polym Compos 498:142–149
6. Nayak D, Mahajan P, Bhatnagar N (2007) Machining studies of UD-FRP composites part 2: finite element analysis. Mach Sci Technol 9:503–528
7. Santiuste C, Soldani X, Miguelez MH (2010) Machining FEM model of long fiber composites for aeronautical components. Compos Struct 92:691–698
8. Calzada KA, Kapoor SG, Devor RE, Samuel J, Srivastava AK (2012) Modelling and interpretation of fiber orientation based failure analysis in machining of CFRP composites. J Manuf Processes 14:141–149
9. Mkadeem A, Demirci I, El Mansori M (2008) A micro-macro combined approach using FEM for modeling of machining of FRP composites: cutting forces analysis. Compos Sci Technol 68:3123–3127
10. Wu M, Gao Y, Cheng Y, Wang B, Huo T (2016) Carbon fiber composite materials finite element simulation analysis of cutting force. Procedia CIRP 56:109–114
11. Ghafarizadeh S, Chatelain JF, Lebrun G (2016) Finite element analysis of surface milling of carbon fiber reinforced composites. Int J Adv Manuf Technol 87:399–409
12. Phadnis VA, Roy A, Silberschmidt VV (2012) Finite element analysis of drilling in carbon fiber reinforced polymer composites. J Phys Conf Ser 382
13. Dhandekar CR, Shin YC (2012) Modelling of machining of composite materials: a review. Int J Mach Tools Manuf 57:101–121
14. Allwin Roy Y, Gobivel K, Vijay Sekar KS, Suresh Kumar S (2017) Impact of cutting forces and chip microstructures in high speed machining of carbon fiber-epoxy composite tube. Arch Metall Mater 62:1771–1777
15. Hashin Z (1980) Failure criteria for unidirectional fiber composites. J Appl Mech 47:329–334
16. Rentsch R, Pecat O, Brinksmeier E (2011) Macro and micro process modelling of the cutting of carbon fiber reinforced plastics using FEM. Procedia Eng 10:1823–1828

Characterization of Mechanical and Thermal Properties in Soda–Lime Glass Particulate Reinforced LM6 Alloy Composites

M. R. Shivakumar and N. V. R. Naidu

Abstract LM6 alloy matrix and soda–lime glass particulate composites were produced by stir casting method. Nine sets of composites with the combinations of 1.5, 3.0 and 4.5 weight per cent of glass particles and 75, 125 and 210 μm glass particle size were developed as test composites. The result showed that, soda–lime glass particles were uniformly distributed and properly wetted with LM6 alloy matrix. Tensile strength of the composites were decreased as the weight per cent and size of the glass particles increases. Hardness increases as the weight per cent and size of the glass particles increases. Wear resistance was improved by increasing the weight per cent of particle. Thermal conductivity of the composites decreased with increase of weight per cent and size of the glass particles.

Keywords Stir casting · Aluminium matrix composite · Glass reinforcement Mechanical and thermal properties

1 Introduction

The nature and characteristics of the reinforcement and the production methodology of composite preparation determines the properties of that composite to a large extent. The properties of composites are controlled by varying the quantity and parameters of reinforcement. Processing cost of composite is moderately higher as compared to the other material processing methods. Aerospace, automobile and electronic packaging areas required specific materials for effective and efficient performance during their usage [1, 2]. The composites are the materials for the critical areas like defence and aerospace, etc., where cost is secondary [3, 4].

Aluminium Matrix Composites (AMCs) are proven as successful 'high-tech' materials in their applications and have economic and environmental benefits [5–7].

M. R. Shivakumar (✉) · N. V. R. Naidu
Department of Industrial Engineering and Management,
M. S. Ramaiah Institute of Technology, Bangalore, India
e-mail: mrshivakumar@msrit.edu

© Springer Nature Singapore Pte Ltd. 2019
A. K. Lakshminarayanan et al. (eds.), *Advances in Materials and Metallurgy*,
Lecture Notes in Mechanical Engineering,
https://doi.org/10.1007/978-981-13-1780-4_16

Usually, particulate reinforcement ceramics such as silicon carbide and alumina are used in AMC. Reinforcement material is expensive and contributes to the cost of AMCs. Low-cost reinforcement materials such as fly ash are usually used to reduce the cost of AMCs. Glass is abundantly available at low price which can also be used as reinforcement material [8].

2 Materials and Methodology

Aluminium with 11.6% silicon alloy (commercially known as LM6 alloy) is the most widely used aluminium cast alloy and having lowest melting point was selected as the matrix material. Soda–lime glass (window glass) powder was used as reinforcement material in the composite preparation. The crushed glass powder was sieved and separated using a series of sieves to obtain the required grain size. In this investigation, stir casting production method was used to produce LM6 alloy/ soda–lime glass particulate composites [9, 10].

The reinforcement particles are usually very poor in getting wetted by a liquid metal. Further, wetting of ceramic particles by molten metal below 1100 °C is poor. If the wettability is inadequate, incorporation of reinforcement becomes incomplete and resulting composite will have less reinforcement than what was intended. A strong interface leads to transfer and distribution of load to the reinforcement from the matrix and promotes enhanced properties of the Metal Matrix Composites (MMCs). To enhance the wettability between glass particles and LM6 alloy pre-treatment of the glass particles [11], addition of very small amount of magnesium to the melt [12] and two-step stir mixing [13–15] techniques were adopted.

3 AMC Test Castings

The methodology of composite production was optimized and maintained same throughout the trial composites by varying glass parameters. The processing parameters such as superheat temperature of melt, pouring temperature, die temperature, preheat temperature of soda–lime glass, amount of magnesium added (as wetting agent), stirring time, stirring speed and processing methodology were kept constant for all the trials. Nine sets of composites with the combinations of 1.5, 3.0 and 4.5 weight per cent of glass particles and 75, 125 and 210 μm glass particle size were developed as test composites.

4 Testing and Results

4.1 Microstructural Analysis

The objective of the microstructural study is to gather information on the quality of the developed composites for interfacial characteristics, bonding and wettability of glass particles with the LM6 alloy matrix and also to analyse the distribution of glass particles in the LM6 alloy matrix [16]. In this investigation, computer-interfaced ZEISS optical microscope and Scanning Electron Microscope (SEM) FEI-QUANTA 200 High-resolution electron microscopy were used. Test samples were drawn from test castings in the as-cast condition.

Figure 1 justifies the successful production of MMCs and the good wettability of the 4.5 wt% glass with LM6 alloy and exhibits occurrence of no reaction between soda–lime glass and LM6 alloy. Figure 2 shows the distribution of 75 μm size glass particles [(a) 4.5 wt% glass, (b) 3 wt% glass and (c) 1.5 wt% glass] in LM6 matrix.

4.2 Tension Test

Tensile test specimens were machined out of the as-cast composite test castings as per the ASTM standard (Designation: E 8/E 8 M—08). The test specimens are shown in Fig. 3.

Tensile tests for three test samples corresponding to different glass particulate sizes and weight contents were conducted results shown in Fig. 3. Figure 4 shows that the increase in addition of glass particles decreases the Ultimate Tensile Strength (UTS) of the composite. Further, as the particle becomes coarser, the

(a)   **(b)**

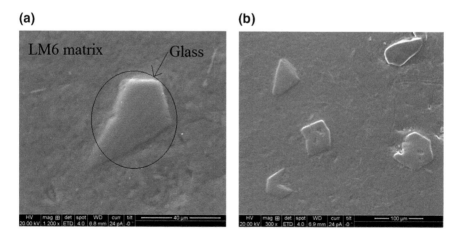

Fig. 1 SEM micrographs of 75 μm particle size and 4.5 wt% glass composites

(a)   **(b)**

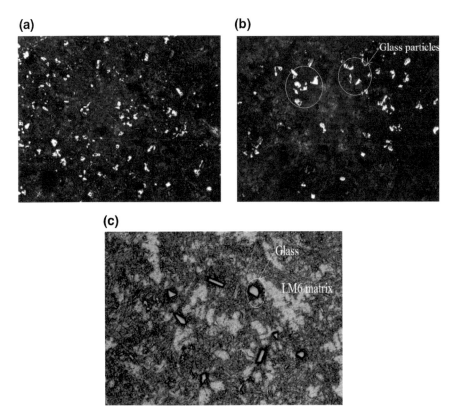

Fig. 2 Optical graphs of composites with varied weight per cent of glass particles **a** 4.5 wt%, **b** 3 wt%, **c** 1.5 wt%

Fig. 3 Test specimens as per the ASTM standard (Designation: E 8/E 8 M—08)

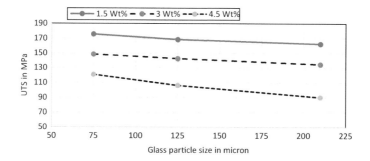

Fig. 4 Mean UTS at different weight per cent and size of glass particles

tensile strength comes down. The UTS at 1.5 wt% and 75 μm was 175.567 MPa and at 4.5 wt% to 210 μm 90.213 MPa.

The difference in thermal expansion values between the LM6 alloy and the soda–lime particles are the main reasons for the thermal mismatch. This leads to the development of elastic stresses, which in turn force the soda–lime glass particles into compression and the LM6 alloy matrix into tension. Material properties are naturally affected by residual stress, and even the values of fracture toughness will be affected. Finally, the composites become more brittle due to these residual stresses.

Increase in the amount of closed pores in composites with increasing amount of soda–lime glass particles which would generate more sites for crack initiation. When the particle size increases, each particle is like a clustering of glass particles, which may result in formation and propagation of cracks. The combination of bigger particle size and de-bonding at the interface between particles and matrix, lead to significant drop in the strength of the composite.

4.3 Hardness Test

Brinell hardness test was conducted on the developed composites. The Brinell hardness tester with a 10 mm diameter of hardened steel ball indenter was used for the tests. The load applied was 500 kgf for duration of 30 s.

Figure 5 shows the hardness of composites with the variations of glass particle size at 1.5, 3.0 and 4.5% weight fractions of glass. The hardness of composites at 1.5%, 75 μm and 4.5%, 210 μm are 50 BHN and 84.3 BHN, respectively. One can observe that the addition of glass powder into the LM6 alloy matrix invariably increases the hardness. Higher value of hardness is mainly due to the positive contribution of the hard glass particles on the soft LM6 alloy matrix. Similar trend is observed with increase in particulate size. Bigger particulates offer higher

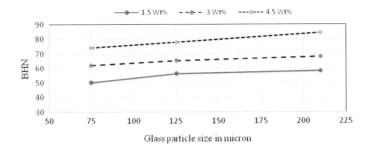

Fig. 5 Variation of hardness with glass particle size

resistance to penetration on the surface, thereby enhancing the hardness. The particle size of reinforcing glass powder contributes to enhance hardness of the composites.

4.4 Wear Test

Samples of LM6 alloy/Soda–lime glass composites were tested using the pin-on-disc apparatus to check the effect of glass reinforcement on wear resistance. Wear test samples (8 mm diameter and 50 mm length) are shown in Fig. 6. During testing, sliding distance and sliding speeds were maintained constant in all the experimental runs. Tests were carried out at a sliding speed of 7 m/s and for a sliding distance of 2100 m at 10 N load to determine weight loss.

Figure 7 shows the effect of reinforcement content and particle size on the weight loss in wear. The wear resistance increases (decrease in weight loss) as the wt% of reinforcement increases; and wear resistance decreases as the particle sizes increases. The weight loss, 0.004 g, is minimum at 4.5%, 75 μm, i.e. the wear resistance is maximum. This is due to the hard glass particles in the composites.

Fig. 6 Specimens used in wear test

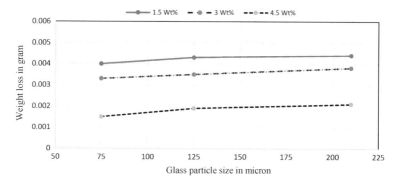

Fig. 7 Variation of wear weight loss with the particle size at different weight fractions of glass

4.5 Thermal Conductivity Test

The well-proven comparative cut bar method, ASTM E1225 Test Method, was used for carrying out the tests; this method has been acclaimed as the best-suited method in axial thermal conductivity tests. Cylindrical specimens were used as per the dimensions of reference specimens of thermal conductivity analyzer. The dimension of the test specimen is 8 mm in diameter and 50 mm in length.

Figure 8 reveals that the thermal conductivity of composite depends on wt% and particle size of reinforcement. The thermal conductivity decreases as the wt% of reinforcement increases and it decreases as the particle sizes increases. At 1.5 wt% and 75 μm, thermal conductivity is 113.267 W/m °C and at 4.5 wt% and 75 μm, is 95.533 W/m °C. This shows the reduction in thermal conductivity by 15.6%. Thermal conductivity at 1.5 wt%, 75 μm it is 113.267 W/m °C and at 1.5 wt%, 210 μm are 113.267 and 108.033 W/m °C.

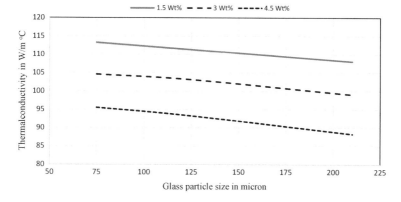

Fig. 8 Variation of thermal conductivity with the particle size at different weight fractions of glass

The reinforced glass has lower thermal conductivity than the LM6 alloy matrix, so thermal conductivity decreases as the glass percentage increases in the composite. It is also observed that the higher particle size glass develops more blockages to the heat flow than the smaller glass particle size in the composite.

5 Conclusions

The stir casting technique is an effective production method for LM6 alloy/glass particulate composite. Better bonding and uniform distribution of glass particles with LM6 alloy matrix are obtained by adopting suitable processing techniques. Tensile tests indicate that the incorporation of soda–lime glass particles into the LM6 alloy does not improve the tensile strength of the composite. Hardness and wear resistance of the composites were improved by the addition of glass particles with the LM6 alloy matrix. The addition of glass particles reduces the thermal conductivity of the LM6alloy/glass composites.

Acknowledgements The authors wish to thank Prof. S. Seshan, Indian Institute of Science, Bangalore for his constant help and encouragement. Financial support received from Visvesvaraya Technological University, Belagavi is gratefully acknowledged.

References

1. Deruyttere A, Froyen L, DeBondt S (1989) Metal matrix composites—a bird's eye view. Bull Mater Sci 12:217–223
2. Surappa MK (2003) Aluminium matrix composites: challenges and opportunities. Sadhana 28:319–334
3. Anthymidis Konstantinos, David K, Agrianidis P, Trakali A (2014) Production of Al metal matrix composites by the stir casting method. Key Eng Mater 592–593:614–617
4. Jayalakshmi S, Satish Kailash V, Seshan S, Kim KB, Fleury E (2016) Tensile strength and fracture toughness of two magnesium metal matrix composites. J Ceramic Process Res 7 (3):261–265
5. Vijayaram TR (2009) Foundry metallurgy of silicon dioxide particulate—reinforced LM6 alloy matrix composites, studies on tensile properties, and fractography. Indian Foundry J 55:21–26
6. Wahab MN, Daud AR, Ghajali MJ (2009) Preparation and characterization of stir cast-aluminum nitride reinforced aluminium metal matrix composites. Int J Mech Mater Eng 4:115–117
7. Ibrahim IA, Mohamed FA, Lavernia EJ (1991) Particulate reinforced metal matrix composites. J Mater Sci 26:1137–1156
8. Rohatgi PK, Asthana R, Das S (1986) Solidification, structures, and properties of cast metal-ceramic particle composites. Int Metal Rev 31:115–139
9. Shobha R, Suresh KR, Niranjan HB (2014) Mechanical and Microstructural evaluation of insitu aluminium titanium boride composite processed by severe plastic deformation. Procedia Mater Sci 5:281–288

10. Madhoo G, Shilpa M (2017) Optimization of process parameters for stir casting technique using orthogonal arrays. Int J Adv Res Methodol Eng Technol 1(2):22–28
11. Canakei Aykut, Arslan Fazli, Yasar Ibrahim (2007) Pre-treatment process of B_4C particles to improve incorporation into molten AA2014 alloy. J Mater Sci 42:9536–9542
12. Saravanan C, Subramanian K, Ananda Krishnan V, Sankara Narayanan R (2015) Effect of particulate reinforced aluminium metal matrix composite—a review. Mech Mater Eng 19 (1):23–30
13. Saravana Bhavan K, Suresh S, Vettivel SC (2013) Synthesis, characterization and mechanical behavior of nickel coated graphite on aluminium matrix composite. Int J Res Eng Technol 2:749–755
14. Eesley GL, Elmoursi A, Patel N (2003) Thermal properties of kinetics pray Al–SiC metal-matrix composite. J Mater Res 18:855–860
15. Alaneme KK, Alukob A (2012) Production and age hardening behavior of borax premixed SiC reinforced Al–Mg–Si alloy composites developed by double stir-casting technique. West Indian J Eng 34:80–85
16. Peng HX, Fan Z, Madher DS, Evans JRG (2002) Microstructure and mechanical properties of engineered short fiber reinforced aluminum matrix composites. Mater Sci Eng A 335:207–216

Tribo Performance of Brake Friction Composite with Stainless Steel Fiber

K. Sathickbasha, A. S. Selvakumar, M. A. Sai Balaji, B. Surya Rajan and MD Javeed Ahamed

Abstract The tribological performance of brake friction composites (FCs) with SSS1140 Stainless steel fiber (Equivalent to EN304) is studied using a pin-on-disk tribometer. Five friction composites, namely SSB_5, SSB_{10}, SSB_{15}, SSB_{20}, and SSB_{25} were developed with 5, 10, 15, 20, and 25% of stainless steel fiber, respectively, by compensating the inert filler $BaSO_4$. The friction and wear characteristics are evaluated at dry sliding condition based on ASTM G99-95. The performance is investigated with three different normal loads (10, 20, and 30 N) and speeds (1, 2, and 3 m/s). It is observed that the increase in fiber content increases the friction coefficient (0.21–0.49). The specific wear rate of the friction composite observed as $SSB_5 < SSB_{10} < SSB_{15} < SSB_{20} < SSB_{25}$. It is observed that the SSB_{20} and SSB_{25} are more aggressive towards rotor wear which is also indicated by the higher hardness of the respective FCs. The physical, chemical, and mechanical properties of the developed friction composites are also studied with IS 2742 and ISO 6315, and the performance values also lies between the prescribed industrial standards.

Keywords Brake pad · Sliding friction · Specific wear rate · Stainless steel fiber

K. Sathickbasha (✉) · A. S. Selvakumar · M. A. Sai Balaji · B. Surya Rajan
M. Javeed Ahamed
Department of Mechanical Engineering, B. S. Abdur Rahman Crescent Institute
of Science and Technology, Vandalur, Chennai 600048, India
e-mail: sathick.basha@bsauniv.ac.in

A. S. Selvakumar
e-mail: selvakumar@bsauniv.ac.in

M. A. Sai Balaji
e-mail: saibalaji@bsauniv.ac.in

B. Surya Rajan
e-mail: suryarajan@bsauniv.ac.in

M. Javeed Ahamed
e-mail: javeedahamed@bsauniv.ac.in

© Springer Nature Singapore Pte Ltd. 2019
A. K. Lakshminarayanan et al. (eds.), *Advances in Materials and Metallurgy*,
Lecture Notes in Mechanical Engineering,
https://doi.org/10.1007/978-981-13-1780-4_17

1 Introduction

Brakes ensure the safety of an automotive system, so care should be taken while selecting appropriate materials or ingredients for effective brake application. Brake pad is the stator component in the brake system which has the cocktail of ingredients in it [1, 2]. The selection of materials is quite complex in satisfying the needs and function of the brake pads. The ingredients selected for the brake friction composite should have high fade resistance, good mechanical strength, stable friction, moderate sliding wear, and compatibility with the work environment [3, 4]. The brake pad materials should possess good mechanical and tribological properties during adverse conditions like high temperature, humidity, and dusty environments [5]. In general, mild steel (K15) and annealed steel fiber used in brake friction composites are affected by the above-said environmental conditions. Stainless steel fibers were proposed to be a good replacement for mild steel fiber with good environment resistivity. Compared to mild steel and annealed steel fibers SSF (SSS1140) is having more hardness. In this work, the tribo performance of the SSF is evaluated using ASTM G99-95 standard under three load and speed conditions. To study the influence of the fiber content in tribo performance fiber brake pad samples was developed with 5, 10, 15, 20, and 25 wt% of SSF content by compensating inert filler barytes (BaSO$_4$).

2 Materials and Methodology

2.1 Material

Five friction material compositions are prepared with different wt% of stainless steel fiber SS1140. SSF supplied by M/S CHIAO YU Friction Co. Ltd. Taiwan. The average diameter and length of fiber is measured as 150 μm and 1 mm, respectively. The chemical composition of the stainless steel fiber is given in Table 1. Scanning electron microscope image and EDAX of the SS1140 fiber is shown in Fig. 1.

2.2 Friction Material Formulation

Five FM formulations were prepared by keeping 62 wt% of materials as a constant parent ingredient. The parent ingredients consist of Kevlar pulp 3%, rock wool and

Table 1 Chemical composition of SS1140 SSF

Chemical elements	C	Si	Mn	Cr	Mo	P	S	Ni
Composition (%)	<0.12	<1	<1	16–18	0.75–1.25	<0.035	<0.03	8–10

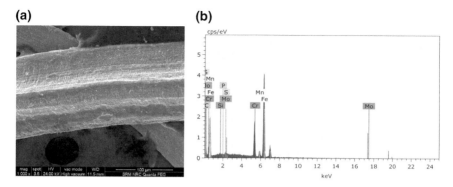

Fig. 1 Scanning electron microscope image with EDAX for stainless steel fiber

glass fiber 11%, Al_2O_3 and boron-modified friction dust 10% [6], Benzoil-modified phenolic resin-12% [7], wollastonite and vermiculite 14%, Graphite and molybdenum disulphide (MoS_2) 10%, and NBR and Crumb rubber particles 2%. The remaining 38% of the material is manipulated with the combination of SS1140 fiber and inter filler barite ($BaSO_4$) as shown in Table 2.

2.3 Fabrication of Friction Composite

The FCs with different steel fiber wt% is manufactured by compression molding technique. The composition containing the parent formulation and ingredients shown in Table 2 are mixed using a shear mixing machine. The perform is made using human hand pressure by pouring 85 g of mixture in the die. The pre-glued backing plate is inserted in the brake pad die of Tata Indica car shown in Fig. 2a.

The perform is cured in the die arrangement along with the backing plate. The curing process takes place at 163 °C for 3 min with six intermittent breathing cycles for volatile removal [8]. The ejected brake pad is post cured in an oven at 120 and 140 °C for 2 hours each. The friction liner is removed from the back plate of the brake pad for creating the pin of 10 mm diameter for the tribo performance test.

Table 2 Brake friction composite formulations

Ingredient	wt%				
	SSB$_5$	SSB$_{10}$	SSB$_{15}$	SSB$_{20}$	SSB$_{25}$
SSS1140 SSF	5	10	15	20	25
BaSO$_4$	33	28	23	18	13

Table 3 FC specimens and tribological test conditions

Material/abbreviation	Load (N)	Speed (m/s)	Sliding distance (km)
SSB$_5$	10	1	15
	20	2	15
	30	3	15
SSB$_{10}$	10	1	15
	20	2	15
	30	3	15
SSB$_{15}$	10	1	15
	20	2	15
	30	3	15
SSB$_{20}$	10	1	15
	20	2	15
	30	3	15
SSB$_{25}$	10	1	15
	20	2	15
	30	3	15

(a)   **(b)**

Fig. 2 **a** Die setup for Tata Indica brake pad, **b** shear load experiment setup 1. Plunger, 2. Die, 3. Backplate, 4. Die base, 5. Heater, 6. Hydraulic power pack, 7. Ram, 8. Shear die, 9. Brake pad, 10. Pressure gage

2.4 Characterization of FC

The friction composites were tested for its physical (density, water absorption, and loss on ignition), chemical (acetone extraction), and mechanical properties (hardness, shear load) based on IS 2742 of 1994 standard. The density of the friction composites was evaluated using the water immersion method follows Archimedes' principle. The hardness of the FCs was evaluated using L scale Rockwell hardness

Table 4 Physical, chemical, and mechanical properties of FCs

Properties	Unit	SSB$_5$	SSB$_{10}$	SSB$_{15}$	SSB$_{20}$	SSB$_{25}$
Density	g/cm^3	1.94	2.07	2.21	2.62	2.85
Hardness	HRL	72–79	73–79	80–88	80–92	87–95
Loss on ignition	%	37	33	27	22	16
Acetone extraction	%	0.89	0.78	0.81	0.80	0.81
Heat swell	mm	0.20	0.25	0.25	0.28	0.3
Water absorption	%	2.7	2.4	2.0	2.0	2.0
Shear strength	Kgf/cm^2	40.80	42.20	48.60	46.00	48.80

tester. The shear strength of the brake composite is calculated using the shearing die setup shown in Fig. 2b. The acetone extraction test is conducted to know the uncured resin percentage in the FC samples.

3 Tribological Testing

The friction and wear performance of the FCs were tested using a pin-on-disk tribometer (Ducom wear tester TR201CL) integrated with data acquisition system and computer shown in the Fig. 3a. The grey cast iron (G3000 with HRB 84) is used as the disk material (diameter 140 mm and thickness 8 mm). The details of composite specimen, sliding speed, and load condition are tabulated in Table 3.

(a)   **(b)**

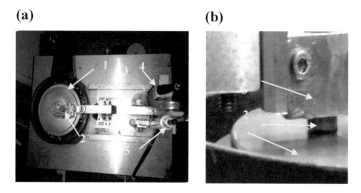

Fig. 3 **a** DOCUM wear tester, **b** pin and sliding disk. 1. Disk, 2. Pin, 3. Loading arm, 4. Load cell, 5. LVDT, 6. Pin holder

4 Result and Discussion

4.1 Physical, Chemical, and Mechanical Properties

The physical, mechanical, and chemical properties are found as per IS 2742 and ISO 6315 standards. The density of FC increases with increase in SSF content. The hardness of the FC ranges between 72 and 95 HRL for the five samples of SSB FC's. The loss of mass due to ignition is decreased because of increase with SSF content. The acetone extraction and heat swell values have negligible difference in all the FC samples. The water absorption values slightly vary from low SSF to high SSF FC's. The shear strength of the friction composite increases slightly with fiber content (Table 4).

4.2 Friction Performance

The friction force is recorded at the periodic interval for the all brake friction material samples (SSB_5, SSB_{10}, SSB_{15}, SSB_{20}, and SSB_{25}) for three different loads (10, 20, and 30 N) and speeds. The friction force plotted for different load for 3 m/s speed is shown in Fig. 4. Friction forces were obtained for the sliding distance of 15,000 m. The friction force value starts with zero value and it reaches a steady value by ensuring the conformal contact between the FC pin and rotor disk. SSB_{25} FC has shown higher friction force at high (30 N) load condition. SSB_5 FC has shown the lowest friction force at low (10 N) load condition. As the wt% of SSF increases the friction force increases.

In general, observations show that the friction force and coefficient of friction increase with the applied load irrespective of the formulation of FCs (Ref. Fig. 5) [9]. The increase in load increases the real area of contact by elastically deforming the contact points, so the friction coefficient tends to increase with load [5]. The variation of friction coefficient is observed with respect to speed. As the speed increases the coefficient of friction decreases. Friction coefficient pertaining to 1 m/s speed is higher than 2 and 3 m/s rotor speeds (Ref. Fig. 5) except SSB_{15}. The speed of rotor causes the rapid temperature rise at the interface which causes the destruction and reformation of tribo layer at the interface which reduces the μ value at higher speed [10]. The reason behind the increase in μ with increases in wt% of SSF is increased abrasive locking and the increased metal–metal contact between the pin and rotor. The increased load also helps to ensure the abrasive locking between the sliding members which subsequently increases the μ with high loads.

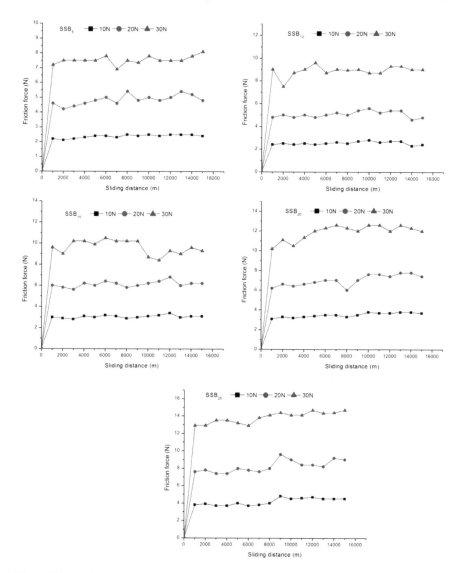

Fig. 4 Friction force (N) versus sliding distance (m) for FCs SSB_5, SSB_{10}, SSB_{15}, SSB_{20}, and SSB_{25}

4.3 Wear Performance

The Specific Wear Rate (SWR) of the composites was plotted for different speeds and loads. It is observed that for the same load and speed condition the SWR increases with increase in fiber content. This is due to the increase in the abrasive nature of the composite with respect to fiber wt%. Irrespective of the composition

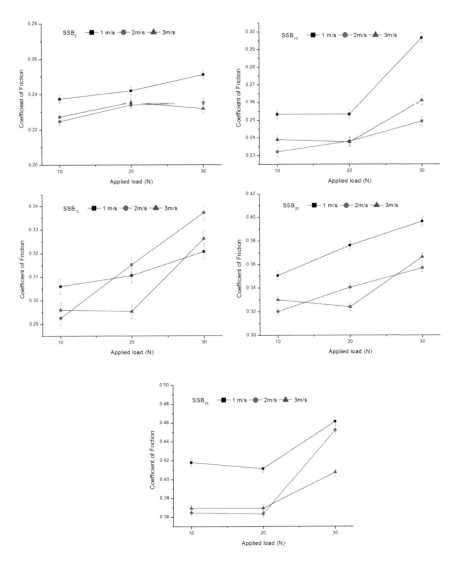

Fig. 5 Coefficient of friction versus applied load (N) for FCs SSB$_5$, SSB$_{10}$, SSB$_{15}$, SSB$_{20}$, and SSB$_{25}$

and the fiber content, the speed of the rotor plays a major role in the SWR. The normal load of on the friction composite has minimal effect with respect to SWR compared to speed of the rotor (Fig. 6).

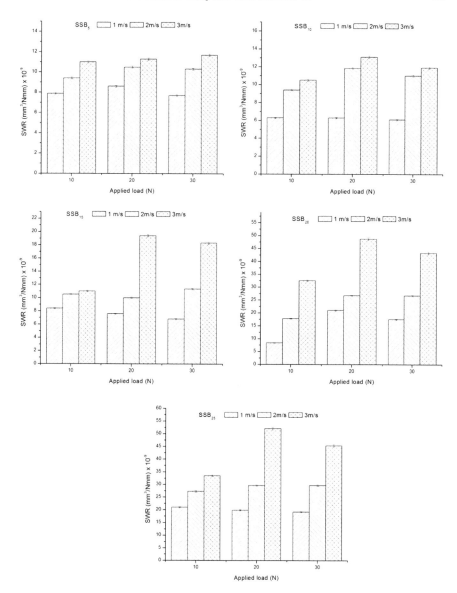

Fig. 6 Specific wear rate (SWR) versus applied load (N) for FCs SSB_5, SSB_{10}, SSB_{15}, SSB_{20}, and SSB_{25}

5 Conclusion

- The physical, chemical, and mechanical properties of the brake friction composites SSB_5, SSB_{10}, SSB_{15}, SSB_{20}, and SSB_{25} were evaluated. The performance variation is as follows:

Density	$SSB_5 < SSB_{10} < SSB_{15} < SSB_{20} < SSB_{25}$
Shear strength	$SSB_5 < SSB_{10} < SSB_{20} < SSB_{15} < SSB_{25}$
Hardness	$SSB_5 < SSB_{10} < SSB_{15} < SSB_{20} < SSB_{25}$
LOI	$SSB_5 > SSB_{10} > SSB_{15} > SSB_{20} > SSB_{25}$
Heat swell	negligible variation
Water absorption	negligible variation
Acetone extraction	negligible variation

- Tribo performance of the FCs is as follows:

Friction	$SSB_5 < SSB_{10} < SSB_{15} < SSB_{20} < SSB_{25}$ [Higher is better]
	$\mu_{1m/s} > \mu_{2m/s} > \mu_{3m/s}$ (overall observation)
	$\mu_{30N} > \mu_{20N} > \mu_{10N}$ (overall observation)
SWR	$SSB_5 < SSB_{10} < SSB_{15} < SSB_{20} < SSB_{25}$ [Lower is better]

References

1. Bijwe J (1997) Composites as friction materials: Recent developments in non-asbestos fiber reinforced friction materials—a review. Polym Compos 18(3):378–396
2. Chan DSEA, Stachowiak GW (2004) Review of automotive brake friction materials. Proc Inst Mech Eng Part D J Autom Eng 218(9):953–966
3. Xiao X, Yin Y, Bao J, Lu L, Feng X (2016) Review on the friction and wear of brake materials. Adv Mech Eng 8(5):1687814016647300
4. Fu Z, Suo B, Yun R, Lu Y, Wang H, Qi S, Jiang S, Lu Y, Matejka V (2012) Development of eco-friendly brake friction composites containing flax fibers. J Reinf Plast Compos 31 (10):681–689
5. Sundarkrishnaa KL (2012) Friction material composites materials prospective. Springer Ser Mater Sci 171. ISBN 978-3-642-33450-4
6. Surya Rajan B, Sai Balaji MA, Velmurugan C, Vinoth B, Selvaraj B (2016) Tribo performance of a NA disc brake pad developed using two types of organic friction modifier. In: Proceedings of the international conference on materials, design and manufacturing process (ICMDM) Chennai, 17–19 Feb, pp 220–227
7. Balaji S, Kalaichelvan K (2012) Optimization of a non asbestos semi metallic disc brake pad formulation with respect to friction and wear. Procedia Eng 38:1650–1657

8. Kumar M, Bijwe J (2010) Studies on reduced scale tribometer to investigate the effects of metal additives on friction coefficient—temperature sensitivity in brake materials. Wear 269 (11):838–846
9. Bajpai PK, Singh I, Madaan J (2013) Tribological behavior of natural fiber reinforced PLA composites. Wear 297(1):829–840
10. Rajan BS, Balaji MS, Velmurugan C (2017) Correlation of field and experimental test data of wear in heavy commercial vehicle brake liners. Friction 5(1):56–65

Influence of Stacking Sequence on Free Vibration Characteristics of Epoxy-Based Flax/Sisal Composite Beams

R. Murugan, N. Nithyanandan and V. Sathya

Abstract Increase in demand for environmental friendly engineered structures make the natural fiber reinforced composites as the best option to synthetic fiber in polymer composite structures. In this study, the influence of stacking sequence of natural hybrid laminates on mechanical and vibration characteristics that are beneficial for structural applications have been focused. To study the effect of stacking sequence efficiently, a high modulus natural fiber, i.e., Flax and a low modulus natural fiber, i.e., Sisal are preferred. The preferred natural hybrid composite laminates were made by hand layup technique. The hybrid laminates were tested for mechanical properties and free vibration characteristics by means of ASTM procedure. The experimental modal frequency values were used for finding the effective elastic constants of natural hybrid composite laminates adopting by simple regression analysis. These effective elastic constants were used for performing theoretical modal analysis of natural composite beams at high frequency level using finite element method. Based on the results of experimental and theoretical modal analysis of Flax/Sisal composite beams, the effective stacking sequence for structural application was suggested.

Keywords Natural fiber · Stacking sequence · Mechanical properties
Material damping · Modal frequency · Mode shapes · Effective
elastic constants

R. Murugan (✉) · N. Nithyanandan · V. Sathya
Department of Mechanical Engineering, Panimalar Institute of Technology,
Chennai 600123, Tamilnadu, India
e-mail: saimurugan1973@gmail.com

N. Nithyanandan
e-mail: nitabi1973@gmail.com

V. Sathya
e-mail: sathya070497@gmail.com

© Springer Nature Singapore Pte Ltd. 2019   171
A. K. Lakshminarayanan et al. (eds.), *Advances in Materials and Metallurgy*,
Lecture Notes in Mechanical Engineering,
https://doi.org/10.1007/978-981-13-1780-4_18

1 Introduction

Fiber-reinforced polymer (FRP) composite structures are commonly used for aerospace and automotive structures since they posses very high strength to weight ratio. But FRP composites made of synthetic fibers which are most widely used today in automotive and aircraft applications are not eco-friendly and non-biodegradable. As a result natural fiber reinforced composites are newly emerging materials that will play important role in next-generation building materials and automotive components. Further these composite structures experience huge vibration during service condition. This can cause durability concerns or discomfort because of the resulting noise and vibration. Evaluation of mechanical properties and vibration characteristics of FRP composite beams with synthetic fibers had been carried out in the past by many authors. There is limited experimental work on vibration characteristics of natural fiber composite materials [1–4].

Hybridization among various available fibers is one of the relatively new concepts for the design of damped composite structures. Akash et al. studied the transverse vibration analysis of Jute/Sisal hybrid composites. They found that modal frequency values of hybrid Jute/Sisal laminate is higher than that of dedicated Jute laminate. Hybrid Jute/Sisal composites possess good damping factor as compared to conventional composites [5]. Vibration damping properties of Flax fiber reinforced composites were characterized and compared with the Glass fiber reinforced composites by Prabhakaran et al. [6]. They observed that the Flax fiber composite have 51% higher vibration damping than the Glass fiber composites. The specific flexural strength and specific flexural modulus for flax fiber reinforced composites also good. These results suggested that the Flax fiber reinforced composites could be a viable candidate for applications which need good sound and vibration properties. In the earlier investigation, Murugan et al. revealed that for a hybrid composite laminate, improved the dynamic stability could be achieved by placing the high modulus fiber as outer layer and low modulus fiber as inner layer [7] and they also established the effective stacking sequence for hybrid composite laminates. In the present study, the effect of stacking sequence of Flax/Sisal hybrid composite laminates on mechanical properties like tensile and flexural strengths and free vibration characteristics like modal frequency, mode shapes and material damping was reported.

Understanding the behavior of natural hybrid composite beam in higher operating frequency range is more important since these structures are exposed to severe vibration in service. Many Authors [8–12] proposed a combined numerical-experimental technique for evaluating the orthotropic elastic constants of composite material. In the present work, the experimental and theoretical frequency results were combinedly used for finding the effective elastic constants of natural composite laminates by simple regression analysis. These effective elastic constants were further used for modal analysis of hybrid composite beams under higher frequency range. The outcome of this analytical study reports the influence of stacking sequence of Flax/Sisal hybrid composite beams on free vibration characteristics.

2 Fabrication of Flax/Sisal Hybrid Composite Laminates

Hand layup technique is used for fabricating the natural fiber reinforced hybrid composite laminates [13]. In the present study, two types of hybrid laminates were fabricated with uniform volume fraction of 0.3. Two types of natural fiber with different elastic modulus values are considered. Flax fiber of elastic modulus of 56 GPa and Sisal fiber with elastic modulus 15 GPa were preferred to study the effect of stacking sequence. Low viscous epoxy resin GY257 is used for increasing the adhesion between the fiber and matrix. After the curing process, the test specimens are cut from the preferred samples as per the standard dimensions. Table 1 shows the symbol used, stacking sequence, dimensions and density of the two hybrid beams considered.

3 Testing of Natural Hybrid Composite Laminates

3.1 Mechanical Testing

ASTM D3039 is the standard test procedure for evaluating the tensile strength of straight-sided rectangular FRP composite coupons. Tensile test specimens were cut to the size of 250 mm × 25 mm × t mm as per ASTM D3039 standard [14]. The tensile tests were conducted on INSTRON 3382® with 5 mm/min of cross head speed at room temperature.

ASTM D 790 is the standard test method to evaluate the flexural properties of composite materials in three point bending mode [15]. This method includes provisions for measuring maximum flexural strength, and flexural modulus. Flexural test specimens were cut to the size of 126 mm × 12.5 mm × t mm. To measure

Table 1 Types of hybrid composite laminate showing stacking sequence, dimensions and density

Symbol	Stacking sequence	Photograph	Dimension l * w * t (mm^3)	Density (kg/m^3)
H1	S F F S		250 * 250 * 3.8	1112.3
H2	F S S F		250 * 250 * 3.9	1114.5

L = length; w = width; t = thickness of laminate

the flexural stresses accurately, the span to depth ratio is controlled as 16:1. The flexural test was conducted on a standard SHIMAD2U-AUTOGRAPH® machine with feed rate of 1.2 mm/min.

3.2 Free Vibration Test Under Fixed-Free Boundary Condition

The preferred hybrid composite specimens were tested under fixed-free end condition to evaluate the modal frequency values and mode shapes [16, 17]. Flax/Sisal hybrid laminates are cut into a size of 250 * 25 * t mm for free vibration study. After fixing one end of the natural composite specimen using a fixture, the span length is set as 200 mm as shown in Fig. 1.

Roving Hammer Method is preferred to obtain the mode shapes of the natural hybrid composite beams. Excitation force is given at various specified points of equal interval of 20 mm distance in composite specimen using an impulse hammer. A tri-axial accelerometer is fixed in the free end of the specimen to capture the response of the composite material for the known excitation. The excitation force and corresponding vibration response of the specimen are fed into data acquisition card where these signals are processed through Fast Fourier Transform software to get the required frequency response function (FRF) plots. By following the similar

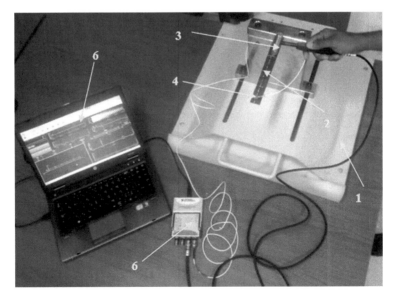

Fig. 1 Image showing Vibration test setup (1) Fixture (2) Composite specimen (3) Impact Hammer (Kistler - 9722A500) (4) Accelerometer (Kistler - 8766A500) (5) Data Acquisition Card (NI-ATA9234) (6) Computer with FFT Software (Dewesoft) showing FRF

procedure, FRF plots for each excitation point and so cumulative FRF plots are obtained to evaluate the mode shapes of composite beams.

3.3 Modal Analysis of Flax/Sisal Hybrid Composite Beam Using FEA

From the constituent material properties using the standard rule-of-mixture equations [18], nine orthotropic elastic constants of woven fabric Flax and Sisal composite laminae were evaluated. Finite element model of natural hybrid beam was developed for the actual beam size of 200 mm span length and 25 mm width using ANSYSv14.5. Layered configuration was done by specifying individual layer properties layer by layer from bottom to top with SHELL 181 element. Figure 2 shows the lay plot arrangement of Flax/Sisal hybrid beams H1 $[0_1/0_2/0_2/0_1]$ and H2 $[0_2/0_1/0_1/0_2]$ in finite element model where 1 and 2 represents Sisal and Flax fabric layers respectively and 0 represents the fiber orientation in terms of degrees considered for the present study. Material properties are assigned to each layer of finite element model according to required stacking sequence by means of the elastic constants evaluated using rule of mixture. Actual density of fabricated samples was given as mass property of composite beam.

4 Results and Discussion

4.1 Mechanical Properties of Natural Hybrid Composite Laminates

Table 2 shows the results obtained by performing tensile and flexural tests on Flax/ Sisal hybrid composite samples. The variation in tensile strength and modulus

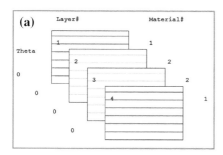 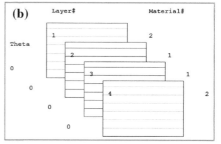

Fig. 2 Lay plots showing different stacking sequence and material property used for finite element model of Flax/Sisal hybrid beams **a** H1 and **b** H2

Table 2 Variation in tensile and flexural properties of Flax/Sisal hybrid laminates

Tensile strength (MPa)		Tensile modulus (GPa)		% of variation in Tensile modulus of H2 against H1	Flexural strength (MPa)		Flexural modulus (GPa)		% of variation in flexural modulus of H2 against H1
H1	H2	H1	H2		H1	H2	H1	H2	
27.3	32.5	2.6	3.1	19.2	47.0	60.9	2.6	3.9	50.0

among hybrid laminates is marginal. The tensile strength of the composite is influenced by strength and modulus of the reinforced fibers. During axial loading, all the four layers of natural fiber hybrid laminates share the tensile load and undergo axial strain with minimal difference in Flax and Sisal fiber layer elongation. This characteristic behavior caused only small difference in tensile modulus. Earlier investigations also showed that there was only minor deviation in tensile modulus of the hybrid laminates with low and high modulus fibers [19, 20].

Table 2 also shows the experimental results obtained by conducting flexural test on hybrid composite samples. Flexural strength of a composite laminate is mainly controlled by the strength of outer layer which is in direct contact with bending load [21]. There is a significant flexural strength variation between natural hybrid laminates H1 and H2. H2 layer arrangement has flexural modulus value 50% higher than that of H1 arrangement as shown in Fig. 3. The presence of high modulus Flax fiber in the outer layer in H2 laminate offers more resistance to flexural loading. Earlier experimental investigation by Ary Subagia et al. [22] and Murugan et al. [23] proved that placing high stiffness fiber away from the neutral axis of the laminate and the low stiffness fiber near the neutral axis, will enhance the flexural modulus significantly and which also confirms the present characteristic study.

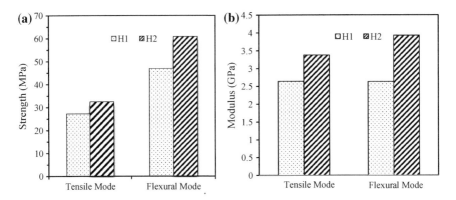

Fig. 3 Comparison graphs of **a** Tensile and **b** flexural properties of Flax/Sisal hybrid laminates

4.2 Free Vibration Characteristics of Flax/Sisal Hybrid Composite Beam

4.2.1 Frequency Response Function Plot

The FRF plots of natural hybrid beams with different stacking sequences were obtained for fixed-free end condition. Comparison of FRF plots of Flax/Sisal hybrid beams, H1 and H2 are shown in Fig. 4. It is revealed from the FRF plots that the amplitude of H1 spectrum is relatively less than H2 beam at all successive resonance sets. This attribute illustrates improved damping performance of the composite beam H1. Loss factor values of the preferred composite beams listed in Table 3 also confirms this attribute. Low amplitude in vibration spectrum represents high damping of composite material tested [24]. Further, free vibration response of hybrid beam H1 shows modified amplitude over the other hybrid arrangement, H2. The variation in natural frequency values of Flax and Sisal fabric layers plied in hybrid beams together supports the increased amplitude in free vibration.

Table 3 shows that there is small increase in natural frequency values between the two different composite beams in fixed-free end condition. It is found from Table 3 that the stacking sequence of Flax/Sisal hybrid beams influences the

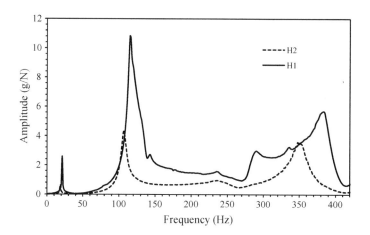

Fig. 4 Comparison of FRF plots of Flax/Sisal hybrid beams in fixed-free boundary condition

Mode No.	Modal frequency (Hz)		Loss factor (η)	
	H1	H2	H1	H2
I	19	21	0.153	0.109
II	106	115	0.087	0.059
III	348	376	0.075	0.061

Table 3 Comparison of modal frequency and loss factor of Flax/Sisal hybrid beams at successive modes

vibration characteristics considerably. Among the two hybrid beams, H2 layer arrangement exhibited improved vibration performance by offering higher resonant frequency level than the other layer arrangement H1 in all successive modes of vibration. It also shows that there is considerable percentage rise in resonant frequency of hybrid beam, H2 against H1 beam at all successive modes. The higher flexural modulus of H2 arrangement than H1 causes this modified performance. Table 3 also shows the loss factor (η) arrived from experimentally recorded FRF function for first three modes of hybrid composite laminates tested. It is observed that there is least variation in loss factor. This attribute is due to the uniform behavior of the common matrix present in both hybrid beams under free vibrating force [25].

4.2.2 Modal Response of Flax/Sisal Hybrid Composite Beams

The mode shapes and corresponding modal frequency values of two types of hybrid sample are reported in Table 4. The mode shapes illustrate the deformed pattern of the composite beams at different modal frequency level. H2 hybrid beam exhibits the first bending mode at 21 Hz whereas the corresponding value of H1 beam is only 19 Hz and similar trend is found for the other successive modes also. Higher flexural modulus value of H2 layering arrangement than H1 arrangement causes this increased modal frequency values at all successive modes. Increased modal frequency values of H2 beam at all modes of vibration reveal the improved dynamic stability.

4.3 Free Vibration Characteristics of Flax/Sisal Hybrid Composite Beams Using FEA

In preliminary stage, the theoretical modal frequency values were evaluated for natural hybrid beams H1 and H2 using the finite element model described in

Table 4 Comparison of experimental mode shapes of Flax/Sisal hybrid composite beams

Mode No.	H1		H2	
	Mode shape	Modal frequency (Hz)	Mode shape	Modal frequency (Hz)
I		19		21
II		106		115
III		348		376

Table 5 Effective elastic constants of flax and sisal lamina by regression analysis

Material	Effective elastic constants of Flax and Sisal fabric lamina					
	$E_{11} = E_{22}$ (GPa)	E_{33} (GPa)	G_{12} (GPa)	$G_{13} = G_{23}$ (GPa)	v_{12}	$v_{13} = v_{23}$
Flax	5.22	5.02	1.92	1.81	0.23	0.30
Sisal	9.48	6.7	2.50	2. 38	0.28	0.37

Sect. 3.3. The effective elastic constants of natural hybrid composite beams were evaluated by comparing the experimental and analytical modal frequency values using simple regression analysis. An error function is defined for the variation between the experimental modal frequency of the real specimen and theoretical modal frequency [26]. Then by minimizing this error function using simple regression analysis the effective elastic constants of hybrid composite beams were evaluated (Table 5).

Further these effective elastic constants are applied in the theoretical modal analysis of the two natural hybrid composite beams by finite element method. The basic mode shapes such as transverse, twisting and shear modes were tapped and corresponding resonant frequency values were noted (Fig. 5). The results obtained from theoretical modal analysis were reported in Table 6. Natural hybrid beams with two different stacking sequences showed a variation in free vibration characteristics. The layering arrangement of two different fibers Sisal and Flax having difference in their stiffness values, together affects the modal frequency of hybrid beams.

Table 7 shows the resonant frequency values of H1 and H2 hybrid beams, for the transverse, shear and twisting mode at higher frequency levels, evaluated by finite element method. Table 7 reveals that hybrid beam H2 exhibited higher resonant frequencies at transverse and twisting mode than H1 arrangement and there is no frequency variation for shear mode. The typical mode shapes of H2 hybrid beam at higher frequency range are shown in Fig. 5. The experimental and theoretical modal analysis of Flax/Sisal hybrid beams reveal that the stacking sequence of hybrid beam significantly influences the free vibration characteristics.

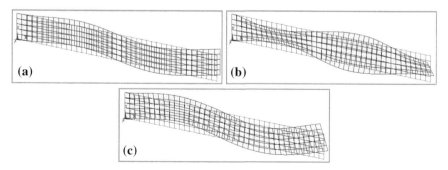

Fig. 5 Mode shapes of H2 hybrid beam under free vibration **a** 3rd transverse mode at 414.7 Hz **b** 3rd twisting mode at 1564.3 Hz **c** 3rd shear mode at 3804.5 Hz

Table 6 Comparision of experimental modal frequency values of Flax/Sisal hybrid beams with FEM results

Mode No.	Mode shape	Modal frequency values (Hz)					
		H1			H2		
		EXP	ANSYS	% of variation	EXP	ANSYS	% of variation
I		19	18.9	0.5	21	22.6	7.6
II		106	116.2	9.6	115	126.5	10.0
III		348	331.1	4.9	376	412.8	9.8

Table 7 Modal frequencies of various modes of Flax/Sisal hybrid composite beams

Mode No.	Modal frequency values of various modes (Hz)					
	Transverse mode		Twisting mode		Shear mode	
	H1	H2	H1	H2	H1	H2
1	18.9	23.6	278.4	302.3	256.9	256.9
2	118.4	147.8	842.8	917.6	1494.5	1494.5
3	332.5	414.7	1429.9	1564.3	3804.5	3804.5

5 Conclusions

The influence of stacking sequence on mechanical properties of woven fabric Flax/ Sisal hybrid composite laminates were experimentally evaluated and analyzed. During axial loading condition, the difference in tensile strength among hybrid laminates, H1 and H2, is very negligible. Under the bending load condition, the desired attribute was observed over the control of stacking sequence. The flexural modulus in the arrangement FSSF, designated as H2 was noticeably high. Since the outermost layers with reference to neutral axis held by Flax fabric carrying high intrinsic modulus of elasticity facilitated to increased resistance to deflection in progressive bending load.

Experimental free vibration characteristics of natural hybrid beam specimens revealed that H2 stacking sequence performed with improved resonance frequency than the other stacking sequence, H1. The increased flexural modulus about 50% hold by H2 stacking sequence than the H1 arrangement confirms this attribute. The stacking sequences with two different elastic modulli of fibers resulted in laminates with different strength arrangement across the thickness and so exhibited modified performance in flexural vibration condition. Free vibration study of hybrid composite beams under fixed-free end condition showed only marginal difference in passive damping characteristics for the control over stacking sequence. Further, finite element results showed that hybrid arrangement H2 offered enhanced vibration stability even at higher operating frequency range than the other layering arrangement H1. Both experimental and theoretical modal analysis of Flax/Sisal hybrid beams were useful in understanding the stacking sequence effect over the mechanical properties and free vibration characteristics and establishing an effective stacking sequence for natural hybrid beam with increased dynamic stability.

References

1. Peters ST (1998) Handbook of composites, 2nd edn. Chapman & Hall
2. Sabeel Ahmeda K, Vijayarangan S (2008) Tensile, flexural and interlaminar shear properties of woven jute and jute-glass fabric reinforced polyester composites. J Mater Process Technol 207:330–335
3. Ramesh M, Palanikumar K, Hemachandra Redddy K (2013) Comparative evaluation on properties of hybrid glass fiber- sisal/jute reinforced epoxy composites. Procedia Eng 51:745–750
4. Ajith Gopinath, Senthil Kumar M, Elayaperumal A (2014) Experimental investigations on mechanical properties of jute fiber reinforced composites with polyester and epoxy resin matrices. Procedia Eng 97:2052–2063
5. Akash DA, Thyagaraj NR, Sudev LJ (2013) Experimental study of dynamic behavior of hybrid jute/sisal fiber reinforced polyster composites. Int J Sci Eng Appl 2(8)
6. Prabhakaran S, Krishnaraj V, Senthilkumar M, Zitoune R (2014) Sound and vibration damping properties of flax fiber reinforced composites. Procedia Eng 97:573–581
7. Murugan R, Ramesh R, Padmanabhan K (2016) Investigation of the mechanical behavior and vibration characteristics of thin walled glass/carbon hybrid composite beams under a fixed-free boundary condition. Mech Adv Mater Struct 23(8):909–916
8. Scida DZ, Aboura ML, Benzeggagh E, Bocherens (1997) Prediction of the elastic behaviour of hybrid and non-hybrid woven composites. Compos Sci Technol 57:1727–1740
9. Araujo AL, Mota Soares CM, Moreira de Freitas MJ, Pedersen P, Herskovits J (2000) Combined numerical-experimental model for the identification of mechanical properties of laminated structures. Compos Struct 50:363–372
10. Hwang SF, Wu JC (2010) Elastic constants of composite materials by an inverse determination method based on a hybrid genetic algorithm. J Mech 26:345–353
11. Pagnotta L, Stiglino G (2010) Elastic characterization of isotropic plates of any shape via dynamic tests: theoretical aspects and numerical simulations. Mech Res Commun 35:351–360
12. Rikards R, Ghate A, Gailis G (2001) Identification of elastic properties of laminates based on experiment design. J Solids Struct 38:5097–5115
13. ASM Hand Book (2001) Composites, ASM International, The Material Information Company
14. ASTM Standard D3039 (2005) Standard test method for tensile properties of polymer matrix composite materials. ASTM International, West Conshohocken, PA
15. ASTM Standard D790 (2003) Standard test method for flexural properties of unreinforced and electrical insulating materials. ASTM International, West Conshohocken, PA
16. Ewins DJ (2000) Modal testing: theory, practice and application. Research Studies Press, Hertfordshire
17. ASTM Standard E756 (2005) Standard test method for measuring vibration damping properties of materials. ASTM International, West Conshohocken, PA
18. Akkerman Renako (2006) Laminate mechanics for balanced woven fabrics. Compos B 37:108–116
19. Bunsell AR, Harris B (1974) Hybrid carbon and glass fiber composites. Composites 5:157–164
20. Pandya KS, Veerraju, Naik NK (2011) Hybrid composites made of carbon and glass woven fabrics under quasi-static loading. Mater Des 32:4094–4099
21. Kretsis G (1987) A review of the tensile, compressive, flexural and shear properties of hybrid fiber reinforced plastics. Composites 18:13–23
22. Ary Subagia I, Kim Y, Tijing LD, Kim CS, Shon HK (2014) Effect of stacking sequence on the flexural properties of hybrid composites reinforced with carbon and basalt fibers. Compos B 58:251–258

23. Murugan R, Ramesh R, Padmanabhan K (2014) Investigation on static and dynamic mechanical properties of epoxy based woven fabric glass/carbon hybrid composite laminates. Procedia Eng 97:459–468
24. Ingle RB, Ahuja BB (2006) An experimental investigation on dynamic analysis of high speed carbon-epoxy shaft in aerostatic conical journal bearings. Compos Sci Technol 66:604–612
25. Kyriazoglou C, Guild FJ (2007) Finite element prediction of damping of composite GFRP and CFRP laminates—a hybrid formulation—vibration damping experiments and Rayleigh damping. Compos Sci Technol 67:2643–2654
26. Murugan R, Ramesh R, Padmanabhan K (2014) Investigation on vibration behaviour of cantilever type glass/carbon hybrid composite beams at higher frequency range using finite element method. Adv Mater Res 984–985

Influence of Cryogenic Treatment on As-Cast AZ91+1.5 wt%WC Mg-MMNC Wear Performance

P. Karuppusamy, K. Lingadurai and V. Sivananth

Abstract In this work, the enhancement of wear performance of cryogenic treated (CT) as-cast AZ91 reinforced with 1.5 wt% WC magnesium metal matrix nano-composite (Mg-MMNC) had been explored using pin-on-disc tribometer and scanning electron microscope (SEM). AZ91 with 1.5 wt% WC reinforcement was prepared with stir casting process and the cryogenic treatment was carried out at -190 °C. The wear test parameters were the applied normal loads of 20 and 40 N, sliding velocities of 1.0, 1.6, 2.1 and 3.1 m/s and a constant slipping distance of 1200 m with tribo-couples of aluminium disc and Mg-MMNC pin at atmospheric conditions. At lower loads, almost all the samples had showed the similar wear loss. But in higher loads, there was significant reduction in wear loss for cryogenic treated Mg-MMNC. The presence of reinforcement and increased $Mg_{17}Al_{12}$ phase particles volume fraction due to CT had significantly enhanced the wear resistance of composite. There was also wear loss in counter-part aluminium disc. The SEM analyses of worn surface indicate that there was abrasion and oxidation. And a changeover occurred from oxidation wear to delamination wear and abrasion wear while the applied load was increased from 20 to 40 N. Adhesion wear took place at the sliding condition of load 40 N and speed 3.1 m/s.

Keywords Mg-MMNC · WC · Cryogenic treatment · SEM · Wear mechanism

P. Karuppusamy (✉)
Department of Automobile Engineering, Dr. Mahalingam College
of Engineering and Technology, Pollachi, India
e-mail: p.karuppusamy@gmail.com

K. Lingadurai
Department of Mechanical Engineering, University College of Engineering,
Dindigul, India
e-mail: lingadurai@gmail.com

V. Sivananth
Engineering Department, Mechanical/Industrial Section,
Ibri College of Technology, Ibri, Oman
e-mail: vsivananth@gmail.com

© Springer Nature Singapore Pte Ltd. 2019
A. K. Lakshminarayanan et al. (eds.), *Advances in Materials and Metallurgy*,
Lecture Notes in Mechanical Engineering,
https://doi.org/10.1007/978-981-13-1780-4_19

1 Introduction

The main drivers for the introduction of new materials into automobiles are Cost, Fuel economy, Environmental (emissions), Performance and Recyclability [1]. Lightweight materials can significantly contribute to enhance the fuel economy and to meet emission norms through weight reduction [2]. Since, the magnesium alloys are having low density ($1.75–1.87$ g/cm^3) compared to other structural metallic materials, they have solid prospective for weight reduction in a widespread range of structural applications. Magnesium possesses other benefits including good castability, better machinability, high specific strength and recyclability [3]. Therefore, in the future, a remarkable usage of magnesium alloys in the automobile industries has been anticipated. On the other hand, magnesium alloys have poor formability, narrow ductility at room temperature, poor wear resistance and high corrosion rate [4–6]. Chen et al. [5] experimented and found that the low hardness of AZ91D alloy resulted in relatively poorer wear resistance than those of the other commercial alloys. These limitations can be improved either by novel alloy development, addition of reinforcements to magnesium matrices [7] or heat/cryogenic treatments [8]. The commonly used reinforcements are SiC, Al$_2$O$_3$, AlN, B$_4$C, TiC, TiB$_2$, CNT, Y$_2$O$_3$ and WC [9]. Gui et al. [10] prepared the magnesium matrix composites using a vacuum assisted stir casting process with Mg-Al9Zn and Mg-Zn5Zr as matrix materials and SiCp as reinforcement. The results from Selvam et al. [11] showed that there was an enhancement in wear resistance of magnesium base alloy considerably, when ZnO nano-particulates added to magnesium matrix. Nguyen et al. [12] employed the pin-on-disc wear test set up to find the wear rate of AZ31B alloy and AZ31B/nano-Al$_2$O$_3$ composites under dry sliding condition and results showed that the minimum value in the wear rates of both base alloy and composites were recorded at a critical speed of 5 m/s for 10 N or 3 m/s for 30 N. The review of Dey et al. [13] on Mg-MMC concluded that Mg-MMC exhibited an improved wear resistance than the base alloy and AZ91 MMC displayed the highest resistance among Mg-MMCs. Further, Cryogenic treatment (CT) also boosted the wear resistance of magnesium alloys. Liu et al. [14] described that the cryogenic treatment featured to refine and increase in volume fraction of secondary phase particles, which in turn improved the wear resistance. Asl et al. [15] described that in AZ91 strengthening effect at room temperature, β phase morphology and the internal microstructure stabilisation were the key factors. Hence, the wear resistance was better at higher loads and sliding speeds. Reinforcement of TiC and WC particles through laser surface treatment increased the hardness values of magnesium alloys [16]. Rockwell method according to scale F was used to find the hardness values. Sanchez et al. [17] and Yildiz et al. [18] reported that WC–Co coating on magnesium alloy contact surface is an effective method to enhance the wear performance. The XRD analyses of AZ91 alloy plasma sprayed with WC–Co showed that the composite layer consisted of Mg$_{17}$Al$_{12}$, WC, W$_2$C, Mg and Mg$_2$C$_3$ phases [19]. Chelliah et al. [20] proposed a schematic model illustrating the wear mechanisms of the AZX915-TiCp composite. Mild wear regime and severe regime were two wear regimes were exposed from SEM analysis on worn surfaces [21]. Oxidation wear, abrasion wear and delamination wear were occurred in the mild wear regime. Wear due to thermal softening and surface melting were in the severe wear regime. A wear transition map was used to summarise the wear mechanisms in each wear regime.

There are limited research literatures available on WC particles as reinforcement in stir casting metal matrix composites. Ravikumar et al. [22] reported that the addition of WC particles enhanced the strength of the Al/WC composites only up to 8 wt% and further increase of WC lead to decrease of strength. Over and above, there are no studies conducted on magnesium and aluminium tribo-couples. In this context, the present study is an attempt to explore the wear characteristics and the related mechanisms of AZ91 reinforced with 1.5 wt% WC Mg-MMNC against aluminium as counter surface. The cost economic and easily adoptable stir casting process was engaged to fabricate the Mg-MMNC.

2 Experimental Details

2.1 Fabrication and Cryogenic Treatment of AZ91+1.5 Wt% WC Mg-MMNC

Since AZ91 MMC displays the highest wear resistance among Mg-MMCs [13], the AZ91 was selected as matrix material in this study. To fabricate AZ91 alloy, commercially available Mg, Al ingots and Zn wire having 99.9% purity were used as alloying elements. WC particulates with 500 nm in size were used as reinforcement. The density of the alloying elements and reinforcement are shown in Table 1.

Vacuum-incorporated stir casting process was chosen for this work to prepare of AZ91+1.5 wt%WC Mg-MMNC. Zhao et al. [23] concluded that the microstructure of AZ91 could be refined by employing the thermal rate treatment (TRT) technique for melting the alloying elements. And $Mg_{17}Al_{12}$ phase precipitations were found to be identical and more dispersive. The casting defects, for instance, hot tears and shot run can be decreased by the TRT technique. Therefore, TRT had been followed for the casing process. The work of Rzychon et al. [24] showed that good castability and mechanical properties were obtained for alloy, which is poured at 780 °C. Hence, the pouring temperature was maintained as 780 °C. The alloying

Table 1 Density of the elements	Elements	Density (ρ) g/cm^3
	Mg	1.73
	Al	2.70
	Zn	7.13
	WC	15.63

elements were placed inside the mild steel crucible. In order to remove the Fe
content present in the raw material, Mn (<0.1%) is to be added to the melt.
Therefore, the separately fabricated master alloy of Al-10%Mn was placed inside
the crucible along with the alloying elements. The elements were superheated to
850 °C in the crucible using an induction furnace. The steel crucible was unin-
terruptedly filled with Argon gas to circumvent the risk of flammability with
magnesium. At the meantime, 1.5 wt% of WC particles was also preheated to 300 °
C in a separate chamber attached with the furnace. Zirconia ceramic coated
mechanical stirrer gently driven at the rate of 500 rpm to create vortex motion in the
molten metal. Then the preheated WC particles were released into the molten AZ91
alloy and the stirring process was continued to 10 min further to disperse WC
particulates uniformly in the matrix. Consequently, the temperature of the molten
alloy was brought down to the pouring temperature 780 °C and maintained. After
5 min, it was released from the crucible to a die which was kept in a die chamber
under Argon gas atmosphere. The schematic diagram of melting technique is given
in Fig. 1. The surface of the fabricated cast samples were in black colour which
indicated the oxidation took place on the surface during pouring.

The wear test specimen should be a cylindrical pin with diameter and length of
10 and 30 mm respectively. Therefore, the fabricated as-cast composite was
machined to the required shape and dimensions. In this study, Aluminium 6061—
T6 alloy was the tribo-pair, i.e. disc material. The chemical compositions of AZ91
and Aluminium 6061 are shown in Table 2. Then the machined sample was placed
in cryogenic treatment chamber where the working fluid was liquid nitrogen. The
temperature was gradually lowered to approximately −190 °C at a rate of 2 K per
minute and the soaking time was 24 h. Then gradually the specimen was led back
to room temperature.

Fig. 1 Sketch map of
melting process: (0–1)
melting, superheating to
850 °C; (1–2) Stirring, adding
WC; (2–3) chilling to 780 °C,
stirring; (3–4) refining,
setting; (4–5) pouring

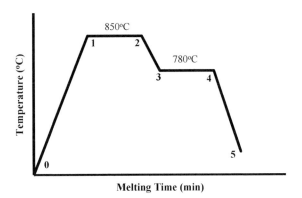

Table 2 Chemical composition of materials

Designation	Chemical composition (wt%)							
Al 6061—T6	Mg	Si	Cu	Mn	Fe	Cr	Ti	Al
	0.92	0.45	0.2	0.1	0.2	0.1	0.05	Balance
AZ91	Al	Zn	Si	Mn	Mg			
	8.7	0.7	0.1	0.1	Balance			

2.2 Dry Sliding Wear Test

Winducom—Pin-on-disc tribometer was used to evaluate the dry sliding wear characteristics of cryogenic treated Mg-MMNC pins. Aluminium alloy disc having a diameter of 170 mm and a thickness of 8 mm was used as disc material. The experiments were performed with the load maintained as constant and varying the sliding speeds as 1.0, 1.6, 2.1 and 3.1 m/s. The tests were repeated for loads 20 and 40 N. The sliding distance was kept as a constant value of 1200 m. The test samples were weighted before and afterwards of each experiment. In order to evaluate the coefficient of friction (COF), it is required to document the friction force and normal force continuously. The in-build load cell in tribometer and data acquisition system monitored the forces. The wear in terms of mass loss had been measured after each test. Wear rate had been represented in terms of the volume loss per unit sliding distance (mm^3/km). To ensure the repeatability of the obtained data, each experiment was repeated and averages of the observations were preceded for the further discussions. Scanning electron microscopy (SEM) is used to identify the principal wear mechanisms take place during the sliding wear. Both the contact surfaces (pin and disc) were studied using SEM.

3 Results and Discussion

3.1 EDS, SEM and Hardness

The energy dispersive spectrometry (EDS) analysis of Mg-MMNC (Fig. 2) confirms that the existence of magnesium (Mg), aluminium (Al), zinc (Zn), tungsten (W) and graphite (C) elements. Cho et al. [25] reported that there was a decomposition of hard WC to less hard W_2C, W and graphite (C). The reaction as follows:

$$2WC \rightarrow W_2C + C \qquad (1)$$

$$WC \rightarrow W + C \qquad (2)$$

The SEM analysis (Fig. 3) showed that there was an increase in secondary β phase precipitation ($Mg_{17}Al_{12}$) in cryogenic treated sample due to the less solubility

Fig. 2 EDS of AZ91+1.5% WC

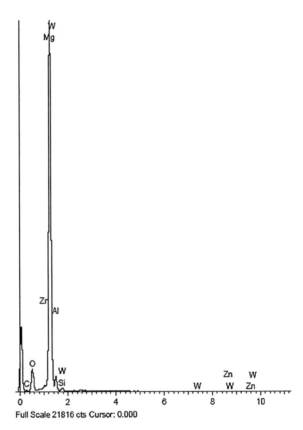

Full Scale 21816 cts Cursor: 0.000

of aluminium in magnesium at low temperature. The existence of hard tungsten carbide also increased the hardness of the composite. The hardness was measured using the Rockwell method according to scale E and the average values of the samples were plotted in Fig. 4. The addition of WC and cryogenic treatment increased the hardness value by 11% compared to the base alloy AZ91.

3.2 Tribological Responses

The volumetric wear rate variation and coefficient of frictional for tribo-couples (Al vs Mg-MMNC) were plotted against the sliding speeds. In order to quantify the changes in the geometry, the mass loss had been converted into volumetric loss. They are presented in Fig. 5. The wear rate greatly depends on the applied load and reinforcement. While increasing the applied load, the wear rate was amplified. At higher load 40 N, the maximum wear rate was obtained for all samples. The frictional coefficient was varying between 0.21 and 0.25.

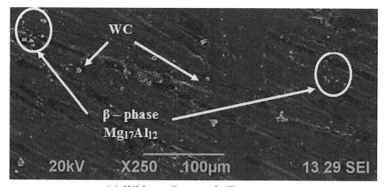

(a) Without Cryogenic Treatment

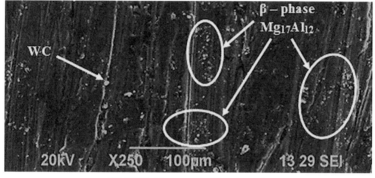

(b) With Cryogenic Treatment

Fig. 3 SEM images of Mg-MMNC

Fig. 4 Hardness value of pin and disc materials

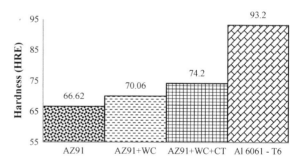

During the start of sliding, there may be uneven contact between the contact surfaces. This would result in minor grooves and pull out certain of the reinforcements from the surface of the pin. Once the speed was increased, the worn debris was compacted on grooves of the pin surface and formed a protective film on the contact surface. The existence of protective layer could be the reason for the reduction of wear rate. For the lower load 20 N, the trend of wear rate was

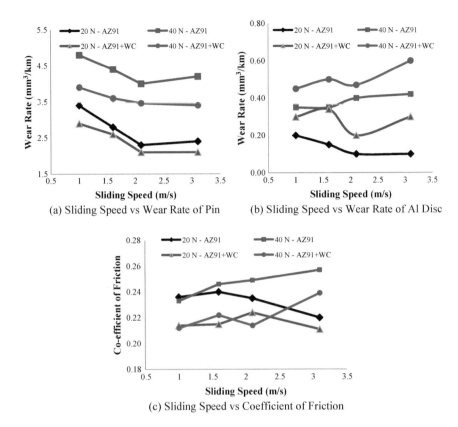

(a) Sliding Speed vs Wear Rate of Pin   (b) Sliding Speed vs Wear Rate of Al Disc

(c) Sliding Speed vs Coefficient of Friction

Fig. 5 Tribological responses

decreasing. But for higher load 40 N, the wear rate was in decreasing trend only up to the critical speed of 2.1 m/s. Afterwards there was an increase in wear rate due the existence of adhesive wear. This describes that the further increase in the speed would result in more weight loss of the pin. The friction coefficient plotted in Fig. 5c which showed that there was an increase in coefficient of friction at higher load 40 N due to adhesiveness of contact surface. Meanwhile compared with AZ91 alloy, the Mg-MMNC showed a better result on wear and friction coefficient. The wear rate of counter face Al disc also plotted and shown in Fig. 5 There was an increasing trend of wear rate when the Mg-MMNC pin was deployed.

3.3   SEM Analysis of Worn Surfaces

Abrasion

Figure 6a–c show that the presence of grooves in the tested specimen in the sliding direction of the pin. The depth of the grooves present in the worn surfaces indicates that the there was a higher penetration with WC particles. The removal of material fragments created the grooves in the pin surface. Hence, it was proposed that the abrasive wear is preceding wear mechanism for the tested sliding conditions. Srinivasan et al. [26] also proposed this kind of feature. It means that the abrasion takes place predominantly in cutting mode at the lesser load 20 N. During the cutting mode, there was a removal of material fragments and formed grooves. The metal to metal contact during the sliding generated the frictional heating. It is directly proportional to the applied load. Consequently, the frictional heating caused a temperature rise on the contact surfaces. The temperature rise on the pin surface resulted in a plastic deformation of the magnesium matrix. There was a slight material shift on either side of the grooves without the detachment of the material from the surface. It was an indication of the mode transition of wear from cutting mode to ploughing mode.

Oxidation

The underlying study revealed that the surfaces of the tested pins under the load 20 N appeared to be dark at speeds of 1 and 1.6 m/s. Then, at the load 40 N and speeds 2.6 and 3.1 m/s, the worn surfaces retain their metallic polish. From Fig. 6a, it was found that the pin surfaces were covered with a thin layer of oxidised fine particles. The temperature rise due to the frictional heating triggered the oxidation on the pin surface. By the removal of these oxide covers, the wear was occurring. Hence, it was concluded that there was a presence of oxidative wear mechanism. By continuous sliding, the valleys on the pin surface was filled with oxide wear debris and squeezed to form as a protective layer. This layer barred the metal to metal contact and so the wear rate was dropped [27]. In the current study, the near-continuous oxide layers were only in moderate sliding condition, meaning that at low load of 20 N at speeds of 1.0 m/s. At the higher applied loads and sliding speed, the stimulated frictional heat becomes sufficient to decay the protective oxide layer. So, the oxide layer was irregular. In addition to the layer, the higher hardness of WC would be responsible for some support to the pin surface that developed the ability to sustain the higher load. From the earlier investigations [28], it was understood that reinforcements might be mixed inside the oxide layer and enhanced the hardness of the sliding pin surface. This enriched the wear resistance of the layer. In this perspective, during the oxidation, the compounded improved wear rate was the benefit of WC reinforcement.

Delamination

The SEM images of worn surfaces showed the tiny cracks almost vertical to the direction of sliding. Because of the intersection in the stimulated cracks, sheet-like wear particles were found on the surface. This was the feature that associated with

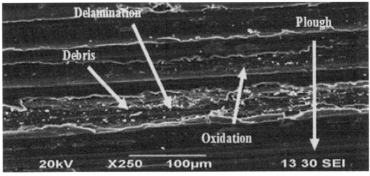

(a) SEM image of Mg-MMNC slided at speed 3.1 m/s and load 20 N

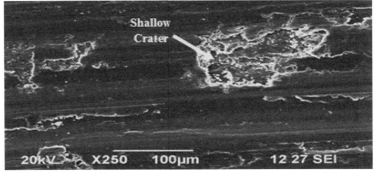

(b) SEM image of Mg-MMNC slided at speed 2.1 m/s and load 40 N

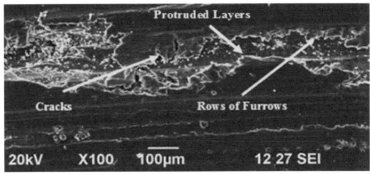

(c) SEM image of Mg-MMNC slided at speed 3.1 m/s and load 40N

Fig. 6 SEM images for analysing the wear mechanism

the process of delamination. Delamination was detected more extensively when the applied load was 40 N. Accordingly the wear rates of the specimens were greater under the higher applied loads. WC–matrix interface provided further void nucleation spots and also favoured the crack propagation tracks. Along with the nucleation sites, growth in delamination caused more wear of contact surface layers of the pin. Figure 6b showed that the delamination had created the shallow crater on

Fig. 7 SEM images of disc surface

the contact surface by the removal of worn-out thin sheet-like materials. The combination of lower speeds and higher loads showed the higher delamination wear.

Adhesion

SEM image of pin tested at higher load (40 N) and speed (3.1 m/s) condition is shown in Fig. 6c. As discussed earlier, the higher load and speed amplified the frictional heating which was sufficient to disintegrate the contact surface layer from the pin. As a result, the disintegrated layers were projected on the surface slightly. The surface also seemed to be quite featureless and wide-ranging transfer of material to the disc surface was observed. From Fig. 7, it was observed that the transferred material was trapped and formed a layer on the sliding pathway of the aluminium disc. There was an increase in the transferring of material with increase in sliding speed which enlarged the groove thickness. These elements imply that the presence of adhesive wear, and its intensity increased with applied load and sliding speed [29].

4 Conclusion

1. Wear testing of cryogenic treated AZ91 alloy and its 1.5% WC-reinforced composite pins against an aluminium counter-face were conducted under dry sliding condition. The wear mechanisms abrasion, oxidation, delamination and adhesion were found to be operated under the tested sliding conditions.
2. Abrasion was the principal wear mechanism that took place in all the sliding condition. Oxidation was the wear mechanism with the mild sliding condition along with abrasion. Since, the composite had the ability to sustain a stable oxide film; it exhibited better wear resistance (10–20% improvement) against the aluminium disc.

3. When the pin was slide under the moderate condition, there was a transition of wear mechanism i.e. delamination was occurred. It created the shallow crater on the pin contact surface.
4. Under severe sliding wear conditions, adhesion was predominant wear mechanism. The composite showed a significant improvement in wear resistance compared to base alloy. Even though, the adhesive wear restricts the Mg-MMNC to use only for the mild sliding conditions.

References

1. Froes FH (1994) Advanced metals for aerospace and automotive use. Mater Sci Eng, A 184:119–133
2. Hirsch J, Al-Samman T (2013) Superior light metals by texture engineering: optimized aluminum and magnesium alloys for automotive applications. Acta Mater 61(3):818–843
3. You S, Huang Y, Kainer KU, Hort N (2017) Recent research and developments on magnesium alloys. J Magnes Alloy
4. An J, Li RG, Lu Y, Chen CM, Xu Y, Chen X, Wang LM (2008) Dry sliding wear behavior of magnesium alloys. Wear 265:97–104
5. Chen TJ, Ma Y, Li B, Li YD, Hao Y (2007) Friction and wear properties of permanent mould cast AZ91D magnesium alloy. Mater Sci Technol 23(8):937
6. Singh Raman RK, Birbilis N, Efthimiadis J (2004) Corrosion of Mg alloy AZ91—the role of microstructure. Corros Eng Sci Technol 39
7. Gupta M, Wong WLE (2015) Magnesium-based nanocomposites: lightweight materials of the future. Mater Charact 105:30–46
8. Yan B, Dong X, Ma R, Chen S, Pan Z, Ling H (2014) Effects of heat treatment on microstructure, mechanical properties and damping capacity of Mg–Zn–Y–Zr alloy. Mater Sci Eng, A 594:168–177
9. Kainer KU (2006) Metal matrix composites: custom made materials for automotive and aerospace engineering. Wiley-VCH, Weinheim, Chichester
10. Gui M, Han J, Li P (2003) Fabrication and characterization of cast magnesium matrix composites by vacuum stir casting process. J Mater Eng Performance 12(2):128–134
11. Selvam B, Marimuthu P, Narayanasamy R, Anandakrishnan V, Tun KS, Gupta M, Kamaraj M (2014) Dry sliding wear behaviour of zinc oxide reinforced magnesium matrix nano-composites. Mater Des 58:475–481
12. Nguyen QB, Sim YHM, Gupta M, Lim CYH (2015) Tribology characteristics of magnesium alloy AZ31B and its composites. Tribol Int 82:464–471
13. Dey A, Pandey KM (2015) Magnesium metal matrix composites—review. Rev Adv Mater Sci 42:58–67
14. Liu Y, Shao S, Xu C, Yang X, Lu D (2012) Enhancing wear resistance of Mg–Zn–Gd alloy by cryogenic treatment. Mater Lett 76:201–204
15. Asl KM, Tari A, Khomamizadeh F (2009) Effect of deep cryogenic treatment on microstructure, creep and wear behaviors of AZ91 magnesium alloy. Mater Sci Eng, A 523:27–31
16. Dobrzański LA, Domagała J, Tański T, Klimpel A, Janicki D (2009) Laser surface treatment of cast magnesium alloys. Arch Mater Sci Eng 35(2):101–106
17. Sanchez E, Bannier E, Salvador MD, Bonache V, Garcia JC, Morgiel J, Grzonka J (2010) Microstructure and wear behaviour of conventional and nanostructured plasma sprayed WC–Co coatings. J Therm Spray Technol 19(5):964–974

18. Yildiz F (2014) Tribological properties of WC–12Co coating on AZ91 magnesium alloy fabricated by high velocity oxy-fuel spray. High Temp Mater Proc 33(1):41–48
19. Mehrjou B, Soltani R, Sohi MH, Torkamany MJ, Valefi Z, Ghorbani H (2016) Laser surface treatment of AZ91 magnesium alloy presprayed with WC–Co. Surf Eng 1–10
20. Chelliah NM, Sing H, Surappa MK (2016) Correlation between microstructure and wear behavior of AZX915Mg-alloy reinforced with 12 wt% TiC particles by stir-casting process. J Magnes Alloy
21. Liang C, Han X, Su TF, Li C, An J (2014) Sliding wear map for AZ31 magnesium alloy. Tribol Trans 57:1077–1085
22. Ravikumar K, Kiran K, Sreebalaji VS (2017) Characterization of mechanical properties of Aluminium/Tungsten Carbide composites. Measurement
23. Zhao P, Gengb H, Wang Q (2006) Effect of melting technique on the microstructure and mechanical properties of AZ91 commercial magnesium alloy. Mater Sci Eng, A 429:320–323
24. Rzychon T, Szala J, Lielbus A (2012) Microstructure, castability, microstructural stability and mechanical properties of ZRE1 Magnesium Alloy. Arch Metall Mater 57(1):245–252
25. Cho TY, Yoon JH, Kim KS, Song KO, Joo YK, Fang W, Zhang SH, Youn SJ, Chun HG, Hwang SY (2008) A study on HVOF coatings of micron and nano WC–Co powders. Surf Coat Technol 202:5556–5559
26. Srinivasan M, Loganathan C, Kamaraj M, Nguyen QB, Gupta M, Narayanasamy R (2012) Sliding wear behaviour of AZ31B magnesium alloy and nanocomposite. Trans Nonferrous Met Soc China 22:60–65
27. Stott FH, Wood GC (1978) The influence of oxides on the friction and wear of alloys. Tribol Int 11:211–218
28. Wilson S, Alpas AT (1997) Wear mechanism maps for metal matrix composites. Wear 212:41–49
29. Mishina H, Hase A (2013) Wear equation for adhesive wear established through elementary process of wear. Wear 308:186–192

Dynamic Mechanical Analysis of Sub-micron Size Flyash Particles Filled Polyester Resin Composites

P. Nantha Kumar, A. Rajadurai and T. Muthuramalingam

Abstract The use of polymer matrix based composites in manufacturing industries tends to increase due to its lowered density and strength to weight ratio. In the present investigation, a trial has been made to analyze the dynamic mechanical behavior of flyash particles filled polyester resin composite. The polyester resin based polymer composite has been synthesized with 2, 3, and 4% of reinforcement. The vibration behavior under operating temperature of the composite have been studied and analyzed. From the DMA analysis, it has been observed that the synthesized polymer matrix composites can withstand higher vibrations within the operating temperature.

Keywords Flyash · Vibration · Polyester · Composites

1 Introduction

The need of flyash-filled polymer composites is rapidly growing in the present scenario. Among various fillers available in the market, byproduct of thermal power plants is being pollutants in the landfill as well atmosphere. The addition of Flyash (FA) can augment the mechanical properties of polymer matrix composites

P. Nantha Kumar (✉)
Department of Mechanical Engineering, Sri Sairam Institute of Technology,
Chennai, India
e-mail: mechnantha@gmail.com1

A. Rajadurai
Department of Production Technology, MIT Campus, Anna University,
Chennai, India
e-mail: rajadurai@mitinidia.edu

T. Muthuramalingam
Department of Mechatronics Engineering, SRM Institute of Science
and Technology, Kattankulathur, India
e-mail: muthu1060@gmail.com

© Springer Nature Singapore Pte Ltd. 2019
A. K. Lakshminarayanan et al. (eds.), *Advances in Materials and Metallurgy*,
Lecture Notes in Mechanical Engineering,
https://doi.org/10.1007/978-981-13-1780-4_20

[1]. In the polymeric composites, the important and basic mechanical properties such as tensile strength, modulus of elasticity, bending strength and bending modulus can be considerably modified by filling FA particles and Aluminum Silicate powders [2]. It has been found that the FA cenosphere particles filled AZ91D based composite can increase the compressive strength than stir cast aluminum alloy. The FA particles have been used in polymeric composites as in metal matrix composites to alter some mechanical properties and reuse the FA particles to some extent [3]. The tensile and flexural strength of the composites have been enhanced with the addition of FA particles. Nevertheless the strength properties of the composites have been lowered for higher accumulation owing to the filler agglomerations at higher filler contents in PET composite [4]. The thermal degradation and ablative rate of cenosphere-added composites can be lowered with higher cenosphere concentration. Hence it has been inferred that cenosphere-added composites could be acted as a thermal inhibitor to prevent material degradation of PP matrix [5]. FA particles could be added to various resins to study the performance of final matrix as compared with plain resin [6]. Optimization techniques could also be used to find the influence of multiple factors based on the input parameter. Among three input parameters ball milling hours contributed more on deciding the experiment for making final polyester resin composite [7]. Thermal property and flexural strength of composite such as degradation and flexural strength could be altered by adding processed FA particles [8]. From the past study, it has been found that only little focus has been given to analyze the effects of flyash addition on the dynamic mechanical behavior of the composites. Hence the present study has been made to investigate the effects of FA addition on vibration behavior of polyester resin composites in the present study.

2 Experiments and Methods

In the present work, the sub-micron level size FA particles have been mixed uniformly in polyester resin using Johnson Photosonic ultra sonic cavitation arrangement. The polyester resin based composite has been made as plates by filling FA particles of three different weight percentages such as 2, 3, and 4%. The pre required size of flyash particles has been reduced using high energy rate ball milling arrangement. The as received FA particles ranging in the size from 580 nm to 3.2 μm have been purchased from Neyveli Lignite Corporation (NLC), Tamil Nadu, India. The as received FA particles have to be processed to alter their original sizes into sub micron level. The size reduction of raw FA particles under different ball milling hours such as 10, 20 and 45 h is shown in Fig. 1. The SEM microstructure image of the developed FA particles reinforced polymer matrix composites is shown in Fig. 2. The calculated percentage of FA filler particles have been added with polyester liquid resin matrix and poured into the mould cavity. In the ultrasonic cavitation setup, the tip of the ultrasonic horn has been made to disperse FA particles inside the liquid resin. The ultrasonic frequency waves have

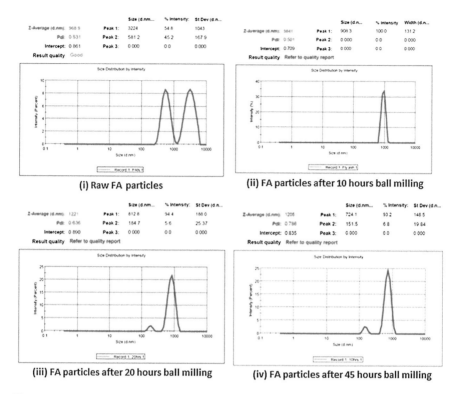

Fig. 1 FA particles under different ball milling hours such as 10, 20 and 45 h

been passed into the liquid with pulse frequency of 20 kHz and 2 kW power for about 30 min. Due to these vibrations, the heat energy has been developed inside the mould. Then the liquid resin with FA particles added compound has been poured in the required shape of mould cavity for curing purpose at ambient temperature. The dynamic behavior has been analyzed using DMA Q800 instrument under temperature sweep mode. The specimen has been placed in a chamber under oscillation frequency of 1 Hz. Then the temperature has been gradually increased from 20 to 200 °C to analyze the performance of the specimens with the temperature increment under dynamic vibration condition. The rate of temperature rise has been chosen as 10 °C per minute.

Fig. 2 SEM image of
synthesized FA particles
reinforced polymer
composites

3 Results and Discussion

3.1 Influence of FA Particles on Dynamic Mechanical Behavior

Figure 3 shows the influence of FA particles on dynamic behavior of the synthesized polymer composites. The Tan Delta curve indicates the temperature at which the polymer composites would lose its ability of solid and transferred into liquid state. That is, the polymer composite losses its property to be an engineering material. From Fig. 3 it has been observed that the plain resin has possessed lower temperature stability under dynamic condition as compared with FA particles filled polymer matrix composites. Since the processed FA particles can restrict the melt flow due to its distinct physical ability, the inclusion of FA particles with resin can increase the transition temperature, i.e., operating temperature. It has been found that addition of 3% of FA particles can considerably increase the operating temperature owing to the uniform distribution of the FA particles in the polyester resin based matrix. Because of decrement of melt flow, the operating temperature has been increased considerably. Nevertheless the further increment in the flyash particles would be resulted in the agglomeration of the particles. This has lead to formation of FA particles cluster. This could also restrict the melt flow. However the restriction has been lower than 3%, since the particles have lost its uniform distribution owing to the cluster formation. Hence the polymer composites with 4% of the FA particles addition has possessed lower safe operating temperature than the lesser addition of the FA particles in the polyester resin.

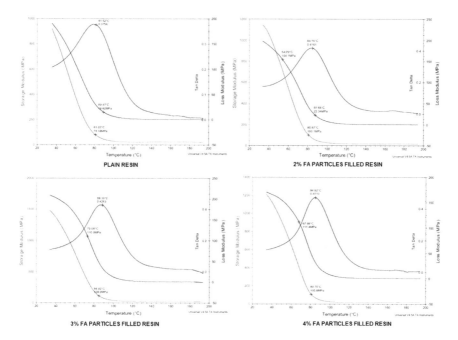

Fig. 3 Effects of FA particles addition on dynamic mechanical behavior

3.2 Influence of Stabilizer on Surface Cracks

During synthesization of polymer composite, the fabrication has been initially performed without accelerator. It has been taken almost 10 days to cure the molten resin. Owing to this slow curing of the composite, the specimen has been affected with surface cracks since the environment heat changes occur. Hence the flyash dispersed resin has been mixed with catalyst in the ratio of 10:1 and 1% of accelerator cobalt octate to reduce the curing period. It has been observed that the curing has been occurred within 24 h. Figure 4 shows the effect of accelerator on surface crack. However on adding 2% of accelerator, the specimens have been observed with cracks due to the excessive addition of the accelerator which produces more heat. This has created a non-homogeneous cooling of the composite which has produced crack. Hence it has been inferred that 1% of accelerator has been optimal value during the polymer composite fabrication.

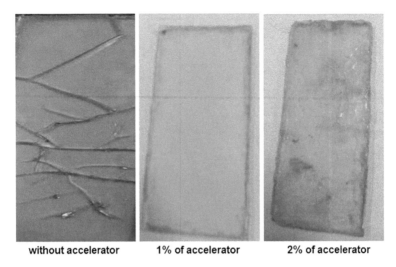

without accelerator   **1% of accelerator**   **2% of accelerator**

Fig. 4 Effects of accelerator on surface cracks

4 Conclusion

The flyash particles added polyester resin polymer matrix composite has been synthesized and the effects of flyash particles addition have been investigated using DMA analysis. Based on the derived experimental results, the following observations have been made.

(a) The addition of 3% of FA particles can considerably increase operating temperature owing to the uniform mixing of the flyash particles in the polyester resin based matrix.
(b) 1% of accelerator is the optimal value for synthesizing the flyash particles filled polymer composite.

References

1. Satheesh Raja R, Manisekar K, Manikandan V (2014) Study on mechanical properties of fly ash impregnated glass fiber reinforced polymer composites using mixture design analysis. Mater Des 55:499–508
2. Tagliavia G, Porfiri M, Gupta N (2010) Analysis of flexural properties of hollow-particle filled composites. Compos Part B 41:86–93
3. Huang Z-Q, Yu S-R, Li M-Q (2010) Microstructures and compressive properties of AZ91D/ fly-ash cenospheres composites. Trans Nonferrous Met 20:458–462
4. Seena Joseph V, Bambola A, Sherhtukade VV, Mahanwar PA (2011) Effect of flyash content, particle size of flyash, and type of silane coupling agents on the properties of recycled poly (ethylene terephthalate)/flyash composites. J Appl Polym Sci 119:201–208

5. Das A, Satapathy BK (2011) Structural, thermal, mechanical and dynamic mechanical properties of cenosphere filled polypropylene composites. Fuel 32:1477–1484
6. Nantha Kumar P, Rajadurai A (2014) Effect of carbon fiber, silica and fly-ash particulate addition on tensile and impact behaviour of polyester and epoxy resin. Appl Mech Mater 592–594:186–191
7. Nantha Kumar P, Rajadurai A, Muthuramalingam T (2018) Multi-response optimization on mechanical properties of silica fly ash filled polyester composites using Taguchi-grey relational analysis. Silicon. https://doi.org/10.1007/s12633-017-9660-8
8. Nantha Kumar P, Rajadurai A, Muthuramalingam T (2018) Thermal and mechanical behaviour of sub micron sized fly ash reinforced polyester resin composite. Mater Res Express 5:045303

Influence of Copper Chills and Pouring Temperature on Mechanical Properties of LM6 Castings

D. M. Wankhede, B. E. Narkhede, S. K. Mahajan
and C. M. Choudhari

Abstract Aluminum silicon alloy castings are highly used in automobile industries due to their excellent casting and machining ability, corrosion resistance and high strength-to-weight ratio. The mechanical properties of aluminum–silicon LM6 alloy are mostly dependent on solidification and casting parameters. This paper presents an experimental investigation made on aluminum–silicon (Al–Si) cast LM6 alloy using sand casting process. The main objective of the research is to investigate the effect of chill thickness and pouring temperature on LM6 sand castings. The casting parameters such as pouring temperature and external copper chills are considered for the experimental work. The simulation work has been carried out with AutoCast-X1 software for obtaining appropriate design parameters. The experimental work has been carried out in the foundry. The pouring temperatures 730, 750 and 770 °C are considered for experiments along with varying chill thickness. The use of end chills during casting not only favors directional solidification but also enhances solidification. Rapid cooling rate helps to get finer structures and improved mechanical properties. In this work, an attempt has been made to obtain the better cooling rate of LM6 castings using copper chills. The mechanical properties such as hardness and ultimate tensile strength (UTS) are analyzed. It is seen that the external chill has a significant influence on the properties of the casting

D. M. Wankhede (✉)
Department of Production Engineering, Veermata Jijabai Technological
Institute, Mumbai 400019, India
e-mail: dmwankhede99@gmail.com

B. E. Narkhede
Industrial Engineering and Manufacturing Systems Group, National Institute
of Industrial Engineering (NITIE), Near Vihar Lake, Powai, Mumbai 400087, India
e-mail: benarkhede@nitie.ac.in

S. K. Mahajan
Technical Education, Mumbai 400072, Maharashtra, India
e-mail: skmahajan@dte.org.in

C. M. Choudhari
Department of Mechanical Engineering, Dwarkadas J. Sanghvi College of Engineering,
Mumbai 400756, India
e-mail: c.choudhari75@gmail.com

© Springer Nature Singapore Pte Ltd. 2019
A. K. Lakshminarayanan et al. (eds.), *Advances in Materials and Metallurgy*,
Lecture Notes in Mechanical Engineering,
https://doi.org/10.1007/978-981-13-1780-4_21

components. The design of experiment has been set up and experiments were conducted as per full factorial array. Castings are made under the constraint of the process and methodical parameters at three different levels. The mathematical models of UTS and hardness have been developed. The micrographs of microstructure of castings with and without applications of copper chills have been compared using optical microscopy. The better mechanical properties such as UTS and hardness were obtained by proper combination of pouring temperature and application of external copper chills.

Keywords External copper chills · Pouring temperature · Mechanical properties Cooling rate · LM6 alloy

1 Introduction

Casting process enables the economical manufacture of products with intricate geometry to near net shape. Almost all metals and alloys are produced from liquids by solidification. The effect of various casting parameters such as pouring temperature and chill thickness has been studied to increase the mechanical properties of castings. The mechanical properties of cast aluminum silicon alloys are very sensitive to composition, metallurgy and heat treatment, the casting process and the formation of defects during filling of mold and solidification. Seah [1] explained the fabrication and testing of Aluminum cast composites in sand molds containing metallic (steel, copper, and cast iron) and non-metallic (silicon carbide) external chills, respectively. The superior mechanical properties of the castings, particularly the ultimate tensile strength (UTS) and fracture toughness, are explained in relation to their microstructure. The mechanical properties of castings can be improved by the addition of chills for local heat transfer and inoculants can also be used to enhance the properties such as ultimate tensile strength and hardness of the castings. The effects of various casting parameters on mechanical properties have been studied. Kanthavel [2] explained the effect of external mild steel chills on steel casting in the sand mold to reduce shrinkage defects. The parameters such as chill distance, chill thickness, pouring temperature and pouring time are studied to reduce the shrinkage defects. Spinelli et al. [3] carried out directional solidification experiments, and interrelations of thermal parameters, microstructure, and tensile properties are established. Linear relationships between the ultimate tensile strength and interfacial heat transfer coefficient have been determined for hypo eutectic Al–Fe and Al–Sn alloys. Chen et al. [4] investigate the effects of pouring temperature on the as-cast structure and mechanical properties of Aluminum–magnesium alloy. With the increase of pouring temperature, the yield strength, tensile strength, and elongation of casted alloy samples are rapidly increased first, and then are slightly declined. Ahmed et al. [5] conducted an experimental study on Al–Si cast alloy (LM6) using green sand casting. The results show that the melt temperature has a significant effect on the quality of green sand casting of LM6 alloy.

The lower pouring temperature provides good microstructure with lesser defects such as porosity. Dehnavi et al. [6] examined the effect of the cooling rate on grain size, secondary dendrite arm spacing, eutectic fraction and thermal analysis characteristic parameters of Al–Cu alloy. The grain size and secondary dendrite arm spacing of alloy determined by microstructural image analysis, show that the both of grain size and SDAS decrease with increasing cooling rate. At varying cooling rates the microstructure does not change fundamentally, but it can be noted that at the lower cooling rate the quantity of trapped liquid in the interdendritic regions is lesser. Anantha Prasad et al. [7] made stir cast hybrid composites of LM13 with garnet and carbon particulate reinforcements. Various external chill materials such as steel, copper, iron and silicon carbide were used to enhance the solidification. Mechanical properties such as hardness and ultimate tensile strength of reinforced composites were examined. Dobrzański et al. [8] investigated that the effect of cooling rate on the SDAS, size of the grains, the size of the precipitation and thermal characteristic results of $AlSi_7Cu_2$ cast alloy have been studied. The work shows that the chemical and thermal modification of the eutectic silicon can be assessed by analyzing thermal characteristics of the cooling curve. Chills are provided in the mold so as to increase the heat extraction capability of the sand mold. A chill normally provides a steeper temperature gradient so that directional solidification as required in a casting can be obtained. The chills are metallic objects having a higher heat absorbing capability than the sand mold. The proper use of chills helps in controlling the localized heat and cooling rate of the cast metal. Chills are metallic inserts which are used into the sand surface to enhance the solidification rates. They are usually made from steel, aluminum, iron, copper and can be machined or cast. The mechanical properties and microstructure of LM6 alloy castings can be improved by using the cooling aids. The external copper chills have been used to improve the cooling rate of castings during solidification. The casting parameters such as pouring temperature and chill thickness have been considered for the experimental work. The design of experiment (DOE) is used to optimize the casting parameters. The faster cooling rate gives a better secondary dendrite arm spacing (SDAS) and grain structure, which improves the mechanical properties of castings. By controlling the casting pouring temperature and chill size, better mechanical properties can be achieved.

2 Experimental Work

The pattern of size $200 \times 30 \times 30$ mm was made of wood. The mold boxes were prepared along with external copper chills. The experiments were carried out on Al–Si (LM6) alloy with and without applications of external copper chills using sand casing method. The temperature of the casting and chill was measured after every thirty seconds by using K-type thermocouple placed at the edge of casting and chills. A data logger was used to measure the temperature with the help of K-type thermocouple. The effect of pouring temperature and copper as an external

(a)   (b)   (c)

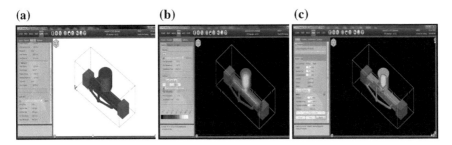

Fig. 1 **a** Casting with chills, **b** simulation process, **c** elimination of shrinkage defect

chill has been considered for the investigation. The design calculations for making a pattern and gating system has been done. The entire casting layout has been simulated in AutoCast-X1 to ensure casting free from solidification related defects. Soundness of the casting has been given the upmost priority and then the subsequent experiments have been performed for the improvement of mechanical properties. The simulation has been done using AutoCast-X1 software get the correct design parameters and identifying the proper placement of riser. This helps in eliminating the major casting defects such as shrinkage and porosity. Few important simulation results are shown in Fig. 1a–c.

The experimental work has been carried out on a rectangular component of Al–Si (LM6) alloy. The ultimate tensile strength and hardness have been measured using universal testing machine (UTM) and Vicker's hardness test method respectively. The sand casted component has been machined and converted as per ASTM- B557 standard for tensile test measurement [9]. The hardness test has been done on casted component as per ASTM-E10:2015 standard. The design of experiment (DOE) has been set up at three different levels. The 30 mm square rectangular bar of length 200 mm is cast as per simulation process. The copper chill is used on both end of casting, which leads to directional solidification and helps in enhancing the mechanical properties of castings. The experiment trials have been carried out in the foundry. Total twenty seven castings were taken out with varying pouring temperature and chill thickness. An experiment was conducted using copper chills with different thicknesses, i.e. 35, 45, and 55 mm. The pouring temperatures were taken 730, 750, and 770 °C. The experimental work is as shown in Figs. 2 and 3a–c.

The levels of casting and methoding parameters have been identified for investigation. Experiments have been arranged for all possible combination of these factors and their levels using design of experiment. Table 1 shows the input factors and their levels.

The orthogonal array selection is based on the number of levels of their various factors. The experimental values of mechanical properties such as UTS and hardness have been obtained using sand casting process is as shown in Fig. 4.

The main effect plots for individual parameters such as ultimate tensile strength are shown in Fig. 5a, b and for hardness as the second output parameter are as shown in Fig. 6a, b. The interaction plot of UTS and hardness is as shown in Fig. 7a, b.

(a)   **(b)**   **(c)**

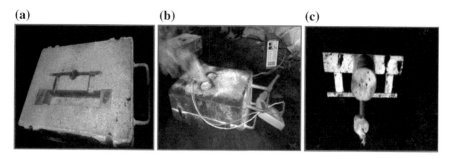

Fig. 2 **a** Casting with chills, **b** casting with milliVoltmeter, **c** casting with a gating system

(a)   **(b)**   **(c)**

Fig. 3 **a** Casting samples, **b** hardness test casting sample, **c** UTS test casting samples

Table 1 Casting input factors and their levels

Input factors	Code	First level	Second level	Third level
Pouring temperature (°C)	X_1	730	750	770
Copper chill thickness (mm)	X_2	35	45	55

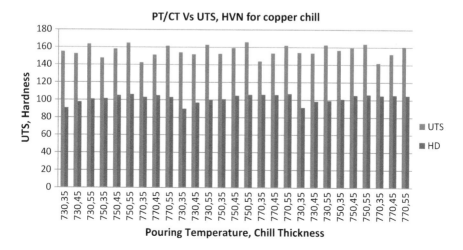

Fig. 4 Experimental values of LM6 castings using copper chills

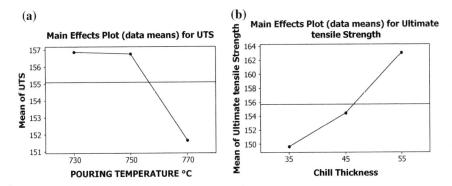

Fig. 5 **a** Main effects plot of pouring temperature versus UTS, **b** main effects plot of chill thickness versus UTS

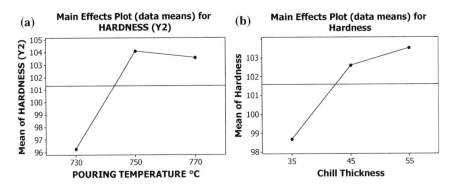

Fig. 6 **a** Main effect plot of pouring temperature versus hardness, **b** main effect plot of chill thickness versus hardness

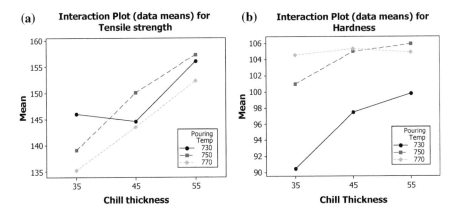

Fig. 7 **a** Interaction plot of pouring temperature, chill thickness versus UTS, **b** interaction plot of pouring temperature, chill thickness versus hardness

(a)   **(b)**

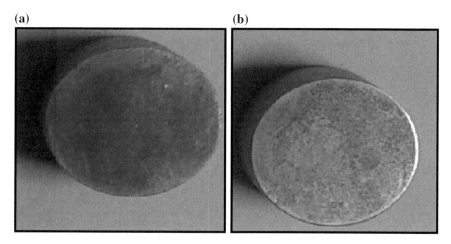

Fig. 8 **a** Sample of casting without using chills, **b** sample of casting using copper chills

The multiple regression analysis is performed using statistical analysis software Minitab 14. It gives the linear relationship between ultimate tensile strength and hardness with chill size and pouring temperature. The regression equations for ultimate tensile strength and hardness in relation with input parameters are shown in Eqs. 1 and 2 given below

The regression equation of tensile strength is

$$Y_1 = 209.64 - 0.1123X_1 + 0.6728X_2 \tag{1}$$

The regression equation of hardness is

$$Y_2 = -78.4874 + 0.2255X_1 + 0.2436X_2 \tag{2}$$

The microstructure analysis has been done on sand casted LM6 alloy samples at 730 °C pouring temperature. The standard metallographic techniques were used to prepare specimens by polishing it on emery paper. The Keller's reagent was used to etch the polished samples. The optical microscope were used to get the micro-graphs. The micrograph has been taken on Olympus GX-51 optical microscope. The casting specimens for microscopic analysis without and with application of copper chills are as shown in Fig. 8a, b. The micrographs (500 × magnification) of casting samples without and with application of copper chills are as shown in Fig. 9a, b.

The coarse primary silicon particles have not seen in casting after application of copper chills. The micrograph shows very less primary silicon particles. The interdendritic network of silicon particles has been observed. The fine grain structure has been obtained which lead to better microstructure as compared to casting without application of chills. The grain structure from the micrograph clearly indicates that the good microstructure gives better mechanical properties.

(a)   **(b)**

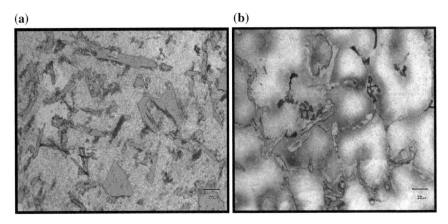

Fig. 9 **a** Microstructure without chill at pouring temp 730 °C, **b** microstructure with chill (copper chills) at pouring temp 730 °C

2.1 Experiment Without External Chills

The simulation work is carried out using AutoCast-X1 software without the application of external chills. The experiment trials have also been carried out on nine castings in the foundry. The pouring temperature has been controlled as per the three specified levels (Table 1). The experiment has been carried out without using external chills. The graphs of pouring temperature versus mechanical properties are as shown in Figs. 10 and 11. The pouring temperature of molten metal and temperature of casting have been measured using data logger.

Fig. 10 Plot for pouring temperature versus UTS

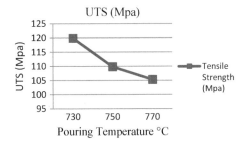

Fig. 11 Plot for pouring temperature versus hardness

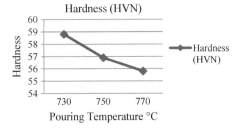

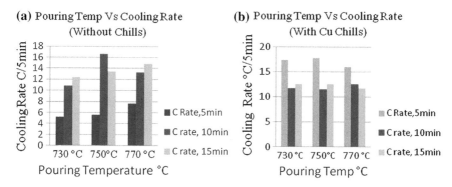

Fig. 12 Graph of pouring temperature versus cooling rate (without chills and with copper chills)

The graph of pouring temperature versus cooling rate for every five minutes during casting solidification of no chills and after using copper chills are as shown in Fig. 12a, b respectively. The cooling rate is drastically enhanced after application of copper chills.

3 Results and Discussion

Total 27 experiments have been conducted in foundry to find out the effect of pouring temperature and external copper chills on mechanical properties of LM6 alloy castings. The graph as shown in Fig. 5 shows that at 730 °C the average UTS value is obtained highest and then it goes on reducing with increased pouring temperature. The UTS has been increasing with increase in chill thickness. The average UTS value was found high at 55 mm chill thickness. The graph as shown in Fig. 6 shows that at 750 °C the average hardness value is obtained highest and then it goes on decreasing with increased pouring temperature. The average hardness value was found high at 55 mm chill thickness. The interaction plot Fig. 7a shows that there is a combined effect of pouring temperature and chill thickness on the UTS, and it has been found highest at 750 °C with 55 mm chill thickness. The hardness value was found highest at 750 °C with 55 mm chill thickness is as shown in Fig. 7b. The micrograph of LM6 alloy with application of copper chill shows a better grain structure. The interdendritic network of silicon particles has been seen after application of external copper chills. The combined effect of controlled input parameters on UTS value was found better at lower pouring temperature with high chill thickness. The hardness value was found higher at moderate pouring temperature with high chill thickness. The cooling rate has been increased by application of external copper chills.

4 Conclusions

- By controlling the casting pouring temperature and chill size, better mechanical properties have been achieved.
- The external copper chills help to get the directional solidification and also have a very good effect on properties of aluminum silicon alloy castings.
- The external copper chills help to improve the cooling rate of solidification of casting. The faster cooling rate gives a better grain structure, which improves the mechanical properties of castings.
- The better UTS and hardness is obtained with application of external chills at 750 °C at 55 mm chill thickness.
- The better values of ultimate tensile strength and hardness were obtained as compared to casting without applications of external chills.
- The regression analysis shows the relationship between ultimate tensile strength and hardness with pouring temperature and chill size.
- The casting with the application of external copper chills has better microstructure compared to casting without external chills.
- Therefore the similar attempts can be made on any casted machine components which will provide improvisation in its mechanical properties of those components.

References

1. Seah KHW, Hemanth J, Sharma SC (2003) Mechanical properties of aluminum/quartz particulate composites cast using metallic and non-metallic chills. Mater Des 24:87–93
2. Kanthavel K, Arunkumar K, Vivek S (2014) Investigation of chill performance in steel casting process using response surface methodology. Procedia Eng 97:329–337
3. Spinelli JE, Cheung N, Goulart PR, Quaresma JMV, Amauri G (2012) Design of mechanical properties of Al-alloys chill castings based on the metal/mold interfacial heat transfer coefficient. Int J Therm Sci 51:145–154
4. Chen S, Chang G, Huang Y, Liu S (2013) Effects of melt temperature on as-cast structure and mechanical properties of AZ31B Magnesium alloy. Trans Nonferrous Met Soc 23:1602–1609
5. Ahmed R, Talib NA, Asmael MBA (2013) Effect of pouring temperature on microstructure properties of Al-Si LM6 alloy sand casting. Appl Mech Mater 315:856–860
6. Dehnavi M, Vafaeenezhad H, Sabzevar MH (2014) Investigation the solidification of Al-4.8 wt.% Cu alloy at the different cooling rate by computer-aided cooling curve analysis. Metall Mater Eng 20(2):107–117
7. Anantha Prasad MG, Bandekar N (2015) Study of microstructure and mechanical behavior of aluminum/garnet/carbon hybrid metal matrix composites (HMMCs) fabricated by chill casting method. Procedia Mater Sci 5:1–8
8. Dobrzański LA, Maniara R, Sokołowski J, Kasprzak W (2007) Effect of cooling rate on the solidification behavior of AC AlSi$_7$Cu$_2$ alloy. J Mater Process Technol 191:317–320
9. ASTM B557-15 Standard Test Methods for Tension Testing, https://www.astm.org/standards/B557.htm

Influence of Deep Cryogenic Treatment on the Wear Behaviour of Different Al–Si Alloys

Bhanuchandar Pagidipalli, Chandan Nashine and Pratik S. Bhansali

Abstract Aluminium and its alloys have been the centre of interest to the engineer's community due to their high strength to weight ratio, high wear resistance, low coefficient of thermal expansion and ease of manufacture. To fulfil the increasing requirement of lighter weight, yet excellent mechanical properties than aluminium alloys, addition of grain refiners with modifiers to the alloys has been proposed. Al–Si alloy was taken as the master alloy, while magnesium and copper were added individually with a tint of modifiers and grain refiner. Two sets of same specimens of varying composition were prepared. One set of samples was annealed and the other underwent deep cryogenic treatment (DCT) followed by tempering. Wear behaviour of Al–Si alloys was analysed before and after the DCT by using the computerized 'pin on disc' wear testing machine. After cryogenic treatment, all the specimens, except those containing grain refiners showed influence on wear resistance.

Keywords Deep cryogenic treatment (DCT) · Wear resistance
Die casting and annealing

1 Introduction

Nowadays, aluminium and its alloys are achieving vast manufacturing implication for their remaining combination of physical as well as tribological properties. Such properties contain improved wear resistance, improved higher temperature strength, and increased damping capacity [1]. The experiment shows that Aluminium Alloy

B. Pagidipalli (✉) · C. Nashine · P. S. Bhansali
Department of Mechanical Engineering, NIT Rourkela, Rourkela, India
e-mail: pagidipallibhanuchandar@gmail.com

C. Nashine
e-mail: Chandan.nashine@gmail.com

P. S. Bhansali
e-mail: bhansalipratiks@gmail.com

© Springer Nature Singapore Pte Ltd. 2019
A. K. Lakshminarayanan et al. (eds.), *Advances in Materials and Metallurgy*,
Lecture Notes in Mechanical Engineering,
https://doi.org/10.1007/978-981-13-1780-4_22

strengthened with approximately 11% SiC composite, which offers identical mechanical properties but with improved thermal conductivity [2]. Therefore, frictional heating such alloys are lower than that of cast irons due to increased thermal conductivity. Hence the demand of Aluminium alloys is grown in the automobile and manufacturing areas where wear and tear are the fundamental difficulties. Components like pistons, turbine blades, cylinder heads, connecting rods, drive shafts and valves are manufactured with their implementation. Al–Si alloys have been used in various industries to a great extent due to their light weight and higher strength. Such alloys are capable of enhancing the service life and decreases the manufacturing cost of the desired products. In this modern age, aluminum alloys are usually widely used in automotive sector particularly due to the real need of weight saving for more reduction of fuel consumption. The alloying elements include copper, magnesium, manganese, silicon, and zinc in the primary applications. Aluminium alloys have an excellent texture in the dry surroundings because of the generation of a protecting layer of aluminum oxide. The alloys of 2xxx, 3xxx and 4xxx series, containing major elemental additives of copper, manganese, and Silicone are widely utilized instead of steel panels in major automobile sectors [2]. Hence these alloys are the subject of centre attraction to various scientific studies in the current research. In modern metallurgical processes, the Deep cryogenic treatment (DCT) is one of unique treatment methods, in which the treatment temperature lower than −123 °C is applied [3, 4]. An optimized implementation of this approach, the hardness, strength, wear resistance and the dimensional accuracy can be improvised [5, 6]. DCT has been regularly utilized for hardening tool steel and Titanium alloys. Though, very limited experiments on the influence of DCT on Al–Si alloys have been carried out. Therefore, in the present work, DCT is being performed to advance the wear resistance for Al–Si cast alloys of varying composition. These varying forms of Al–Si alloys with magnesium and copper plays a major role in automobile and aerospace engineering; hence the study of the influence of DCT over these casted alloys becomes significant. In this experiment, the influence of DCT on Al–Si alloys cast with varying composition of copper and magnesium is being investigated. A complete study has been achieved to observe the microstructure and wear resistance in the alloys.

2 Experimental Work

Aluminium–silicon–magnesium and aluminum–silicon–copper alloys were prepared with different weight percentage of Mg and Cu by stir casting route in an induction heating furnace (melting furnace). Samples of wear test standard dimensions were cut. Wear behaviour of different compositions samples were studied by conducting several wear tests on computerized Ducom friction and wear monitor pin on disc wear test machine. The Al–Si–Cu and Al–Si–Mg alloys with varying Cu and Mg composition in pure aluminum (99.4%) along with Al–20Si master alloy in a clay graphite crucible in an electrical resistance furnace under a cover flux (45% NaCl + 45% KCl + 10% NaF) and the melt held at 800 °C. After degassing with 1% solid hexachloroethane, master alloy chips duly packed in aluminum foil were added to the melt for grain refinement. For modification, 0.02% pure strontium alloy was used. The melt was stirred for 30 s with the coated iron rod after the addition of grain refiner and modifier. Melts were held for 5 min and poured into a cylindrical graphite mold (25 mm diameter and 120 mm height) surrounded by fireclay brick. The details of the alloys, grain refinement and modification treatment are given in Tables 1 and 2 (Figs. 1 and 2).

The cast samples are of 120 mm length and 25 mm diameter. After casting the specimens were cut into samples as per ASTM standard for wear specimen (35 mm length and 8 mm diameter) by turning and facing operations to conduct wear test. Since the casted specimen is of 25 mm diameter, the sample is reduced to a diameter of 8 mm which is required for wear analysis on disc wear testing machine. The samples obtained from the above procedure possesses wear and tear on their surfaces resulting in weaker mechanical strength. This problem can be overcome by

Table 1 Test specimens Al–Si–Cu series

S. No.	Alloy composition
1	Al–9Si–2.5Cu
2	Al–9Si–2.5Cu–0.02Sr
3	Al–9Si–2.5Cu–0.02Sr–M51

Table 2 Test specimens Al–Si–Mg series

S. No.	Alloy composition
4	Al–9Si–0.3Mg
5	Al–9Si–0.3Mg–0.02Sr
6	Al–9Si–0.3Mg–0.02Sr–M51

Fig. 1 Raw material in the crucible

Fig. 2 Melting furnace

heat treating the specimens above its recrystallization temperature followed by DCT for relieving internal stresses and improving mechanical properties. As per Al–Si phase diagram, the recrystallization temperature is around 350 °C. The specimens are heated to that temperature and maintained for one hour. After that annealing process is done, DCT has been applied to A1–Si alloy that has already been heat treated, i.e., Slow cooling without any thermal shock to approximately −196 °C and soaking of the specimen at around −196 °C for 24 h. After DCT the specimens were again reheated slowly without thermal shock to ambient temperature. The test temperatures were attained from ambient to −196 °C using liquid nitrogen (LN$_2$) to significantly slow atomic and molecular activity in the material. Gradual temperature changes cause the change in microstructure from the core to the surface, releasing residual stresses and homogeneously stabilizing the alloy. This process may take 24 h or longer to keep the entire mass in equilibrium throughout temperature cycling. All processes have been done under control, and separate equipment was used for the individual processes (Figs. 3 and 4).

Fig. 3 Specimen after deep cryogenic treatment

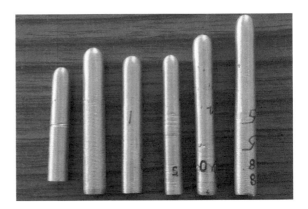

Fig. 4 Disk type wear testing machine

In the present experiment, the Computerized Pin on Disc wear tester instrument has been appropriated for the wear tests.

3   Results and Discussion

3.1   Microstructure Analysis

Cryogenic treatment of samples were done for the individual sets of specimens, i.e., Al–Si–Cu and Al–Si–Mg series. The microstructural analysis of both the sets of samples were done using Field Emission Scanning Electron Microscope (FESEM). The microstructure of both untreated and cryogenically treated sets of specimen for both the series at 5000× magnification are shown in Fig. 5.

From the figure it can be observed that, there has been a significant influence of cryogenic treatment on the grain boundaries of the test specimen. Finner grains are observed on the cryogenically treated micrographs as compared to the untreated sample. Grains were influenced in both the sets of series after cryogenic treatment of 24 h.

3.2   Wear Analysis

Wear tests of all the specimen samples were done individually. Information collected from the experiments are reported and further analysed. Wear test on the specimens is performed using wear and friction monitor. The tests were carried out at applied load of 25 N, sliding speed of 300 rpm, and time Taken is 10 min. Although the test is conducted in the dry conditions, therefore no lubrications were utilized. All required measures were taken regarding the cleaning of test samples

(a)   **(b)**

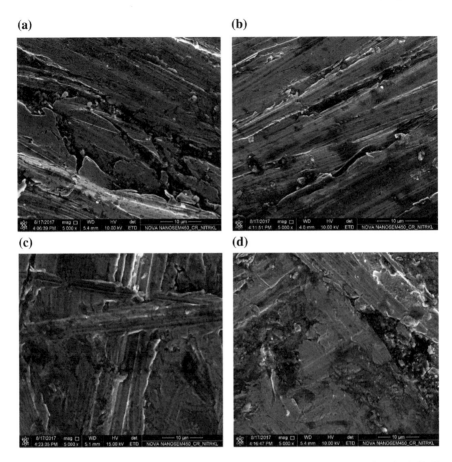

(c)   **(d)**

Fig. 5 **a** Untreated sample of Al–Si–Cu series; **b** cryogenically treated sample of Al–Si–Cu series; **c** untreated sample of Al–Si–Mg series; **d** cryogenically treated sample of Al–Si–Mg series

under this analysis to withdraw the capture of wear trash. This will accomplish uniformity in the experiments (Tables 3 and 4; Figs. 6, 7 and 8).

From the above figures, it can be observed that on implementation of DCT the wear resistance for both Al–Si–Cu and Al–Si–Mg series increases. All the alloys responded to DCT. It is understood that cryogenic treatment for Al–Si–Cu series showed a better improvement in wear resistance than Al–Si–Mg series.

Table 3 Wear behaviour of specimen before cryogenic treatment

Alloy no.	Alloy composition	Weight (n)	Speed (rpm)	Wear (μm)	Wear loss (gm)
01	Al–9Si–2.5Cu	25	300	508	0.008
02	Al–9Si–2.5Cu–0.02Sr	25	300	478	0.007
03	Al–9Si–2.5Cu–0.02Sr–1M51	25	300	402	0.005
04	Al–9Si–0.3Mg	25	300	399	0.006
05	Al–9Si–0.3Mg–0.02Sr	25	300	382	0.005
06	Al–9Si–0.3Mg–0.02Sr–1M51	25	300	338	0.004

Table 4 Wear behaviour of specimen after cryogenic treatment

Alloy no.	Alloy composition	Weight (N)	Speed (rpm)	Wear (μm)	Wear loss (gm)
01	Al–9Si–2.5Cu	25	300	430	0.005
02	Al–9Si–2.5Cu–0.02Sr	25	300	375	0.003
03	Al–9Si–2.5Cu–0.02Sr–1M51	25	300	604	0.007
04	Al–9Si–0.3Mg	25	300	367	0.005
05	Al–9Si–0.3Mg–0.02Sr	25	300	375	0.005
06	Al–9Si–0.3Mg–0.02Sr–1M51	25	300	816	0.007

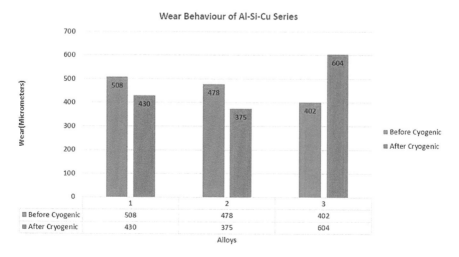

Fig. 6 Graph showing variation in wear rate of Copper series before and after cryogenic treatment

B. Pagidipalli et al.

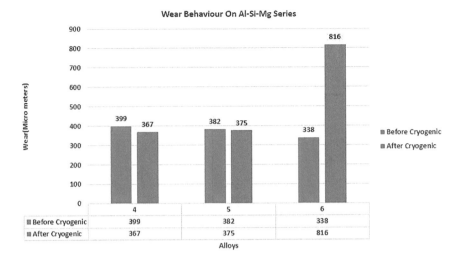

Fig. 7 Graph showing variation in wear rate for magnesium series before and after cryogenic treatment

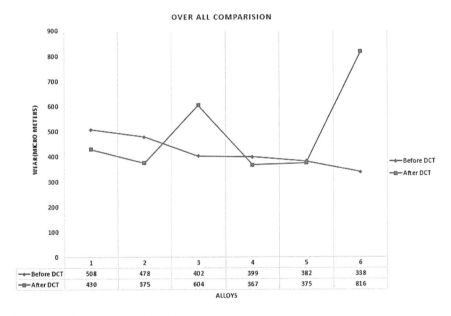

Fig. 8 Overall comparison of wear rate

4 Conclusions

1. The addition of modifier and grain refinery increases the wear resistance for both Al–Si–Cu and Al–Si–Mg series.
2. The combined addition of modifier and grain refiner (1%wt Al–5Ti–1B + 0.02% wt Sr) improved wear resistance by 20% for Al–9Si–2.5Cu alloy.
3. The combined addition of modifier and grain refiner (1%wt Al–5Ti–1B + 0.02% wt Sr) improved wear resistance by 15% for Al–9Si–0.3Mg alloy.
4. Cryogenic treatment improved wear resistance by 15% for Al–9Si–2.5Cu alloy and 21% for Al–9Si–2.5Cu–0.02Sr alloy.
5. Cryogenic treatment improved wear resistance by 8% for Al–9Si–0.3Mg alloy and 2% for Al–9Si–0.3Mg–0.02Sr alloy.
6. Cryogenic treatment has a significant influence on alloys of both series. Fine grains are observed in the cryogenically treated sets of specimen.

References

1. Barron RF (1982) Cryogenic treatment on metals to improve wear. Cryogenics 22:409–413 (Elsevier)
2. Hany A (2010) Influence of metallurgical parameters on the mechanical properties and quality indices of Al–Si–Cu–Mg and Al–Si–Mg casting alloys, PhD thesis, Universite du
3. Baldissera P, Delprete C (2008) Deep cryogenic treatment: a bibliographic review. Open Mech J 2:1–9
4. Amini K, Akhbarizadeh A, Javadpour S (2017) Cryogenic heat treatment—a review of the current state. Metall Mater Eng 23(1):1–10
5. Torabian H, Pathak JP, Tiwai SN (1994) Wear characteristics of Al–Si alloys. Wear 172:49–58
6. Mónica P, Bravo PM, Cárdenas D (2017) Deep cryogenic treatment of HPDC AZ91 magnesium alloys prior to aging and its influence on alloy microstructure and mechanical properties. J Mater Process Technol 239:297–302

Inverse Estimation of Interfacial Heat Transfer Coefficient During the Solidification of Sn-5wt%Pb Alloy Using Evolutionary Algorithm

P. S. Vishweshwara, N. Gnanasekaran and M. Arun

Abstract The study of the interfacial heat transfer coefficient (IHTC) is one of the major concerns during solidification of casting. In order to find out the IHTC at the metal–mold interface, a one dimensional transient heat conduction model is numerically investigated during horizontal directional solidification of Sn–5wt%Pb alloy. The forward model is solved using explicit finite difference method to obtain the exact temperatures for the known boundary conditions. The estimation of the unknown IHTC is attempted using Particle Swarm Optimization (PSO) as an inverse approach along with Bayesian framework. In order to prove the robustness of the proposed methodology, the estimation is accomplished for the simulated measurements. The simulated measurements are then added with noise to replicate the experimental data. The present approach not only minimizes the difference between simulated and measured temperatures but also takes in to account "a priori" information about the unknown parameters.

Keywords Swarm · Optimization · Inverse · Solidification · Interfacial heat transfer coefficient · Bayesian framework

1 Introduction

The study on heat transfer during casting solidification gives a complete understanding of the solidification process. During solidification, the metal shrinks and there will be a formation of air gap due to the contraction of metal and expansion of mold. Air gap at this mold metal interface creates a resistance to heat transfer from molten metal to the mold and the measure of this resistance is referred as interfacial heat transfer coefficient (IHTC). The value of IHTC depends on the geometry, direction of solidification, applied pressure, superheat of metal, chill material and

P. S. Vishweshwara · N. Gnanasekaran (✉) · M. Arun
Department of Mechanical Engineering, National Institute of Technology Karnataka,
Surathkal, Mangalore 575025, Karnataka, India
e-mail: gnanasekaran@nitk.edu.in

© Springer Nature Singapore Pte Ltd. 2019   227
A. K. Lakshminarayanan et al. (eds.), *Advances in Materials and Metallurgy*,
Lecture Notes in Mechanical Engineering,
https://doi.org/10.1007/978-981-13-1780-4_23

temperature, etc., Hence, modeling of IHTC at the mold metal interface is very important in understanding the solidification process. Few analytical solutions are available which are solved assuming perfect contact between the mold and cast metal and neglecting the superheat. The procedure for analytical solutions from the work of [1] shows the mathematical complexity involved in solving equations for different methods of solidification [2]. This complexity can be reduced by adopting numerical methods to solve the governing equations. In order to determine the IHTC, the surface temperature of the both cast and mold are necessary. Due to the moving boundaries inside the casting, the location of thermocouple at these boundaries distorts the thermal field at the interface and the properties of the materials are temperature-dependent [3] hence the estimation of IHTC is carried by non linear estimation method developed by Beck [4]. Early studies in the estimation and understanding of IHTC for upward and downward casting, flat and cylindrical casting is available in [5, 6]. Santos et al effectively highlighted the effect of casting parameters on IHTC. The values of the IHTC have a major role in obtaining temperature distribution in casting [7]. Commercial simulation softwares use IHTC as one of the inputs for solving the numerical simulation. In industries, it is common practice to use these IHTC as trial-and-error methods. As IHTC values depends on various factors as mentioned above, it is essential to determine the correct value of IHTC for a particular cast and mold system. Vasileiou et al conducted different simulations for varying casting geometry using ProCAST with an assumption of IHTC correlation with respect to temperature as stepwise and exponential as well as stepwise function of time [8, 9]. The estimation of IHTC values for continuous casting of steel billets by using Differential Evolution Algorithm and the results were validated with the industrial data [10]. In the present study, PSO is used as an inverse method to estimate the a and b values for IHTC correlation obtained from the literature data for horizontal directional solidification of Sn–5wt%Pb alloy against carbon steel mold. The exact temperatures are added with noise in order to mimic the actual experiments. The unknown interfacial heat transfer coefficient is retrieved by calculating the posterior distribution function in the objective function based on the Bayesian framework.

2 Direct Problem

Direct problem is solved based on one dimensional unsteady heat conduction equation for solidification process to obtain the exact temperatures for the assumed value of IHTC. Figure 1 shows the schematic view of the unidirectional solidification problem. The fluid flow effects during solidification are neglected and phase change problem is solved only regards with conduction. Carbon steel mold of length 110 and 60 mm width with water cooled system is considered for solving horizontal directional solidification of Sn–5wt%Pb alloy casting. Insulation was provided to maintain no heat is transfer through the walls. In order to obtain the temperature data, three sensors are located in the domain. In the casting cavity, two

Fig. 1 Schematic view of metal and mold system

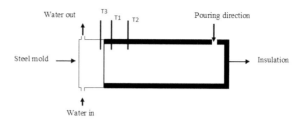

Fig. 2 Discretisation of mold and metal system

sensors T_1 and T_2 at a distance of 6 mm and 12 mm, respectively from mold metal the interface are located. One more sensor T_3 is located inside the mold at a distance of 2 mm from the mold metal interface. The discretised form of the solidification domain is as shown in Fig. 2.

2.1 Modeling of Interfacial Heat Transfer Coefficient

The interfacial heat transfer coefficient 'h_i' (IHTC) between the metal and mold surface is given by,

$$h_i = \frac{q}{(T_C - T_M)},$$

(1)

where q is the average heat flux across the metal–mold interface in W/m^2 and T_C and T_M are the casting and mold surface temperature in °C, respectively.

2.1.1 Governing Equations for the Heat Transfer in the Mold

The unsteady state one-dimensional conduction heat transfer is given by,

$$\frac{\partial^2 T}{\partial^2 x} = \frac{\rho C}{k} \frac{\partial T}{\partial t},$$

(2)

where T is the temperature in °C, k is thermal conductivity (W/mK), ρ is the density (kg/m^3) and C is the specific heat capacity (J/kgK) respectively.

2.1.2 Governing Equation for the Heat Flow in the Casting

The governing equation for heat transfer in the casting contains heat source term added to the left hand side of Eq. (2). The heat source term includes latent heat, fraction rate terms which should be solved by Schiel's equation. The final form of the equation is given by Eq. (3).

$$\frac{\partial^2 T}{\partial^2 x} = \frac{\rho C'}{k} \frac{\partial T}{\partial t}, \tag{3}$$

where $C' = C - l\frac{\partial f_s}{\partial T}$ and f_s is fraction of solid, l is latent heat.

2.1.3 Boundary Conditions

At $x = 0$, $\frac{\partial T_C}{\partial x} = 0$; At $x = i_g$ (casting surface), $-k_C \frac{\partial T_C}{\partial x} = h_i(T_C - T_M)$
At $x = i_{g+1}$ (mold surface), $-k_M \frac{\partial T_M}{\partial x} = h_i(T_C - T_M)$; At $x = L$,
$-k_M \frac{\partial T_M}{\partial x} = h_a(T_M - T_\alpha)$,

where k_M is thermal conductivity of mold (W/mK), k_C is thermal conductivity of casting (W/mK), h_a mold-environment is heat transfer coefficient of air (W/m^{2}K).

The thermophysical properties of the materials are mentioned in the Table 1. The initial temperature of the melt and mold are taken as 248 and 27 °C respectively. The IHTC (h_i) is time dependent hence in this study, it is assumed to vary as a power function with time with a and b as the unknown parameters as shown in Eq. (4).

$$h_i = at^{-b} \tag{4}$$

Here, in the present study, in order to solve direct model, IHTC values are presumed based on literature data [11] which is as shown in Eq. (5).

$$h_i = 6000t^{-0.33} \tag{5}$$

The simulated temperatures are obtained by solving direct model. Generally, experimental temperatures are prone to errors. In order to mimic the experiments, noise of $0.01T_{max}$, $0.02T_{max}$ and $0.03T_{max}$ are added as shown in Eq. (6)

Table 1 Thermophysical properties of the Sn–5%wtPb and steel materials [11]

Properties	k_s	k_l	σ_s	σ_l	C_s	C_l	T_f	T_s	T_l	k_p	l
Alloy	65.6	32.8	7475	7181	217	253	232	183	226	0.0656	59,214
Steel	46		7860		527						

$h_a = 5.7t^{0.15}$ (SI units: T—°C, σ—kg/m^3, C—J/KgK, K—W/mK, l—J/kg), suffix l—liquidus, s—solidus, f—fusion

$$Y_{iM} = T_{\text{exact}}(t_i, \text{Sensor}_M) + \varepsilon\sigma, \tag{6}$$

where M is the number of sensors, σ is the standard deviation of the temperature measurements and ε is the random numbers varying between -2.576 to 2.576 for normally distributed errors with zero mean and 99% confidence bounds [12]. The maximum temperature is considered is the initial temperatures of the molten metal and mold. The noisy data is now considered as measured data Y_{iM}, which in turn used to solve the inverse problem.

3 Particle Swarm Optimization as Inverse Method

In 1995, Kennedy and Eberhart developed Particle Swarm Optimization (PSO) as a robust technique which can be implemented to various inverse problems [13]. In comparison with the existing biological algorithms, PSO uses lesser parameters to solve the objective function. PSO is based on the swarm of birds moving in the direction of food or shelter and the searching is accomplished by mutual communication and learning. In a N dimensional search space with M particles moving in a swarm, the position and the velocity vectors of the i-th particle are given by $X_i = (x_{i1}, x_{i2}, ..., x_{iM})$ and $V_i = (v_{i1}, v_{i2}, ..., v_{iN})$. In every iteration, the addition of displacement to the current location X_i^n will give a new position X_i^{n+1} which is given by,

$$X_i^{n+1} = X_i^n + V_i^{n+1} \tag{7}$$

The corresponding updated velocity, V_i^{n+1} from the previous velocity V_i^n of the each particle in the direction of $P_{best.i}$ and $g_{best.i}$ is given as

$$V_i^{n+1} = W_a V_i^n + c_1.r_1.(P_{best.i} - X_i^n) + c_2.r_2.(g_{best.i} - X_i^n) \tag{8}$$

where c_1, c_2 are cognition learning coefficient, W_a is inertial coefficient, r_2 and r_2 are random numbers. In the present case, the values of c_1 and c_2 are taken to be 1.43 respectively. The overall representation of the present work is as shown in Fig. 3.

The objective function is expressed based on Bayesian framework. For every iteration in PSO, the fitness function as given in Eq. (9) is calculated for the best fitness value.

$$-lnPPDF = \sum_{m=1}^{M} \sum_{i=1}^{N} \frac{[Y_{im} - T_{im}(h_i)]^2}{2\sigma_x^2} + \frac{(a - \mu_a)^2}{2\sigma_a^2} + \frac{(b - \mu_b)^2}{2\sigma_b^2}, \tag{9}$$

Where h_i is the unknown parameter, Y_{im} is the i-th observation from the m-th measurement; M and N are the number of measurements and observations, respectively. $T_{im}(h_i)$ is the simulated temperature obtained from the forward model.

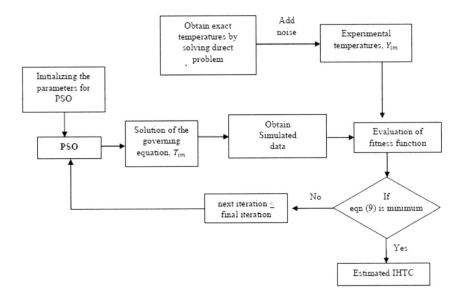

Fig. 3 Representation of the present work

a and b are the range of values generated by PSO, $\mu_a = 6000$, $\mu_b = 0.33$, σ_x, σ_a and σ_b are takes as $0.1T_{max}$, $0.01\mu_a$ and $0.01\mu_b$ respectively.

4 Results and Discussion

The simulated transient temperatures after solving direct model for three different sensors are as shown in the Fig. 4. A thin solidified skin is first formed when liquid metal moves in contact with the mold which will not allow the incoming molten metal. As the latent heat is continuously released, there is an increase in temperature of mold T_3. The sensor T_1 shows lesser value than T_2 as it its closer to the mold metal interface. In order to solve the above mentioned inverse problem, the input parameter of the PSO are initially set as; number of particles = 15, number of iterations = 50. From Eq. (4), a and b are the unknown parameters to be estimated. The values of a and b are assumed in the range of [1000 10000] and [0 0.6] respectively. The range of a and b was chosen from the data available based on the work of Silva et al. [11]. Firstly, the inverse estimation is performed for exact temperatures later the analysis is conducted for measured temperature with added noise. The inverse estimation is accomplished using in-house codes developed using MATLAB and executed in computer with configuration of 12 GB RAM, INTEL i5 Core, 1.70 GHz. Within the specified range, the particles are randomly generated. PSO uses each particle represents a and b values and the objective function given by Eq. (9) is solved and the error is minimized using the Bayesian

Fig. 4 Temperature values of the sensors T_1, T_2 and T_3 respectively

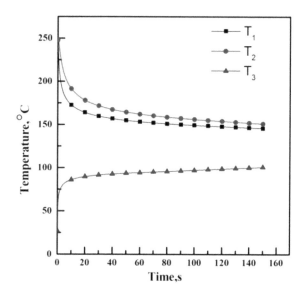

framework. Later the estimation of a and b values are carried out for temperature data with Gaussian white noise of $0.01T_{max}$, $0.02T_{max}$ and $0.03T_{max}$ respectively. The computational time taken for each run to converge to the solution was found in between 7 and 8 h.

For exact temperature measurement two runs with same initial guess is performed and the retrieved values of a and b are noted as shown in Table 2. It can be noted that PSO estimates the unknown values very close to the original values of $a = 6000$ and $b = 0.33$ with error less than 0.4. The convergence of a and b values and corresponding *-lnPPDF* are plotted as shown in Figs. 5 and 6 respectively. The variation of the results with the change in the number of particles is also checked as noted in Table 3. It has been observed that the time taken for the evaluation is higher compared to the 15 particles. But the % error for 15 particles is less compared to 30 number of particles and PSO was able to estimate good results hence further estimations are carried out for 15 particles. Sometimes increase in PSO gives more satisfactory results as a large search space is provide in the process of estimation but it is time consuming [14] (Fig. 6).

Now the estimation is conducted using noisy data. Table 4 shows estimated values of a and b for $0.01T_{max}$. Even for noisy temperature data, PSO algorithm was able to estimate a and b values within a error of 2%. The aim of the present study is

Table 2 Retrieval of a and b values for exact temperature values

Runs	Actual value of a	Actual value of b	Retrieved value of a	Retrieved value of b	% error for a	% error for b
1	6000	0.33	6002	0.33	0.033	0
2	6000	0.33	5980.3	0.329	0.328	0.3
Average	6000	0.33	5991.1	0.329	0.148	0.3

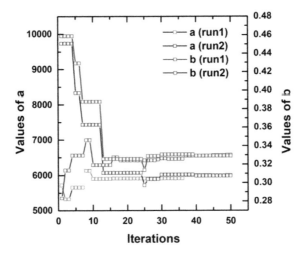

Fig. 5 Convergence of *a* and *b* values for two runs for exact temperature data

Table 3 Effect of number of particles on retrieval of a and b values for exact temperature values

Particles	Actual value of a	Actual value of b	Retrieved value of a	Retrieved value of b	% error for a	% error for b
15	6000	0.33	6002	0.33	0.033	0
30	6000	0.33	5967.75	0.328	0.537	0.606

Fig. 6 Convergence of *-lnPPDF* values for exact temperature data

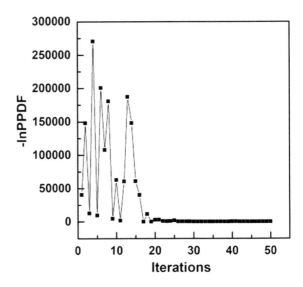

Table 4 Retrieval of a and b values for noisy data of $0.01T_{max}$

Runs	Actual value of a	Actual value of b	Retrieved value a	Retrieved value b	% error for a	% error for b
1	6000	0.33	6032.31	0.332	0.538	0.60
2	6000	0.33	5903.462	0.335	1.608	1.51
Average	6000	0.33	5967.88	0.333	0.535	0.90

Fig. 7 Comparison of cooling curves for $0.01T_{max}$ noisy temperature data

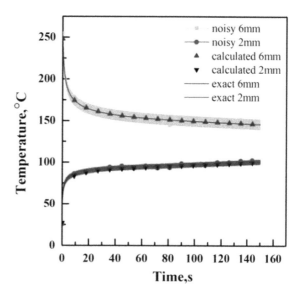

to prove the ability of PSO to retrieve the unknown parameters for different noise levels. From the estimated values, the temperature distribution is calculated and from the Figs. 7 and 8 it can be observed that the calculated temperatures using estimated a and b values matches well with exact, noisy temperature data. Thus, PSO has a capability of handling noisy data and the results of the estimation of a and b were found satisfactory with an error less than 2% as seen from the Table 5. Similar result was found in [12] where GA has been adopted for the solidification of Al–4.5w%Cu.

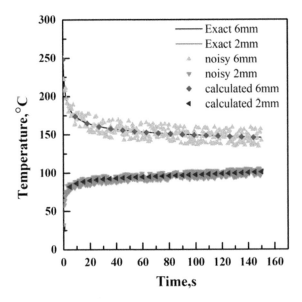

Fig. 8 Comparison of cooling curves for $0.02T_{max}$ noisy temperature data

Table 5 Retrieval of a and b values using PSO for different noisy data

Noise	Actual value of a	Actual value of b	Retrieved value of a	Retrieved value of b	% error for a	% error for b
0.01 (T_{max})	6000	0.33	6032.31	0.332	0.538	0.60
0.02 (T_{max})	6000	0.33	6008.3	0.33	0.138	0
0.03 (T_{max})	6000	0.33	6116.5	0.335	1.94	1.51

5 Conclusions

- Interfacial heat transfer coefficient for horizontal directional solidification of Sn–5wt%Pb alloy was successfully estimated using PSO Algorithm.
- Initially, the estimation of unknown parameters was carried out for exact temperature measurements. Later, noise added simulated measurements were used in the estimation process.
- It was concluded that even for the noisy data, PSO was able to estimate values of a and b close to the actual values, hence the proposed methodology can effectively be used for solving the inverse heat transfer problem in solidification.

References

1. Davey K (1993) An analytical solution for the unidirectional solidification problem. Appl Math Model 17(12):658–663
2. Stefanescu DM (2015) Science and engineering of casting solidification. Springer, Berlin
3. Jonathan WW, Woodbury K (2008) Accounting for sensor errors in estimation of surface heat flux by an inverse method. In: Proceedings of the ASME summer heat transfer conference, HT
4. Beck JV, Blackwell B, Clair C (1985) Inverse heat conduction, ill-posed problems. Wiley, London
5. Ho K, Pehlke RD (1985) Metal-mold interfacial heat transfer. Metall Mater Trans B 16 (3):585–594
6. Nishida Y, Droste W, Engler S (1986) The air-gap formation process at the casting-mold interface and the heat transfer mechanism through the gap. Metall Mater Trans B 17(4): 833–844
7. Santos CA, Siqueira CA, Garcia A, Quaresma JM, Spim JA (2004) Metal–mold heat transfer coefficients during horizontal and vertical unsteady-state solidification of Al–Cu and Sn–Pb alloys. Inv Prob Sci Eng 12(3):279–296
8. Vasileiou AN, Vosniakos GC, Pantelis DI (2015) Determination of local heat transfer coefficients in precision castings by genetic optimisation aided by numerical simulation. In: Proceedings of the Institution of Mechanical Engineers, Part C: J Mech Eng Sci, vol 229, issue 4, pp 735–750
9. Vasileiou AN, Vosniakos GC, Pantelis DI (2017) On the feasibility of determining the heat transfer coefficient in casting simulations by genetic algorithms. Procedia Manuf (11):509–516, ISSN 2351-9789
10. Yu Y, Luo X (2017) Identification of heat transfer coefficients of steel billet in continuous casting by weight least square and improved difference evolution method. Appl Therm Eng 114:36–43
11. Silva JN, Moutinho DJ, Moreira AL, Ferreira IL, Rocha OL (2011) Determination of heat transfer coefficients at metal–mold interface during horizontal unsteady-state directional solidification of Sn–Pb alloys. Mat Chem Phys 130(1):179–185
12. Ranjbar AA, Ghaderi A, Dousti P, Famouri M (2011) A transient two-dimensional inverse estimation of the metal-mold heat transfer coefficient during squeeze casting of AL-4.5 wt% CU. Int J Eng-Trans A Basics 23(3 and 4):273
13. Kennedy J, Eberhart RC (1995) Particle swarm optimization. In: Proceedings of IEEE international conference on neural network, pp 1942–1948
14. Vakili S, Gadala MS (2009) Effectiveness and efficiency of particle swarm optimization technique in inverse heat conduction analysis. Num Heat Trans Part B: Fundamentals 56:119–141

Tractable Synthesis of Graphene Oxide by Electrochemical Exfoliation Method

Azmeera Srinivasanaik and Archana Mallik

Abstract The aim of this work is electrochemical exfoliation of pyrolytic graphite for mass production of few-layer oxygen-functional graphene, commonly known as graphene oxide (GO). It is synthesized by intercalation of graphite sheets in the 1 M concentration of nitric acid electrolyte by application of positive bias. The voltage is gradually increased with an increment of 0.5 V up to 8 V and an interval of 3 min. The X-ray diffraction peaks corresponding to GO ((001) plane) and graphene sheet ((002) plane) were observed at 2θ positions of 26.35° and 13.56° respectively. The morphology of as-synthesized GO is characterized by field emission scanning electron microscopy. The transparent layers of GO are observed in transmission electron microscopy. AFM topography revealed that the thickness of the few-layer GO nanosheets are in the range of 3–5 nm only. The hexagonal ring structure of GO sheets was identified by selected area diffracted pattern. Through FTIR studies, the presence of functional groups of O–H and C–O has been identified. The synthesized material can be used as a base material for the future applications such as desalination of sea water, supercapacitors, sensors, solar cells, and coatings.

Keywords Pyrolytic graphite · Electrochemical · Intercalation
Exfoliation · Graphene oxide nanosheets · AFM · TEM

A. Srinivasanaik · A. Mallik (✉)
Department of Metallurgical and Materials Engineering,
National Institute of Technology Rourkela, Rourkela 769008,
Odisha, India
e-mail: archanam@nitrkl.ac.in

A. Srinivasanaik
e-mail: azmeera91@gmail.com

© Springer Nature Singapore Pte Ltd. 2019
A. K. Lakshminarayanan et al. (eds.), *Advances in Materials and Metallurgy*,
Lecture Notes in Mechanical Engineering,
https://doi.org/10.1007/978-981-13-1780-4_24

1 Introduction

Graphene is one of the allotropes of carbon consisting of single layer thick carbon atoms in a hexagonal lattice. It has the association of both the thermal and electric conductivities at remarkably great [1]. Some of the remarkable characteristics of graphene like thermal conductivity makes astonished values i.e. 4.9×10^3 to 5.35×10^3 W/mK [2], Young's modulus of $E = 1.0$ Tpa [3], intrinsic nobilities in the level of 200,000 cm^2 V^{-1} s^{-1} at an electron density of 2×10^{11} cm^{-2} [4, 5]. Graphene has very low electrical resistance with an optical transmittance of about 85%, so graphene sheets are being used as thin transparent conducting electrodes [6, 7]. Based on aforementioned characteristics graphene finds applications in display technology and medical science as well.

Graphene can be synthesized by different techniques, but at first, it was synthesized by the famous scotch tape method. High-quality samples of graphene were delivered by above-mentioned technique for initial investigations. Later several methods have been developed with different approaches and technologies to synthesize few-layer graphene nanosheets. These methods can be categorized into two approaches (1) top-down and (2) bottom-up approaches. Based on the practices they have some advantages and disadvantages as well. In order to select a particular method, it is important to have quantity and quality of the graphene produced.

Gas-phase microwave production and chemical vapor deposition, epitaxial growth techniques and thermal annealing of silicon carbide were embraced for the bottom-up approach for the synthesis of graphene nanosheets(GNSs) [8]. Nanomechanical cleavage of graphite was one of the classical methods, it derives high-quality graphene but yields small films in order of a few tens of micrometers and hence is not a productive one [9]. High amounts of reduced graphene oxide (RGO) sheets have been produced by chemical reduction of exfoliated graphite oxide layers, but it resulted in damage of graphene lattice and defects. Due to defects in the lattice, the synthesized GNSs electrical properties were reduced.

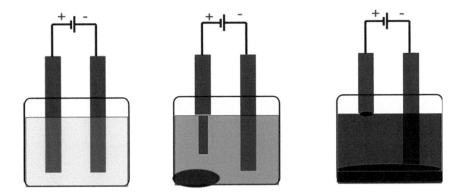

Fig. 1 Schematic diagram of electrochemical intercalation and exfoliation process

RGOs were further reduced to GNSs by dispersion technique [10], but productivity was low, and also films produced were not continuous. Hence they do not meet the requirement for many specialized applications.

In competition with these classical approaches electrochemical exfoliation is one such method, which can offer several advantages, like the ease of experiment, cost-effective, eco-friendly and can operate at ambient temperature. Electrochemical exfoliation technique is still under developing stage because of generation of defects in the final graphene lattice. In this method, pure graphite is used as a working electrode with protic ionic electrolytes at an appropriate DC bias. Due to applied potentials (may be positive/negative) to the graphite electrode, anionic/cationic intercalate from the electrolyte start intercalate the flaky graphite by breaking the weak van der walls force between the two adjacent layers. Hence corresponding redox reactions primes to the evolution of gases at the electrode outer surfaces, leading to in situ stress formation. This phenomenon could also create a surface blistering of the flaky graphite substrate [11]. Subsequently, single to few-layer thin sheets of flakes gets spread in the solution. Many findings with different anionic intercalants like polystyrene sulfonate ions, perchlorate ions, tetrafluoroborate ions, dodecyl sulfate ions, sulfate ions, nitrate ions, hexafluorophosphate ions have been explored and reported. The rate of intercalation and quality of sheets finally produced depends on these anionic intercalates primarily. Cathodic intercalation has also played a major role in synthesizing of graphene, which is a conventional technique than cathodic intercalation. Due to intercalation process, some of the gases evolve around the electrode (Graphite). Mainly oxygen and carbon dioxide gases were evolved during anodic intercalation, whereas hydrogen gas was evolved in cathodic intercalation. It is believed that with an increase in the volume of gas evolved during the anodic process [12], exfoliation is expected to be higher. But there is a high chance of functional group (epoxy, carboxylic and hydroxyl) attachment onto graphene nanosheets which formed freshly. These could be minimized by proper optimization of parameters (applied potentials, electrolyte).

Following the above observations, in the present study, a fast and easy approach is used to the synthesize graphene oxide nanosheets in an electrolyte, from pyrolytic graphite sheets by applying suitable potentials by anionic intercalation. Also, the structure and topography of GO were characterized by XRD, FESEM, AFM, and TEM.

2 Experimental Works

Analytical grade HNO_3 had been used as an electrolyte, the double distilled water was used as a solvent for all the experiment. Super fine grain and high-quality pyrolytic graphite sheet of full dimension (10 × 15 × 3) mm (by Asbury Graphite Mills), IPG 15 was employed as both the counter and working electrodes in this experiment. A regulated DC power supply by APLAB, model number 7103, was engaged in applying DC potential between electrodes. All the experiments were done at atmospheric pressure and temperature only.

The electrodes were kept vertically and facing each other at a separation of 1 cm. This setup was held in a beaker with the help of a stand. Initially, the WE was connected as cathode and CE as an anode. After the proper connection, a DC bias of 10 V for 30 s was applied, this treatment will serve a dual purpose, i.e., to clean the surface of the electrodes by removing the impurities and create the availability of surface and lattice pores of graphite electrode for effective intercalation.

After the cathodic pre-treatment, the electrodes were switched, i.e., WE electrode was connected as anode and CE were connected as a cathode. Subsequent exfoliation was started by raising the DC bias from 0 to 8 V with an increment of 0.5 V and an interval of 3 min. As a result of electrolysis, in an acidic electrolyte the hydrogen ions (H^+) and NO_3^- ions were redounded. The H^+ ions are materially adsorbed onto graphite lattice surface (Eq. 1) and successively released in the form of molecular hydrogen gas (Eq. 2) (Fig. 1).

$$C_{lattice} + H^+ \rightarrow C_{lattice}H_{adsorbed} \tag{1}$$

$$C_{lattice}H_{adsorbed} + H_2O + e^- \rightarrow C_{lattice}(expanded) + H_2 \uparrow + OH \tag{2}$$

The oxidation reactions will occur at anode due to the intercalation process, which was carried out by NO_3^- ($d = 0.358$ nm) intercalants. It will result in exfoliation of a mixed phase multilayer of graphite to few layers of graphene nanosheets. These exfoliated graphite layers were then dispersed in the electrolyte and also large particles were slowly deposited at the bottom of the beaker. After that, the electrolytic bath which consists of exfoliated graphite flakes was thoroughly washed with double distilled water to remove any acid contents, ionic forms and undissipated particles. The obtained mixed phase of graphene oxide nanosheets further fragmented by the ultrasonification process at a frequency of 20 kHz for 20 min [13]. The resulted colloidal was centrifuged at 4500 rpm for 30 min. Due to centrifuging the large particles were collected at bottom of the flask and GO were in colloidal form. The attained colloidal was dried in hot air oven at 100 °C for 24–48 h. The dried flakes were collected for additional characterizations.

The as-synthesized flakes were then subjected to characterization. The Ultima IV system using a monochromatic Cu Kα radiation ($\lambda = 0.154$ nm) was used for the X-ray diffraction (XRD) spectrum, from a 5°–60° range of 2θ. FEI Nova NanoSEM 450 FESEM along with EDS and a JEM 1200 JEOL operated at 200 kV TEM were used to investigate surface Morphology. The Shimadzu IR Prestige-21, FTIR instrument, was used to investigate the details about the various functional groups bonded to the exfoliated GO in the range of 4000–750 cm^{-1}. The SPM lab programmed Veeco diInnova MultiMode Scanning Probe Microscope with tapping (non-contact) mode was used for AFM topographical analysis of GO.

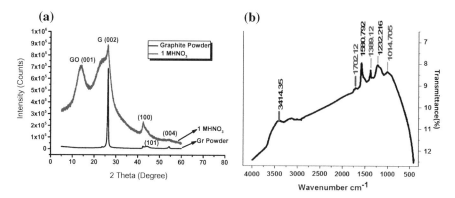

Fig. 2 **a** XRD spectrum of Graphite powder and as synthesized GO. **b** FTIR transmittance spectrum of as-synthesized GO

3 Results and Discussion

3.1 XRD Analysis

The XRD spectrum of as-synthesized GO nanosheets is shown in Fig. 2a. In the figure, the blue line corresponds to Graphene oxide nanosheets and black one corresponds to pristine graphite in powder form. The peak at $2\theta = 26.25°$ refers to the (002) orientation with the lattice space (d) equal to 3.392 A° which confirms a highly oriented layer structure in both the Graphite powder as well as in GO. The size of intercalant (NO_{3-}) is very close to the interlayer spacing of graphene sheets in the graphite electrode. Therefore, these anions should effectively interstice the graphite lattice. The effective intercalation leads to the high expansion of the graphite lattice along with higher oxidation. It has been observed that the sharp basal plane peaks centered at 26.35 for PGr has been broadened and shifted to 25.99. This shows the increase in interlayer spacing from 3.378 to 3.426 nm in the (002) planes. Due to intercalation, the lattice space of graphite is increased and broad peak could be observed at $2\theta = 26.35°$. The diffraction peak at $2\theta = 13.56°$ refers to the (001) plane which confirms the oxidation of graphite. The broad peak at (001) plane is due to the existence of further functional groups within the lattice space, hence the corresponding lattice space (d) at the plane (001) is 6.53 A°. The remaining diffraction peaks at 42.40°, 44.37°, and 54.44° (2θ values) correspond to (100), (101), and (004) reflections respectively as shown in Fig. 2a. The above-said planes are common in both XRD spectrum of the graphite powder and GO, [14] so it may be confirmed that some unexfoliated graphite particles were present in as-synthesized samples.

3.2   FTIR Spectral Analysis

The electrochemically exfoliated GO nanosheets are expected to be endowed with several oxygen-functional groups with the carbon atoms that causing from the protic solvent media with applied potential. Furthermore, the carbon structure and different functional groups attached to the GO nanosheets have been analyzed by fourier transform infrared spectroscopy (FTIR). The extent of these functional groups has been studied by FTIR spectrum of GO nanosheets dispersed in KBr. Figure 2b, shows the FTIR spectra of GO nanosheets, which illustrates the infrared transmittance peak at 1015 cm^{-1} corresponding to carbonyl (C–O), and the peak at 1232.21 cm^{-1} corresponding to epoxy group C–O–C. The peak at 1389.12 cm^{-1} is corresponding to COO–H/CO–H group. The peak at 1580.79 cm^{-1} was recognized as rising from C–C vibrations of the graphitic province. A peak at 1702.12 cm^{-1} is observed as carboxylic acid group C=O. The pretty broad peak at 3414 cm^{-1} may be due to the adsorbed water on the surface of the GO nanosheets. Followed by the above observations it is confirmed that the quality of graphene is good but have partial functionalization by oxygenation has also occured during the exfoliation process.

3.3   Morphological Studies

To acquire the morphological and topographical evidence of the GO nanosheets the FESEM, TEM and AFM investigations were carried out. The FESEM observations revealed the structure of graphite and confirmed the stratified nature of the planner graphitic crystal planes as shown in Fig. 3a (inset). A number of layers stacked as bundles in a lateral direction with small pores in between the grains of graphitic layers. These pores act as easy intercalate sites for ions/intercalants for uniform intercalation throughout the electrodes. The structure of GO nanosheets, that a few layers of sheets stacked together with curl edges are observed and are shown in Fig. 3a, b. Same as its parent material graphite, but here the layers looks like tot fragments. Also found that unexfoliated layers of graphite of size 1–3 μm are stacked with each other. TEM images of GO, which are deposited on the copper grid shown in Fig. 3c. The TEM image reveals that as prepared GO transparent sheets have a lateral size of 0.5–1.5 μm. The small unexfoliated graphite particles are also combined with the transparent GO nanosheets. The SAED pattern of GO is confirming the XRD data. The ring-like structure is analyzed and found that the planes (001), (100) and (002), which are already discussed in the XRD spectrum shown in Fig. 3d. The crystallographic structure of the as-synthesised GO could be confirmed from the above results and discussions.

The exfoliation of graphite to Graphene takes place in the vicinity of the anode along with the rapid evolution of H_2 gas at an anode and oxygen at the cathode. During intercalation, the exfoliated graphene oxidizes to graphene oxide due to the presence of dissolved oxygen. The presence of single to few layers of graphene

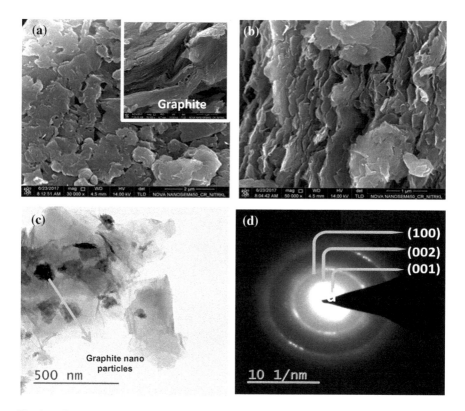

Fig. 3 **a**, **b** FESEM images of GO. (inset) Graphite powder. **c** TEM image of GO. **d** SAED Pattern of GO

oxide is confirmed by the TEM image analysis. Along with single layer (SL) and few-layer GO can be observed, also the presence of some exfoliated graphite nanoparticles as a by-product. The formation of graphite nanoparticles (≈15 nm) might be due to the defects induced during the exfoliation process resulting in the formation of nanodimension graphite particles. In most of our experiments, we observed the presence of graphite nanoparticles. Hence it is conclusive that the formation of SLGO/FLGO and graphite nanoparticles takes place simultaneously during exfoliation.

3.4 Atomic Force Microscopy Observations of GO on Mica Substrate

Broad AFM investigations are done to characterize the GO nanosheets, which are deposited onto mica surface. The AFM sample preparation is made by mixing the

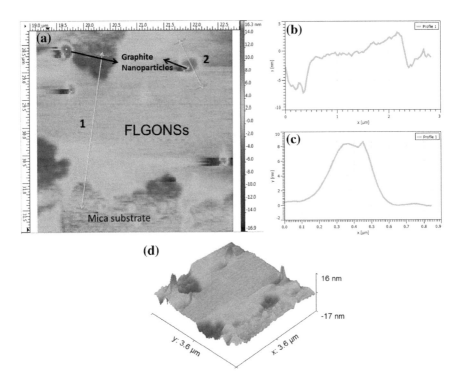

Fig. 4 **a** AFM topography of well dispersed GO on the surface of freshly cleaved mica. **b** Height profiles respected to line profile 1. **c** Height profile respected to line profile 2. **d** Three-dimensional topography of GO and Graphite nanoparticles

GO powder with 10 ml of double distilled water at the rate of 1 mg/mL. The solution is sonicated for 45 min at room temperature by using bath sonicator. As-synthesized and well dispersed GO nanosheets are drop cast on to an atomically flat freshly cleaved mica sheet so that evenly distributed GO nanosheets are obtained and dried in ambient conditions [14]. The AFM observations are performed in tapping (non-contact) mode of action using quadrilateral silicon cantilevers with a spring constant (k) of ~ 40 N m^{-1} and general resonance frequencies between 300 and 370 kHz. The typical Imaging is abled by tip-sample interaction, recording topography (height) and phase images simultaneously.

The AFM topography shows that a 1–3 μm dimensioned GO nanosheet is detected and is shown in Fig. 4a. The lateral thickness of as-synthesized GO nanosheets is observed in the range of 2–5 nm as shown in the height images of Fig. 4b, c. Two height profiles were observed at two different locations, first height profile across the GO it showing flat few layers of nanosheets of thickness in the range of 3–5 nm only (Fig. 4b). The second height profile showing the thickness of the graphite nanoparticles of the size 10 nm (Fig. 4c) i.e. the same characterization also observed in the TEM characterization as well. The AFM topography has also revealed an evidence for the presence of unexfoliated Graphite nanoparticles. The three-dimensional histogram is also shown in Fig. 4d, it shows the size of graphite particles are in the range of 10 nm to 16 nm.

4 Conclusions

An efficient, green and scalable method for synthesis of GO nanosheets by electrochemical intercalation and exfoliation of pyrolytic graphite sheets has been discussed in this paper. The as-synthesized GO has low quality due to impurities, like graphite nanoparticles.

1. The exfoliation of graphite in HNO_3 electrolyte has found very rapid, due to the small size of intercalates (NO_3^-), so they can easily perforate into the lattice of the graphite and increases the rate of exfoliation.
2. The FESEM observations revealed that few layers of graphene oxide nanosheets in the range of 1–3 µm.
3. FTIR studies confirm that the existence of the functional groups, but they could be removed by using a simple chemical reduction process.
4. A 1–2 transparent layers Graphene oxide can be observed in the TEM analysis.
5. The TEM observations and AFM observations are showing that the existence of both the GO and graphite nanoparticles.

Acknowledgements The authors are very thankful for the financial and infrastructure support from National Institute of Technology, Rourkela, India.

References

1. Novoselov KS, Fal'ko VI, Colombo L, Gellert PR, Schwab MG, Kim K (2012) A roadmap for graphene. Nature 490(7419):192–200
2. Balandin AA et al (2007) Extremely high thermal conductivity of graphene: experimental study. Mater Sci, 1–16
3. Lee C, Wei X, Kysar JW, Hone J (2008) Measurement of the elastic properties and intrinsic strength of monolayer graphene. Science 321:385–388
4. Bolotin KI et al (2008) Ultrahigh electron mobility in suspended graphene. Solid State Commun 146(9–10):351–355
5. Zhu Y et al (2010) Graphene and graphene oxide: synthesis, properties, and applications. Adv Mater 22(35):3906–3924
6. Cai W, Zhu Y, Li X, Piner RD, Ruoff RS (2009) Large area few-layer graphene/graphite films as transparent thin conducting electrodes. Appl Phys Lett 95(12):2–5
7. Li X, Zhu Y, Cai W, Borysiak M, Han B, Chen D, Piner RD, Colombo L, Ruoff RS (2009) Transfer of large-area graphene films for high-performance transparent conductive electrodes. Naon Lett 9:4359–4363

8. Bhuyan MSA, Uddin MN, Islam MM, Bipasha FA, Hossain SS (2016) Synthesis of graphene. Int Nano Lett 6(2):65–83
9. Novoselov KS et al (2005) Two-dimensional gas of massless Dirac fermions in graphene. Nature 438(7065):197–200
10. Blake P et al (2008) Graphene-based liquid crystal device. Nano Lett 8(6):1704–1708
11. Sahoo SK, Mallik A (2015) Simple, fast and cost-effective electrochemical synthesis of few layer graphene nanosheets. NANO 10(2):1550019
12. Morales GM et al (2011) High-quality few layer graphene produced by electrochemical intercalation and microwave-assisted expansion of graphite. Carbon N Y 49(8):2809–2816
13. Sahoo SK, Ratha S, Rout CS, Mallik A (2016) Physicochemical properties and supercapacitor behavior of electrochemically synthesized few layered graphene nanosheets. J Solid State Electrochem 20(12):3415–3428
14. Pandey D, Reifenberger R, Piner R (2008) Scanning probe microscopy study of exfoliated oxidized graphene sheets. Surf Sci 602(9):1607–1613

Fact-Finding on Physical and Mechanical Properties of 3Y-TZP Toughened Alumina (ZTA) Composites Incorporation of Functionalized Multi-walled Carbon Nanotubes

D. Manikandan and S. Renold Elsen

Abstract Zirconia-toughened alumina (ZTA) has been the famous composite utilized for the fabrication of articulating components in a hip joint prosthetics. The demand for longer life and better performance the material characteristics of the articulating components have to be enhanced. In recent literatures it has been described that the addition of multi-walled carbon nanotubes (MWCNT) into an alumina matrix of zirconia-toughened alumina, ZTA to improve the flexural strength, fracture toughness, and fatigue resistance. The intent of the current work is to establish and authenticate that the material's toughness and hardness could be significantly tailored by preparing 3Y-TZP toughened alumina (ZTA) composites by the combination of functionalized MWCNT using conventional sintering method. For this method, homogenous spreading of CNTs in ceramic matrix has been reached from 0.5 wt% up to 1.8 wt% CNTs using ball milling then compacted and finally sintered. The density and micro hardness were studied related to the experimental runs established using box-behnken design. A clear enhancement in the physical properties was achieved after the adding MWCNTs at the range of 0.5 to 1.8 wt% and sintering temperature varied from 1500 to 1600 °C. The addition of MWCNT in the matrix exhibited the better porosity, density over 3Y-TZP toughened alumina (ZTA) sintered at the same temperature. This results designates that the properties of Zirconia-toughened alumina (ZTA) with MWCNT reinforcements based composites are strongly rely on the process of adding CNT and sintering. The optimized process parameter were also identified form the studies.

Keywords Hip joint prosthetics · ZTA · MWCNT · Box-Behnken design
Ball milling · Conventional sintering · Density and hardness

D. Manikandan
Department of Mechanical Engineering, M.I.E.T Engineering College,
Tiruchirappalli, Tamil Nadu, India
e-mail: mani08mech08@gmail.com

S. Renold Elsen (✉)
School of Mechanical Engineering, VIT, Vellore, Tamil Nadu, India
e-mail: renoldelsen@gmail.com

© Springer Nature Singapore Pte Ltd. 2019
A. K. Lakshminarayanan et al. (eds.), *Advances in Materials and Metallurgy*,
Lecture Notes in Mechanical Engineering,
https://doi.org/10.1007/978-981-13-1780-4_25

1 Introduction

Zirconia-Toughened Alumina remains one of the often used biomaterials in the field of orthopedic in particular hip arthroplasty. Regardless of the excellent performance of ZTA composites, still few issues associated to their failure requires a strong focus to offer resolution [1]. For several years, the adding strong particles/ whiskers (alumina whiskers [2], graphene platelets [3], etc.,) incorporate better mechanical and physical properties in the ZTA composite matrices. Also, high-performance material requires improved mechanical and physical properties of the ZTA composite achieved by the developed by ZTA fibers [4]. The prospective for nano particles reinforced ceramics with improved mechanical and physical properties over monolithic ceramic materials has been reported by many researchers [5, 6]. Adding Single-Wall and Multi-Wall CarNano-Tubes to the monolithic ceramics promised excellent strength and stiffness of the ceramics [7, 8]. CNTs had higher had tensile strength 200 GPa and Young's modulus 1 TPa higher than whiskers. The MWCNTs were reported to have a maximum Young's modulus of 950 GPa in its outermost layer and maximum tensile strength of 63 GPa [9]. Many researches focus on carbon nano tubes enclosed on Polymer Matrix Composites (PMCs) to enhanced electric, optical and mechanical characteristics [10]. However, only minimal studies were reported on Ceramic Matrix Composites (CMCs) reinforced with CNTs [11]. The additions of Zirconia in any ceramic matrix develops improved fracture toughness owing to the transformation from tetragonal phase to monoclinic phase induces increase in volume. This increase in volume improves stresses in the ceramic matrix, which will result into hindrance for crack propagation [12]. Spear et al., reported the microstructure, hardness and fracture ZTA reinforced by MWCNTs and pure monoclinic and nanosized ZrO_2 particles. They studied the characteristics of new ceramic nano-composite before and after HIP sintering which offered exceptional chemical and mechanical properties [13].

2 Experimental Work

Response Surface Methodology (RSM) is an approach used to model a particular behaviour and assess its performance and in addition to optimize the factors affecting the process [14]. Box-Behnken Design (BBD) has been selected to study the effect of composition of Alumina in ZTA, composition of Multi-walled Carbon Nanotubes (MWCNT) and sintering temperature on density and hardness of a composite. Since, relatively minimum combinations of process parameters are required BBD is used often [10]. In a variety of experiments, it is probable to state the independent factors concerned into a quantitative form as in the Eq. (1).

Table 1 Controlling parameters and their levels

Symbol	Variables	Units	Experimental level		
			Low level (−1)	Mid level (0)	High level (1)
A	Composition of Al$_2$O$_3$ in ZTA	wt%	70	80	90
B	Composition of MWCNT with ZTA	wt%	0.5	1.15	1.80
C	Sintering Temperature	°C	1500	1550	1600

$$R = \theta(f_1, f_2, \ldots, f_n)e, \tag{1}$$

where, the R is the response, $f_1, f_2, \ldots, f_k$ are the n quantitative factors and e assesses the errors from the experiments. The mathematical form of Θ is unknown and it can be approximately acceptable within the experimental region by polynomial.

The alpha phase Alumina (α-Al$_2$O$_3$) and Zirconia (ZrO$_2$) as well as Multi-walled CNTs commercial of grade were purchased from IENT Salem, Tamilnadu, India. Alumina powders of weight ratio (70, 80 and 90%) are blended together with Zirconia. The ratio of zirconia is selected based on the wt% of the Alumina and MWCNT added to the ZTA composite. And then the powders of MWCNTs of different weight ratios (0.5, 1.15, and 1.80 wt%). Using Ball mill for four hours with charge to powder ratio of 1:1 for blending ZTA composite powders [15] is done initially for alumina and Zirconia and then finally the MWCNT is blended to the powder mixture. The powder mixture is poured into the die cavity with 10 mm diameter and then compacted using a ram. ZTA green compacts of cylindrical profile with diameter 10 mm with thickness 10 mm were obtained at load of 160 N/mm^2.

Initially 1500 °C is selected as the lower limit as the composite is found to have developed good density at 1560 °C [16] and 1600 °C, as the upper limit a maximum density is reported for ZTA composite [10]. The green compacts are then sintered in the box furnace at 1500, 1550 and 1600 °C temperatures with a raise in temperature of 5 °C per minute with a soaking period of 6 h. The composites were fabricated according to the experimental runs were developed by the DESIGN EXPERT based on the BBD given in Table 1.

3 Results and Discussion

3.1 Results

The Box–Behnken Technique based on the control parameters and levels specimens were prepared. The selected variables and levels for this work along with the responses observed were provided in Table 2. The density of the composite varies

Table 2 Process design layout using Box–Behnken design & Test results

Run	Variables			Response	
	Composition of Al$_2$O$_3$ in ZTA	Composition of MWCNT with ZTA	Temperature	Density	Hardness
	Weight Percentage	Weight Percentage	(°C)	(g/mm^3)	(H$_V$)
1	80	0.5	1500	3.63	1555
2	90	1.15	1500	3.31	1583
3	80	1.8	1500	3.59	1544
4	70	1.15	1500	3.71	1482
5	70	1.8	1550	4.19	1583
6	80	1.15	1550	4.05	1767
7	80	1.15	1550	4.08	1765
8	90	0.5	1550	3.69	1848
9	80	1.15	1550	4.03	1761
10	70	0.5	1550	4.26	1594
11	90	1.8	1550	3.64	1835
12	80	1.15	1550	4.07	1769
13	80	1.15	1550	3.98	1766
14	90	1.15	1600	4.12	1989
15	70	1.15	1600	4.47	1768
16	80	0.5	1600	4.38	1939
17	80	1.8	1600	4.32	1913

between a minimum of 3.31 to a maximum of about 4.47 g/mm^3. Also, the hardness of the composite varies between a minimum of 1532 H$_V$ and 1989 H$_V$ with a variation of about 22.97%.

3.2   Influence of the Process Parameter on Density

The density of the composite varies between 3.7 and 4.52 g/mm^3 from 1500 to 1600 °C at 70 wt% of Alumina and 0.5 wt% of MWCNT as shown in Fig. 1. The density of the composite is found not to be influenced by the addition of the MWCNT as observed in Fig. 2. The density of the composite varies between 4.5 and 4.06 g/mm^3 from 70 to 90 wt% of Alumina in Fig. 3. This is due to the addition of Zirconia which has higher density compared to alumina which make the composite much denser. The 90 wt% of Alumina with 1.15 wt% of MWCNT had the best relative density of 97.33% of the theoretical density.

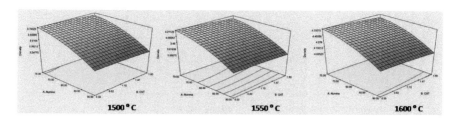

Fig. 1 Density variation for Weight percentage of Alumina in ZTA versus CNT at 1500, 1550 and 1600 °C sintering temperature

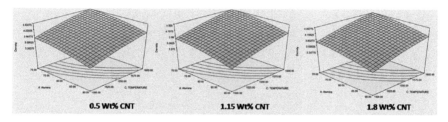

Fig. 2 Density variation for Weight percentage of Alumina in ZTA versus Sintering Temperature at 0.8, 1.15 and 1.8 wt% of MWCNT

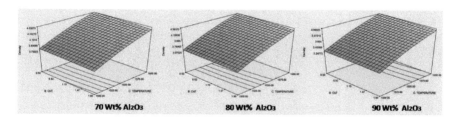

Fig. 3 Density variation for Weight percentage of CNT with ZTA versus Sintering Temperature at 70, 80 and 90 wt% of Alumina

3.3 Influence of the Process Parameter on Hardness

Hardness is increased from 1438 to 1599 H_V, 1604 to 1829 H_V, 1732 to 2023 H_V at 1500, 1550, 1600 °C respectively with the addition of Alumina (90–70 wt%) to the ZTA and the minimal limit addition of MWCNT 0.5 wt% to the composite as shown in Fig. 4. The hardness of the composite is found not to be influenced much by the addition of the MWCNT as observed in the Fig. 5 and the maximum variation is found to be 13%. It shows the minimal variation of hardness and density of composites as compared with sintering through spark plasma sintering (variation 17%) [17]. The hardness of the composite varies between 1532 H_V and 1923 H_V from 70 to 90 wt% of Alumina shown in Fig. 6. The improvement in hardness is

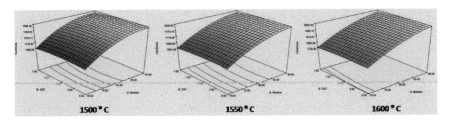

Fig. 4 Hardness variations for Weight percentage of Alumina in ZTA versus CNT at 1500, 1550 and 1600 °C

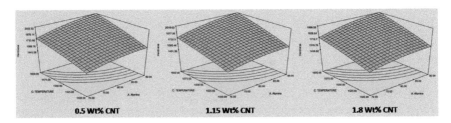

Fig. 5 Hardness variation for Weight percentage of Alumina in ZTA versus Sintering Temperature at 0.8, 1.15 and 1.8 wt% of MWCNT with ZTA

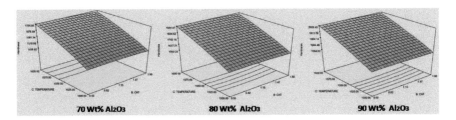

Fig. 6 Hardness variation for Weight percentage of CNT with ZTA versus Sintering Temperature at 70, 80 and 90 wt% of Alumina in ZTA

about 20.33% and this is due to the reduction of Zirconia and also the addition of MWCNT is minimal which make the composite much denser, similar results were observed [10, 18].

3.4 ANOVA Response for Density and Hardness

The "Model F value" 52.59 for density and 58.99 for hardness, indicates the significance of the model as observed from Table 3. Also, merely a 0.01% possibility that "Model F-Value" this huge can happen is due to presence of noise for both density and hardness. Further, the values of "Prob > F" is less than 0.0500 indicate that the model terms are significant for both. From the table further it can be assessed that weight percentage of ZTA and temperature is found to influence the density and hardness of the composite. But, the weight percentage of MWCNT (B) and other interactions have no impact on the density and hardness of the developed composite. The "Pred R-squared" of 0.8208 for density and 0.7931 for hardness has reasonable harmony with the "Adj R Squared" of 0.9666 for density and 0.9702 for hardness. Furthermore "Adeq Precision" of 27.13 for density and 28.43 for hardness, which gives the signal to noise ratio which is greater than 4 indicates an sufficient signal. The Lack of Fit is also found to be insignificant for the model from ANOVA. The correlation in terms of coded factors and actual factors for density and hardness is given by Eqs. (2), (3), (4) and (5) respectively.

$$
\begin{aligned}
\text{Density} = {} & +4.04 - 0.23 \times A - 0.028 \times B + 0.38 \times C - 0.087 \\
& \times A^2 - 9.750 \times 10^{-3} \times B^2 - 0.052 \times C^2 \\
& + 5 \times 10^{-3} \times AB + 0.013 \times AC - 5 \times 10^{-3} \times BC
\end{aligned} \tag{2}
$$

$$
\begin{aligned}
\text{Density} = {} & -58.7883 + 0.0765 \times \text{Alumina} + 0.1876 \times \text{CNT} + 0.0705 \times \text{Temperature} - 8.725 \times 10^{-4} \\
& \times \text{Alumina}^2 - 0.0231 \times \text{CNT}^2 - 2.09 \times 10^{-5} \times \text{Temperature}^2 + 7.6923 \times 10^{-4} \\
& \times \text{Alumina} \times \text{CNT} + 2.5 \times 10^{-5} \times \text{Alumina} \times \text{Temperature} - 1.5384 \times 10^{-4} \\
& \times \text{CNT} \times \text{Temperature}
\end{aligned} \tag{3}
$$

$$
\begin{aligned}
\text{Hardness} = {} & +1765.60 + 103.50 \times A - 7.62 \times B + 180.62 \times C \\
& - 41.42 \times A^2 - 9.18 \times B^2 - 18.68 \times C^2 \\
& - 0.50 \times AB + 30.00 \times AC - 3.75 \times BC
\end{aligned} \tag{4}
$$

$$
\begin{aligned}
\text{Hardness} = {} & -18047.628 - 16.2815 \times \text{Alumina} + 223.2159 \times \text{CNT} + 22.1022 \times \text{Temperature} \\
& - 0.4142 \times \text{Alumina}^2 - 21.7159 \times \text{CNT}^2 - 7.47 \times 10^{-3} \times \text{Temperature}^2 - 0.0769 \\
& \times \text{Alumina} \times \text{CNT} + 0.06 \times \text{Alumina} \times \text{Temperature} - 0.1153 \times \text{CNT} \\
& \times \text{Temperature}
\end{aligned} \tag{5}
$$

3.5 Optimization Studies

Optimization is a process which uses various procedures to find the location of the best response within the defined constraints and confinements in a specific design domain [19]. The three constraints taken are (i) the weight percentage of Alumina with 70–90% as the limit, (ii) the weight percentage of MWCNT with 0.5–2.0% as the limit and (iii) 1500–1600 °C of sintering temperature were taken as the limits.

Table 3 ANOVA for density and hardness surface quadratic model

Source	Sum of squares		Degrees of Freedom		Mean square		F-Value		p-value Prob > F	
	D^*	H^{**}	D^*	H^{**}	D^*	H^{**}	D^*	H^{**}	D^*	H^{**}
Model	1.6538	360548.6	9	9	0.1838	40060.95	52.5979	58.9943	<0.0001	<0.0001
A-Alumina	0.4371	85.698	1	1	0.4371	85.698	125.1191	126.2001	<0.0001	<0.0001
B-CNT	0.0061	465.125	1	1	0.0061	465.125	1.7318	0.6850	0.2296	0.4352
C-Temp	1.1628	261003.1	1	1	1.1628	261003.1	332.8435	384.3570	<0.0001	<0.0001
AB	0.0321	7225.392	1	1	0.0321	7225.392	9.1748	10.6402	0.019	0.0138
AC	0.0004	354.4447	1	1	0.0004	354.4447	0.1146	0.5220	0.7449	0.4934
BC	0.0115	1468.445	1	1	0.0115	1468.445	3.2903	2.1625	0.1126	0.1849
A^2	1E-04	1	1	1	1E-04	1	0.0286	0.0015	0.8704	0.9705
B^2	0.0006	3600	1	1	0.0006	3600	0.1789	5.3014	0.6850	0.0548
C^2	1E-04	56.25	1	1	1E-04	56.25	0.0286	0.0828	0.8704	0.7818
Residual	0.02446	4753.45	7	7	0.0035	679.0643				
Lack of fit	0.0182	4718.25	3	3	0.0061	1572.75	3.858811	178.7216	0.1125	0.0001
Pure error	0.0063	35.2	4	4	0.0016	8.8				
Cor total	1.6782	365.302	16	16						

Density(D^*): Std. Dev.—0.059106, R-Squared—0.9854, Mean—3.9717, Adj R-Squared—0.9667, C.V.—1.4882, Pred R-Squared—0.8209, PRESS—0.3006 and Adeq Precision—27.1328

Hardness(H^{**}): Std. Dev.—0.059106, R-Squared—0.9854, Mean—3.9717, Adj R-Squared—0.9667, C.V.—1.4882, Pred R-Squared—0.8209, PRESS—0.3006 and Adeq Precision—27.1328

Table 4 Constraints and limits for density and micro hardness

Constraints		Limits	
Name	Goal	Lower limit	Upper limit
A: Alumina	Is in range	70 (wt% of Alumina)	90 (wt% of Alumina)
B: CNT	Is in range	0.5 (wt% of MWCNT)	1.8 (wt% of MWCNT)
C: Temperature (°C)	Is in range	1500	1600
Density (g/mm^3)	Maximum	4.47	
Vickers Hardness (H$_V$)	Maximum	1984	

Table 5 Comparison of confirmation experiments with the results

Exp. No	Density (g/mm^3)			Vickers Hardness (H$_V$)		
	Obtained	Predicated	Error (%)	Obtained	Predicated	Error (%)
1	4.0254	4.34711	7.4	1910.99	1953.98	2.2
2	4.0693	4.34711	6.39	1864.09	1953.98	4.6
3	3.9211	4.34711	9.8	1897.31	1953.98	2.9

The goal is set for maximum density and hardness given in Table 4. The predicted levels of optimized responses were 81.89 wt% Alumina in ZTA, 0.53 wt% MWCNT with ZTA and 1600 °C sintering temperature with 0.912 desirability value resulting the density 4.34711 g/mm^3 and hardness 1953.9 H$_V$.

The confirmation experimental studies were conducted with 82 wt% of Alumina and 0.5 wt% of with CNT with a 160 MPa and sintering temperature of 1600 °C. The results from the experimental methods were validated with the predicted value and are given in Table 5. The experimental results are found to have minimal error percentage of 6.39 for Density and percentage of 2.2 for Hardness with predicted value.

4   Conclusion

The density and hardness of ZTA with MWCNT composite prepared following the BBD using powder forming method were characterized in this work. A quadratic model was developed using regression to explain the relationship amongst the density, hardness with the process parameter. Temperature is the influential factor in the selected design space, the increase in temperature improves the density and Vickers hardness of the ZTA with MWCNT composite. The addition of MWCNT showed minimal variation in the density and Vickers hardness. The reduction of weight percentage of alumina from 90 to 70 wt% improves the density by 10.42% and reduces the micro hardness of the composite by 18%. The process parameter which affects the density and hardness were identified and the optimized process parameters with limits were found by using RSM with minimal error.

The following conclusions were made from the investigation.

- The significant Process parameters were identified from the ANOVA for density and hardness as sintering temperature and followed by composition of alumina added to the composite with a better R-Squared value of 82.09 and 79.33% respectively.
- The confirmation test conducted was found to have error percent close to the predicted values with 6.39% for Density and 2.2% for Hardness.

References

1. De Mattia JS, Castiello E, Affatato S (2017) Clinical issues of ceramic devices used in total hip arthroplasty. In: Advances in ceramic biomaterials, pp 313–328
2. Nevarez-Rascon A, Aguilar-Elguezabal A, Orrantia E, Bocanegra-Bernal MH (2011) Compressive strength, hardness and fracture toughness of Al_2O_3 whiskers reinforced ZTA and ATZ nanocomposites: Weibull analysis. Int J Refract Met Hard Mater 29(3):333–340
3. Liu Jian, Yan H, Reece MJ, Jiang K (2012) Toughening of zirconia/alumina composites by the addition of graphene platelets. J Eur Ceram Soc 32(16):4185–4193
4. Pfeifer S, Demirci P, Duran R, Stolpmann H, Renfftlen A, Nemrava S, Niewa R, Clauß B, Buchmeiser MR (2016) Synthesis of zirconia toughened alumina (ZTA) fibers for high performance materials. J Eur Ceram Soc 36(3):725–731
5. Rincón A, Moreno R, Chinelatto ASA, Gutierrez CF, Salvador MD, Borrell A (2016) Effect of graphene and CNFs addition on the mechanical and electrical properties of dense alumina-toughened zirconia composites. Ceram Int 42(1):1105–1113
6. Lupo F, Kamalakaran R, Scheu C, Grobert N, Ruhle M (2004) Microstructural investigations on zirconium oxide–carbon nanotube composites synthesized by hydrothermal crystallization. Carbon 42(10):1995–1999
7. Sun J, Gao L, Jin X (2005) Reinforcement of alumina matrix with multi-walled carbon nanotubes. Ceram Int 31(6):893–896
8. Sun J, Iwasa M, Nakayama T, Niihara K, Gao L, Jin X (2004) Pressureless sintering of alumina carbon nanotubes composites in air atmosphere furnace and their mechanical properties. J Ceram Soc Jpn Supplement 112–1, PacRim5 Special Issue
9. Yu M-F, Lourie Oleg, Dyer MJ, Moloni K, Kelly TF, Ruoff RS (2000) Strength and breaking mechanism of multiwalled carbon nanotubes under tensile load. Science 287(5453):637–640
10. Renold Elsen S, Ramesh T, Aravinth B (2014) Optimization of process parameters of zirconia reinforced alumina by powder forming process using response surface method. Adv Mater Res 984:129–139
11. Fan J, Zhao D, Wu M, Xu Z, Song J (2006) Preparation and microstructure of multi-wall carbon nanotubes-toughened Al_2O_3 composite. J Am Ceram Soc 89(2):750–753
12. Santos C, Maeda LD, Cairo CAA, Acchar W (2008) Mechanical properties of hot-pressed ZrO_2–NbC ceramic composites. Int J Refract Met Hard Mater 26(1):14–18
13. Rose L, Spear RE (2008) Carbon nanotubes for orthopaedic implants. IntJ Mater Form 1 (2):127–133
14. Renold Elsen S, Ramesh T (2016) Analysis and optimization of dry sliding wear characteristics of zirconia reinforced alumina composites formed by conventional sintering using response surface method. Int J Refract Met Hard Mater 58:92–103
15. Renold Elsen S, Jegadeesan K, Ronald Aseer J (2017) X-Ray diffraction analysis of mechanically milled alumina and zirconia powders. Nano Hybrids Compos 17:96–100

16. Renold Elsen S, Ramesh T (2015) Optimization to develop multiple response hardness and compressive strength of zirconia reinforced alumina by using RSM and GRA. Int J Refract Met Hard Mater 52:159–164
17. Ipek AKIN (2015) Investigation of the microstructure, mechanical properties and cell viability of zirconia-toughened alumina composites reinforced with carbon nanotubes. J Ceram Soc Jpn Supplement 123–5:405–413
18. Renold Elsen S, Ramesh T (2016) Shrinkage characteristics studies on conventional sintered zirconia toughened alumina using computed tomography imaging technique. Int J Refract Met Hard Mater 54:383–394
19. Aseer JR, Sethupathi PB, Chandradass J, Renold Elsen S (2017) Taguchi based analysis on hole diameter error of drilled glass/BahuniaRacemosa fiber polymer composites. In: SAE Technical Paper. Issue 2017-28-1983

Effect of Carbon Nano Tubes (CNT) on Hardness of Polypropylene Matrix

R. Ashok Gandhi, V. Jayaseelan, K. Palani Kumar, B. K. Raghunath and S. Krishnaraj

Abstract Hardness is an important mechanical property which determines the applicability of polymer composites. Carbon Nanotubes (CNTs) invented by Iijima by arc-discharge technique. It possesses some unique properties like Young's modulus, the values varies from 0.42 to 4.15 TPa, tensile strength of 1 TPa, density varies between 1.3 and 3 g/cm^3 which is comparatively lower than commercial carbon fibers. This makes CNT as a potential reinforcement with metal and polymers for enhancement of properties. This work describes about preparation of PP-CNT composites with different ratios. Hardness of the composites were measured using Nanoindentation method and found that hardness of the PP-CNT system increases significantly with the increase of CNT proportion in the PP matrix.

Keywords Hardness · Corbon Nano Tubes (CNT) · Poly Propylene (PP) Nanoindentation

R. Ashok Gandhi (✉) · S. Krishnaraj
Mechanical Engineering, Sri Sai Ram Engineering College, Chennai, India
e-mail: ashokgandhi.mech@sairam.edu.in

S. Krishnaraj
e-mail: krishnaraj.mech@sairam.edu.in

V. Jayaseelan
Mechanical Engineering, Prathyusha Engineering College, Chennai, India
e-mail: jaiseelanv@gmail.com

K. Palani Kumar
Sri Sai Ram Institute of Technology, Chennai, India
e-mail: principal@sairamit.edu.in

B. K. Raghunath
Manufacturing Engineering Department, Annamalai University, Chidambaram, India
e-mail: bkrau@rediffmail.com

© Springer Nature Singapore Pte Ltd. 2019
A. K. Lakshminarayanan et al. (eds.), *Advances in Materials and Metallurgy*,
Lecture Notes in Mechanical Engineering,
https://doi.org/10.1007/978-981-13-1780-4_26

1 Introduction

Carbon nanotubes (CNTs) are sheets of graphite hollow cylinders and it is used as the most promising modifiers of the conventional polymers. This causes the material matrix system to increase its multifunctional properties. Ashok Gandhi et al. [1] have proved that inclusion of nanomaterials increases the wear resistance of the whole system. Wear and hardness are interrelated. To have a wear-resistant material, then it should possess better hardenability. The invention of carbon nanotubes opened the way for numerous researches related to the nanotubes. This also resulted in a dramatic increase in their co-related composites. The challenge with the single-walled and multi-walled carbon nanotubes is determination of its true mechanical properties. Different researchers found the Young's modulus of SWCNT. Tu and Ou-yang [2] reported as 4.7 TPa. However, computational studies done by Belytschko et al. [3] and Maiti et al. [4] expressed that the properties of nanotubes is governed by dislocation of carbon atoms. The accuracy of the calculation is highly dependent on the initial boundary condition, which is applied to the simulated models and sizes of the system. The important reason behind reduction of mechanical strength of nanocomposites is caused by the weak van der Waals interaction between layers of multi-walled nanotubes.

Bauhofer and Kovacs [5] observed that agglomeration of CNTs caused by strong van der Waals attraction between different CNTs. This could be avoided by oxidation or introduction of other nanoparticles. Kashiwagi et al. [6] indicated that reinforcement of montmorillonite (MMT) into the system will enhance dispersion. Hou et al. [7], Pujari et al. [8], and Wu et al. [9] reported that melt compounding is the most convenient and efficient method to produce PP/CNT composites. Using such a method, Pujari et al. [8] proved that CNT has the highest thermal conductivity. Choi et al. [10] reported that the CNTs, present in the PP matrix to align perpendicularly to a crack, are able to slow down their propagation by bridging the crack faces. Thus, these materials may be used to improve the out-of-plane and interlaminar properties of advanced composite structures by increasing the matrix strength. Pradhan and Iannacchione [11] showed the higher thermal conductivity of CNT/polymer composites can be achieved using suitable dispersion methods and higher quality nanofiller materials. The thermal conductivity carbon material performs because of its atomic vibrations or phonons. Lopez Manchado et al. [12] reported that the thermal conductivities of composites are controlled by filler concentration, filler conductivity, filler geometry, and interface conductance between the filler and polymer and homogeneity of the filler dispersion.

Kanagaraj [13] observed the possibility of good load transfer because of this network of CNTs. They have observed more surface area, per unit volume, which results in a more uniform stress distribution load transfer in the matrix that will increase the resistance to wear of the composite.

Fig. 1 Granules of PP-CNT blend

Table 1 Composition of PP-CNT blends

S. No.	Weight % of PP	Weight % of CNT
1	100	0
2	99	1
3	97	3
4	95	5
5	93	7

2 Preparation of PP/CNT Blend

Melt blending of PP, PP-g-MAH (10 wt%), and the CNT of 1, 3, 5, and 7 wt% was carried out in an intermeshing counter-rotating twin screw extruder (ctw-100, Haake Germany) having barrel length of 300 mm and an angle of entry 90° prior to extrusion, the matrix polymer and the CNT were dehumidified in a vacuum oven at 60 °C for a period of 6 h. The PP was fed at the rate of 5 kg/h and the CNT was subsequently introduced at the melting zone. The process was carried out at a screw speed of 150 rpm and a temperature difference of 160, 170, and 180 °C between feed zones and die zone, followed by granulation (length 3–5 mm and diameter 3 mm) in a pelletizer (Fission, Germany) and drying as shown in Fig. 1. The compositions of samples are shown in Table 1.

3 Preparation of Test Samples

Die has been prepared with a tolerance of 0.05 mm as per the die design procedure as shown in Fig. 2, which is used to shape the mixed granules into testing specimens as per ASTM G99 standards.

Now, the prepared die is used in the hand injection molding machine as shown in Fig. 3 for further processing. It consists of two main parts, an injection unit and

Fig. 2 Die

Fig. 3 Injection molding

Fig. 4 Test specimen

a clamping unit. The objective of using hand molding is to produce the test specimens at low cost and to have the same characteristics as the original compounds (Fig. 4).

4 Nanoindentation Testing

Hardness measurement of polymeric materials creates significant challenges. The major reason is polymers have poor mechanical properties and the response of hardness cannot be measured at all with conventional methods. Load–displacement method is used in nanoindentation tests to extract the hardness of the specimen material from measurements. Conventional hardness measurement methods measure the size impression in the specimen by the indenter during loading. This is a measure of the contact area of indenter in the specimen. In case of nanoindentation test, the size of the residual impression would be a few microns and it is viewed as a great sign of difficulty by direct measurement.

This nanoindentation testing gives quantitative, absolute measurements of hardness of polymer material with nanoscale spatial resolution. So, this can be a key to understand the mechanical behavior of technologically important material systems. The polypropylene, and the CNT composites with variations in compositions were tested in TI 950 TriboIndenter as shown in Fig. 5. The tests were conducted on the samples at various locations. The value of hardness obtained from the testing is plotted. The CNT is reinforced in PP with various percentages (1, 3, 5, and 7%) and its effect on the hardness of matrix is shown in Table 2.

Figure 6 shows the comparison of hardness of PP-CNT composites with variations in weight percentages of the CNT reinforcement. The X-axis represents the various locations of testing and the Y-axis represents the values of hardness in GPa. It is observed that the hardness of PP+1% CNT is very less when compared to the other compositions. Lau and Hui [14] observed that the hardness of carbon nanotube/polymer composite increased with increasing weight fraction. They also

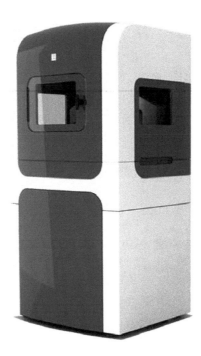

Fig. 5 TI 950 TriboIndenter

Table 2 Comparison of hardness of polypropylene and CNT composite at various locations

S. No.	H (GPa)	H (GPa)	H (GPa)	H (GPa)
Composition	PP+1% CNT	PP+3% CNT	PP+5% CNT	PP+7% CNT
A1	0.039292	0.382635	0.097093	0.361362
A2	0.147477	0.319819	0.311016	0.404103
A3	0.108873	0.282372	0.248817	0.287997
A4	0.046057	0.237653	0.345612	0.251833
A5	0.044553	0.218505	0.227845	0.235015
A6	0.208124	0.206667	0.079464	0.212338
A7	0.024463	0.196641	0.162885	0.208511
A8	0.013742	0.137806	0.114412	0.186198

found that the hardness was found dropped at low nanotube weight fraction samples because of weak bonding between the nanotube and the polymer matrix.

From Table 2, it is observed that the hardness of the composite increases with the increase in the percentage of CNT and the maximum hardness of 0.404 GPa was obtained in PP+7% CNT composite. This occurs due to the size of the reinforcement, which can easily occupy the molecular and nanovoids of the matrix and increasing the CNT weight fraction that would result in forming a mesh-like networking structure with high aspect ratio of CNTs. Hence, it is assured that more the

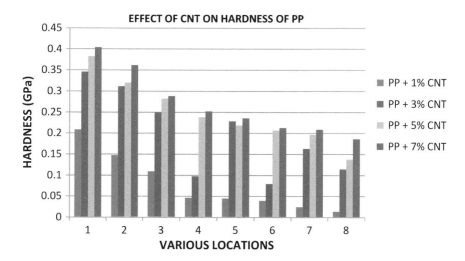

Fig. 6 Comparison of hardness of PP-CNT composites with variations in weight percentages of CNT reinforcement

reinforcement, more will be the resistance to penetration, besides increase in the hardness. If the hardness of the composite increases, naturally, it will enhance the wear properties.

Figure 7 shows the SEM image of Polypropylene without reinforcement, the surface is free from irregularities like crack, flaw and molding defects. Figure 8 shows the image of PP-1% CNT composites test specimen in this image, uniform distribution of CNTs in the PP matrix is noticed, and a good amount of surface interaction between PP and CNT. In order to induce the enhanced properties in the matrix materials, uniform dispersion of reinforcement elements is essential. The CNTs are extremely hard to disperse and align it into a polymer matrix because the CNTs form stabilized

Fig. 7 PP-0% CNT

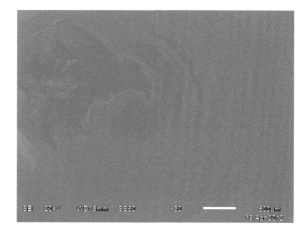

Fig. 8 PP-1% CNT

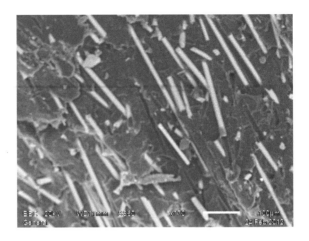

bundles due to van der Waals interactions. Melt mixing of CNT in PP matrix has been performed with twin screw extruder machine, and it has prevented agglomeration of CNT and its uniform distribution in the PP matrix.

Figure 9 shows the image of PP-3% CNT composite specimen, small cavities, and debris formation are observed when CNTs started to disintegrate from the PP matrix. Lau and Hui [14] showed the local deformations of the matrix found adjacent to stiff nanotubes and deformed the surrounding matrix, thus enlarging the holes from where they came. Choi et al. [10] reported that the CNTs, present in the PP matrix to align perpendicularly to a crack, are able to slow down their propagation by bridging the crack faces. Thus, these materials may be used to improve the out-of-plane and interlaminar properties of advanced composite structures by increasing the matrix strength. From this image, we can also infer that the CNTs are good load carriers, and the applied load on the matrix has been transferred to the reinforced carbon nanotubes this tendency is the reason for increase of hardness of

Fig. 9 PP-3% CNT

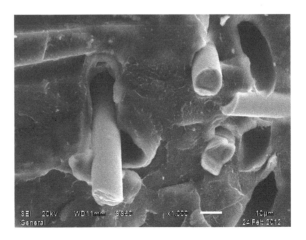

Fig. 10 PP-7% CNT

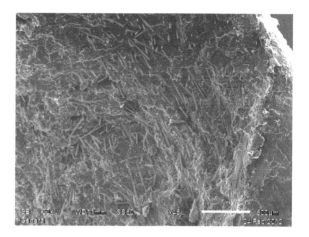

the system with an increase in CNT weight percentage. Figure 10 shows PP-7% CNT image which clearly shows the distribution of CNT in the polypropylene matrix.

5 Conclusion

Nanoindentation hardness testing was performed on PP-CNT composites at different locations of the test specimen with variations in the weight percentage of CNT reinforcement and the following conclusions were made.

1. Hardness of the composite is more if the CNT present in the test location.
2. Hardness of composite increases with increase in CNT weight percentage.
3. For optimum results, more tests has to be performed.

References

1. Ashok Gandhi R, Palani Kumar K, Raghunath BK, Paulo Davim J (2013) Role of carbon nano tubes (CNT's) in improving wear properties of polypropylene (PP) in dry sliding condition. Mater Des 48:52–57
2. Tu ZC, Ou-yang ZC (2002) Single-walled and multiwalled carbon nanotubes viewed as elastic tubes with effective Young's moduli dependent on layer number. Phys Rev B 65:233–407
3. Belytschko T, Xiao SP, Schatz GC, Ruoff RS (2002) Atomistic simulations of nanotube fracture. Phys Rev B 65(23):235430
4. Maiti A, Svizhenko A, Anantram MP (2002) Electronic transport through carbon nanotubes: effects of structural deformation and tube chirality. Phys Rev Lett 88(12):126–135, 235–430

5. Bauhofer W, Kovacs JZ (2009) A review and analysis of electrical percolation in carbon nanotube polymer composites. Compos Sci Technol 69:1486–1498
6. Kashiwagi T, Grulke E, Hilding J, Groth K, Harris R, Butler K (2004) Thermal and flammability properties of polypropylene/carbon nanotube nanocomposites. Polymer 45:4227–4239
7. Hou ZC, Wang K, Zhao P, Zhang Q, Yang CY, Chen DQ, Du RN, Fu Q (2008) Structural orientation and tensile behaviour in the extrusion-stretched sheets of polypropylene/ multi-walled carbon nanotubes composite. Polymer 49:3582–3589
8. Pujari S, Ramanathan T, Kasimatis K, Masuda J, Andrews R, Torkelson JM, Brinson LC, Burghardt WR (2009) Preparation and characterization of multi-walled carbon nanotube dispersions in polypropylene: melt mixing versus solid-state shear pulverization. J Polym Sci Polym Phys 47:1426–1436
9. Wu DF, Sun YR, Zhang M (2009) Kinetics study on melt compounding of carbon nanotube/ polypropylene nanocomposites. J Polym Sci Polym Phys 47:608–618
10. Choi ES, Brooks JS, Eaton DL, Al-Haik MS, Hussaini MY, Garmestani H, Li D, Dahmen K (2003) Enhancement of thermal and electrical properties of carbon nanotube polymer composites by magnetic field processing. J Appl Phys 94:6034–6049
11. Pradhan NR, Iannacchione GS (2010) Thermal properties and glass transition in PMMA +SWCNT composites. J Phys D Appl Phys 43:342–344
12. Lopez Manchado MA, Valentini L, Biagiotti J, Kenny JM (2005) Thermal and mechanical properties of single-walled carbon nanotubes–polypropylene composites prepared by melt processing. Carbon 43:1499–1505
13. Kanagaraj S (2007) Mechanical properties of high density polyethylene/carbon nanotube composites. Compos Sci Technol 67:3071–3077
14. Lau KT, Hui D (2002) Effectiveness of using carbon nanotubes as nano-reinforcements for advanced composite structure. Carbon 40:1605–1606

Dissimilar Friction Welding of AISI 304 Austenitic Stainless Steel and AISI D3 Tool Steel: Mechanical Properties and Microstructural Characterization

Sangathoti Haribabu, Muralimohan Cheepu,
Lakshmipathi Tammineni, Naresh Kumar Gurasala,
Venkateswarlu Devuri and Venkata Charan Kantumuchu

Abstract In recent years, the continuous demand for dissimilar joining combination of various materials increasing in manufacturing sector for various applications such as power plants, nuclear, and aerospace applications. The joining of dissimilar metals using conventional fusion welding methods is exhibited in unsatisfactory joint strength. The use of solid-state welding methods is most suitable for joining of dissimilar combination in the current scenario. In this study, dissimilar joining of 304 austenitic stainless steel and D3 tool steel are joined using friction welding process to investigate the properties and joint interface characteristics. To identify the feasibility of joining dissimilar materials using friction welding process, the experiment is performed at different input welding conditions such as friction

S. Haribabu · L. Tammineni · N. K. Gurasala
Department of Mechanical Engineering, Sri Venkatesa Perumal College
of Engineering and Technology, Puttur 517583, Andhra Pradesh, India
e-mail: haribab204513@gmail.com

L. Tammineni
e-mail: lakshmipathi4761@gmail.com

M. Cheepu (✉)
Department of Mechatronics Engineering, Kyungsung University, Busan 48434,
Republic of Korea
e-mail: muralicheepu@gmail.com

V. Devuri
Department of Mechanical Engineering, Marri Laxman Reddy Institute
of Technology and Management, Hyderabad 500043, Telangana, India
e-mail: dvriitr@gmail.com

V. C. Kantumuchu
Department of Industrial and Manufacturing Engineering and Technology, Bradley
University, Peoria, IL 61625, USA
e-mail: charan1102srikar@gmail.com

V. C. Kantumuchu
A Division of Methode Electronics Malta Ltd., Hetronic USA,
Oklahoma City, OK 73112, USA

© Springer Nature Singapore Pte Ltd. 2019
A. K. Lakshminarayanan et al. (eds.), *Advances in Materials and Metallurgy*,
Lecture Notes in Mechanical Engineering,
https://doi.org/10.1007/978-981-13-1780-4_27

pressure and upset pressure were varying from minimum to maximum values to obtain the reliable joint strength. The friction welded joints were characterized using microscope observations at the weld interface and failure modes are discussed.

Keywords Friction welding · Austenitic stainless steel · D3 high carbon steel Microstructure · Mechanical properties · Optimization technique Joint interface

1 Introduction

Nowadays, the joining of materials became an essential requirement of manufacturing industries. The most important needs of present fabrication industries are focusing on the manufacturing of cost-effective materials and the use of less expensive procedures other than the safety and quality issues [1, 2]. The joining of dissimilar metals has several challenges unlike similar metals welding such as metallurgical drawbacks, interface formation which can cause in-service failure. The most often pronounced welding failures are solidification cracks owing to the inadvertent deposition of improper selection of welding electrodes, among them carbon steel electrodes are the primary metal for failures [3–5]. The failures such as fragile and crack susceptible brittle phases occur in the stainless steel interface in case of dissimilar welds, those appear along center of the fusion zone [3]. Such kind of brittle phases along the fusion zone may depict dissimilar metal joints amenable to the formation of hydrogen embrittlement, localized presence of pitting corrosion, and stress-induced cracks. This frequently occurs in the soft zone of the welds of heat-affected zone of the dissimilar metal joints of the ferrite materials. The previous studies reported that the dissimilar joining of conventional fusion welding combinations resulted in the formation of very brittle and low ductility intermetallic compounds owing to the metallurgical reactions and thermal mismatch during solidification [6–8]. To avoid the difficulties of using conventional joining methods for dissimilar materials combinations, solid-state welding techniques have been contemplated. Solid-state welding methods possess the absence of melting of base materials during welding and limit the extent of weld fusion zones or intermixing zones. There are various solid-state welding methods available for dissimilar joining such as friction welding, friction stir welding, diffusion bonding, explosive welding, and roll bonding. Each and every process has its own advantages and applications based on the joint configurations. Among them, friction welding process has many benefits and a wide range of applications, especially for the incompatible materials [9–12]. Using friction welding method, many dissimilar materials are successfully joined. The dissimilar combination of titanium to stainless steel, stainless steel to aluminum, and aluminum to ceramic materials are not feasible to weld by conventional fusion welding methods. Friction welding was successfully applied for these combinations and exhibits excellent outcome of the joint properties [13–19]. To improve the mechanical properties and strength of dissimilar material combinations, interlayer material techniques have been implemented to avoid the direct contact between the substrates [20–23].

Moreover, this method is characterized by minimal heat input levels which are generated by the transformation of rotating mechanical energy into heat energy at the weld center owing to the rotation under external pressure on the substrates. Friction welding is widely used for joining of various materials and shapes such as plates to plates, rod to rod, and rod to plates. Austenitic stainless steels (ASSs) have specific properties of strength and ductility, and its complete austenite phase leads for different applications in various fields of nuclear, chemical, cryogenic, and food industries [19]. Therefore, ASS of 304 type steels are most often used for the applications of corrosion resistance and it can be replaced for the aggressive corrosion environments of marine applications due to their superior qualities as like titanium alloys. ASS has excellent weldability due to their low alloy contents, whereas the fusion welding of these alloys have several defects and can destroy the austenite phase and formation of solidification cracks. In addition, the joining of 304 ASS with D3 tool steel exhibits several issues of intermetallics formation and changes in formation of microstructures during fusion welding. Therefore, the coarser grains formation near to the weld interface which can deteriorate the mechanical properties and corrosion properties of the welds [24]. Murti et al. [25], investigated the thermal behavior of the microstructural changes during friction welding of austenitic and ferritic transition joints. Their further studies on dissimilar combinations of medium carbon steel to high-speed steel friction welding joints behavior evaluated with the experimental conditions which are appropriate to the proposed service applications. The friction welding was carried out using optimized welding conditions and the produced joints were exposed to heat treatment conditions as per the schedule that is usual for intended applications then the joint properties were evaluated in dynamic torsion methods [26]. The properties of D3 tool steel are quite similar to 304 ASS materials and its combination of joints applications are widely used in many products. The similar studies have been studied by Akata et al. [27] on friction welding between stainless steel and medium carbon steel. The strength of the joints was evaluated and proposed the necessary conditions for enhancing the mechanical properties. In their extended work, the notch impact test and fatigue properties were evaluated. The other studies on friction welding of quenched and tempered grade A517 steels joint properties had been evaluated and a uniform pressure and fatigue cycles were applied to identify the joint properties under different cyclic loads [28, 29]. However, the properties of a 304 austenitic stainless steel (304 SS) and D3 tool steel dissimilar combinations have not yet been friction welded to characterize the interface microstructural behavior and mechanical properties. This research work is aimed to understand the weldability behavior of the 304 austenitic stainless steel (304 SS) and D3 tool steel dissimilar combinations using friction welding technique. The results, hence, would provide feasibility of the welding between these alloys and its microstructural formation and mechanical properties at different welding conditions. The welding conditions are also studied and found the range of major welding conditions of upset force and friction force on tensile strength of the friction welded joints.

2 Experimental Work

The round bars of AISI 304 austenitic stainless steel and D3 tool steel with a diameter of 16 mm were cut to the length of 80 mm for the friction welding joints. The chemical composition of the base substrates of AISI 304 austenitic stainless steel and D3 tool steel are provided in Table 1. Mechanical properties of the 304 stainless steel and D3 tool steel materials were tensile strength 548 and 651 MPa, and the elongation 67 and 20% (heat-treated condition), respectively. To get the accuracy of the joint strength of the all welds before welding of base, metal substrates faying surfaces are machined and is followed by polish with 800 grit emery papers. The samples were polished until to get the required surface roughness, which has more contribution on the enhancing bond strength of the welds [30]. Before welding, all the samples were cleaned with acetone and dried to remove the oil, grease, dirt, etc. To produce the joints, continuous drive friction welding method was used with the machine capacity of 200 kN of ETA make. To measure the axial force on the substrates and for its control, a load cell was used in this machine. In order to understand the welding conditions effect on joints, all the important friction welding conditions such as axial load, spindle position, and spindle speed are read and plotted online during friction welding. The variation of the welding conditions of process chart helps to monitor the process behavior and to calculate the heat input of the joints. Figure 1 illustrates the present study process chart of the friction welding conditions. In friction welding process, the process parameters have a significant effect on the mechanical and metallurgical properties of the welds. The welding conditions of upset force, upset time, friction force, friction time, and spindle speed have direct relation to the joint formation and its strength. To achieve the high-quality joints, a new set of welding parameters were designed after the several experimental trails of the welds. A new set of welding conditions which are found as upset force in the range of 4–8 tons, upset time 6 s, friction force in the range of 3–7 tons, friction time 5 s, and spindle speed of 1500 rpm was used. The resultant of the friction welded joint using these new set of welding conditions is exhibited in Fig. 2.

Table 1 Chemical composition of the base metals used in the present study (wt%)

Materials	C	Si	Mn	P	S	Cr	V	W	Ni	Bal
D3 tool steel	2.32	0.441	0.362	0.014	0.025	11.025	0.068	0.032	–	Fe
304 stainless steel	0.02	0.04	1.73	0.03	0.03	18.07	–	–	8.06	Fe

Fig. 1 The process chart illustrates the variation of welding conditions with time

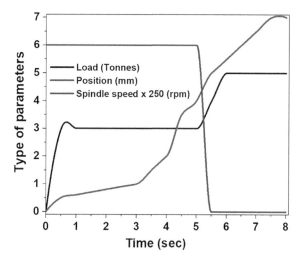

Fig. 2 The photograph of the friction welded joint showing the weld flash at interface

After friction welding, samples were cut to make the cross sections for metallographic samples preparation, and the cross section of the samples were prepared as per the metallographic procedure for microstructural analysis and hardness measurements. The tensile samples were prepared as per the ASTM E8 standard for the evaluation of mechanical properties [31]. The joint tensile strength was carried out by machining the weld flash using a universal testing machine TFUC-400 India. In addition, Vickers hardness test was conducted for the joints to clarify the joint properties and intermetallics formation across the weld interface. The microstructural observations were made using optical microscope after etching the samples with 2% nitol for D3 tool steel and aqua regia solution of 304 austenitic stainless steel for the joint interface and microstructural formations before and after welding.

3 Results and Discussion

3.1 Microstructural Characterization

The joining of dissimilar metals using any fabrication techniques is still challenging with the several problems of the welded joints of stainless steel to tool steels which have excellent properties and their wide range of applications. The analysis of the conventional fusion weld interfaces between 304SS and D3 tool steel exhibits solidification problems and improper bonding, thus resulted in deterioration of the mechanical properties [32, 33]. To make use of these alloys, an alternative welding technique is found to be successful in producing sound welds and the results satisfy the intended service applications. Solid-state welding process of friction welding is one of the prominent welding processes for producing joints between incompatible materials. The welding of these two materials is very difficult to weld due to their different physical and metallurgical properties. The microstructures of the substrates are illustrated in Fig. 3. The microstructure of the tool steel contains the martensitic structure with presence of carbides and ferrites. The equiaxed grains with the twin boundary and the full austenite structure exhibits in the 304 austenitic stainless steel microstructure. The two distinct microstructures of two different alloys makes the weldability poorer and subjected to several solidification problems in the fusion zone, therefore mechanical properties of the welds deteriorate across the brittle phases.

During friction welding, faying surfaces of the substrates frictionally rub each other and generates the heat between the two surfaces and subjected to deformation. The weld flash formation due to deformation on the D3 tool steel side is higher than the 304 SS side. The physical properties of the D3 tool steel changes with temperature, due to this, most of the applications are restricted below 250 °C. Whereas, the deformation on stainless steel side is very less when compared to D3 tool steel.

(a)   **(b)**

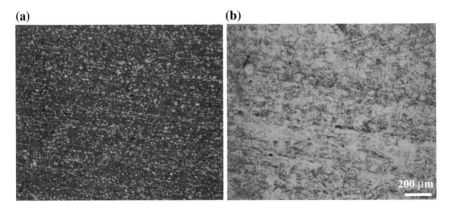

Fig. 3 Optical micrographs showing the base metal microstructures of **a** D3 tool steel and **b** 304 austenitic stainless steel

(a) **(b)**

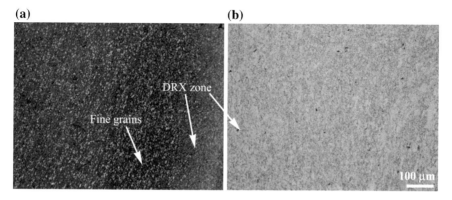

Fig. 4 Microstructures shows cross section of the weld interface with different zones at **a** D3 tool steel side and **b** 304 austenitic stainless steel side

The microstructures of the cross section of the joints at weld interface are depicted in Fig. 4. Figure 4a clearly shows the occurence of plastic deformation adjacent to the weld interface which resembles as unetched region with the presence of finer grains. During welding, the material is subjected to heat and pressure, thus the softened region of the material experienced to the formation of dynamic recrystallization zone (DRX) and the grains are refined to form as fine grain structure. The material flow in this region is inclined to the weld interface and its structure is modified from the base metal. The microstructure on the stainless steel exhibits the very small region of dynamic recrystallization unlike D3 tool steel (see Fig. 4b). The weld interface at center regions is smooth and the formation of very thin interfacial layer has been observed. Whereas, the weld interface at the periphery of the joints exhibits different from the center region with the formation of several width of interface. The relative speed and its effect at the periphery region are higher than the center region thus the width of intermixing zone formation is higher at the periphery region. The interface at the center region is free from the defects formation, whereas at the periphery region the presence of micro cracks has been observed, this is due to the formation of wider intermixing zone.

3.2 Mechanical Properties

Figure 5 shows the microhardness across the welds at center and periphery region of the friction welded joints. It is found that the peak hardness value has been recorded at the weld interface for both the conditions of center and periphery regions. The hardness values are gradually increased towards interface from the base metal, and the highest hardness at adjacent to the weld interface subjected to the strain hardening affect and the formation of finer grains in the dynamic recrystallization zone. The hardness profiles at the center and periphery regions

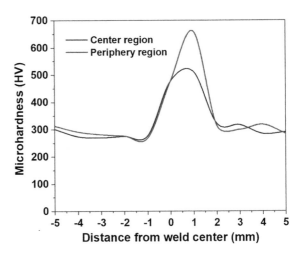

Fig. 5 Mircohardness across the weld interface of the joints at center and periphery regions

show the significant difference that the highest profile recorded for the periphery regions. The reason for the highest hardness values at periphery regions is due to the formation intermixing zone. It is expecting that the intermixing zone consists of formation of hard and brittle phases. As discussed earlier in microstructural section, the weld interface at periphery region owing to the cause for recording highest hardness. The large difference in hardness values of center and periphery regions shows that the failure region and crack initiation occurring from the intermixing zone at periphery of the weld joints. The tensile strength of the joints is evaluated after removing the flash from weld interface. In order to achieve the highest tensile strength, several welding condition combinations are tried for initial experiments. The welding conditions of upset force and friction force have been identified as the most influencing parameters for the selected dissimilar combination of friction welds. The tensile strength of the joints at lower values and improper combination of welding conditions failed at drop test. To achieve the highest tensile strength friction force and upset force varied in several ranges and the corresponding results of the joint strength is illustrated in Fig. 6. The strength of the joints increases with the increasing of friction force and upset force initially and starts to decrease after reaching the maximum strength of the joints. The strength at minimum level of friction force (3 tons) and upset force (4 tons), joint strength is 210 MPa, and it is increased to 388 MPa at the friction force (6 tons) and upset force (6 tons). Whereas, it is decreasing to 290 MPa with the further increasing of friction force to 7 tons and upset force to 8 tons as illustrated in Fig. 6. The joints at the higher upset and friction force combinations are resulted in the formation of large amount of weld flash, which leads to weaken the strength of the joints. From the joints tensile tests, it is observed that the weld failure took place in the weld interface with the mixed regions of fracture surfaces. The fracture surfaces clearly indicted the crack initiation from the periphery region where the wider region of intermixing zones

Fig. 6 Tensile strength of the friction welded joints at different friction force and upset force welding conditions

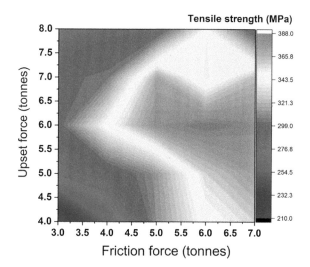

formed and formation of brittle and ductile mode of mixed failure regions has been identified.

4 Conclusions

Friction welding of dissimilar materials between D3 tool steel and 304 austenitic stainless steel has been successfully joined. The feasibility of the joining of these materials using friction welding and welding conditions are studied extensively. The joint microstructural characteristics, weld interface and mechanical properties are evaluated. Based on the experimental results, some conclusions can be drawn as follows:

- Friction welding of D3 tool steel to 304 austenitic stainless steel produced joints have higher strength than other welding methods.
- Friction welding process successfully avoided the formation of brittle phases at the weld interface and solidification problems with the formation smooth weld interface.
- The weld interface is irregular and wider at the periphery region of the welds compared to center of the joints.
- Microhardness of the joints increased towards weld interface from base metal due to the strain hardening effect and the presence of fine grains on the weld interface.
- The highest hardness values are recorded at the periphery of the joints owing to the formation of wide range of hard interfacial layers.

- The tensile strength of the joints increasing with the increasing of upset force and friction force initially and decreasing after reaching the maximum value of 388 MPa.

Acknowledgements The authors are grateful to Professor G. D. Janaki Ram, Materials Joining Laboratory, Department of Metallurgical and Materials Engineering, IIT Madras for providing the Friction Welding facility to carry out this research experiments.

References

1. Özdemir N (2005) Investigation of the mechanical properties of friction-welded joints between AISI 304L and AISI 4340 steel as a function rotational speed. Mater Lett 59:2504–2509
2. Muralimohan CH, Haribabu S, Reddy YH, Muthupandi V, Sivaprasad K (2015) Joining of AISI 1040 steel to 6082-T6 aluminium alloy by friction welding. J Adv Mech Eng Sci 1 (1):57–64. http://dx.doi.org/10.18831/james.in/2015011006
3. Arivazhagan N, Singh S, Prakash S, Reddy GM (2008) An assessment of hardness, impact strength, and hot corrosion behaviour of friction-welded dissimilar weldments between AISI 4140 and AISI 304. Int J Adv Manuf Technol 39:679–689
4. Muralimohan CH, Haribabu S, Reddy YH, Muthupandi V, Sivaprasad K (2014) Evaluation of microstructures and mechanical properties of dissimilar materials by friction welding. Procedia Mater Sci 5:1107–1113
5. Cheepu M, Haribabu S, Ramachandraiah T, Srinivas B, Venkateswarulu D, Karna S, Alapati S, Che WS (2018) Fabrication and analysis of accumulative roll bonding process between magnesium and aluminum multi-layers. Appl Mech Mater 877:183–189. http://dx.doi.org/10.4028/www.scientific.net/AMM.877.183
6. Devireddy K, Devuri V, Cheepu M, Kumar BK (2018) Analysis of the influence of friction stir processing on gas tungsten arc welding of 2024 aluminum alloy weld zone. Int J Mech Prod Eng Res Dev 8(1):243–252. https://doi.org/10.24247/ijmperdfeb201828
7. Meshram SD, Mohandas T, Reddy GM (2007) Friction welding of dissimilar pure metals. J Mater Process Technol 184:330–337
8. Venkateswarulu D, Cheepu M, Krishnaja D, Muthukumaran S (2018) Influence of water cooling and post-weld ageing on mechanical and microstructural properties of the friction-stir welded 6061 aluminium alloy joints. Appl Mech Mater 877:163–176. http://dx.doi.org/10.4028/www.scientific.net/AMM.877.163
9. Lalam SV, Reddy GM, Mohandas T, Kamaraj M, Murty BS (2009) Continuous drive friction welding of Inconel 718 and EN24 dissimilar metal combination. Mater Sci Technol 25:851–861
10. Cheepu M, Muthupandi V, Srinivas B, Sivaprasad K (2018) Development of a friction welded bimetallic joints between titanium and 304 austenitic stainless steel. In: Pawar PM, Ronge BP, Balasubramaniam R, Seshabhattar S (eds) Techno-Societal 2016, International conference on advanced technologies for societal applications ICATSA 2016, Springer, Cham, Chapter 73, pp 709–717. https://doi.org/10.1007/978-3-319-53556-2_73
11. Muralimohan CH, Ashfaq M, Ashiri R, Muthupandi V, Sivaprasad K (2016) Analysis and characterization of the role of Ni interlayer in the friction welding of titanium and 304 austenitic stainless steel. Metall Mater Trans A 47:347–359
12. Fuji A, North TH, Ameyama K, Futamata M (1992) Improving tensile strength and bend ductility of titanium/AISI 304L stainless steel friction welds. Mater Sci Technol 8(3):219–235

13. Dey HC, Ashfaq M, Bhaduri AK, Rao KP (2009) Joining of titanium to 304L stainless steel by friction welding. J Mater Process Technol 209:5862–5870
14. Muralimohan CH, Muthupandi V, Sivaprasad K (2014) The influence of aluminium intermediate layer in dissimilar friction welds. Inter J Mater Res 105:350–357
15. Sahin M (2009) Joining of stainless-steel and aluminium materials by friction welding. Int J Adv Manuf Technol 41:487–497
16. Taban E, Gould JE, Lippold JC (2010) Dissimilar friction welding of 6061-T6 aluminum and AISI 1018 steel: properties and microstructural characterization. Mater Des 31:2305–2311
17. Cheepu M, Ashfaq M, Muthupandi V (2017) A new approach for using interlayer and analysis of the friction welding of titanium to stainless steel. Trans Indian Inst Met 70:2591–2600. https://doi.org/10.1007/s12666-017-1114-x
18. Fauzi MNA, Uday MB, Zuhailawati H, Ismail AB (2010) Microstructure and mechanical properties of alumina-6061 aluminum alloy joined by friction welding. Mater Des 31:670–676
19. Cheepu MM, Muthupandi V, Loganathan S (2012) Friction welding of titanium to 304 stainless steel with electroplated nickel interlayer. Mater Sci Forum 710:620–625
20. Muralimohan CH, Muthupandi V, Sivaprasad K (2014) Properties of friction welding titanium-stainless steel joints with a nickel interlayer. Procedia Mater Sci 5:1120–1129
21. Meshram SD, Reddy GM (2015) Friction welding of AA6061 to AISI 4340 using silver interlayer. Defence Technol 11:292–298
22. Cheepu M, Muthupandi V, Che WS (2018) Improving mechanical properties of dissimilar material friction welds. Appl Mech Mater 877:157–162. http://dx.doi.org/10.4028/www.scientific.net/AMM.877.157
23. Muralimohan CH, Muthupandi V (2013) Friction welding of type 304 stainless steel to CP titanium using nickel interlayer. Adv Mater Res 794:351–357
24. Murugan S, Rai SK, Kumar PV, Jayakumar T, Raj B, Bose MSC (2001) Temperature distribution and residual stresses due to multipass welding in type 304 stainless steel and low carbon steel weld pads. Int J Press Vessels Pip 78:307–317
25. Murti KG, Sundaresan S (1985) Thermal behavior of austenitic-ferritic transition joints made by friction welding. Weld J 64(12):327–334
26. Murti KG, Sundaresan S (1986) Structure and properties of friction welds between high-speed steel and medium-carbon steel for bimetal tools. Mat Sci Tech 2:865–870
27. Akata HE, Sahin M (2001) Friction welding of different composition steels II, Makina Malzemesi ve Imalat Teknolojisi Sempozyumu, 7–9 Kasim 2001, Manisa, pp 595–602 (in Turkish)
28. Sahin M, Akata HE (2004) An experimental study on friction welding of medium carbon and austenitic stainless steel components. Ind Lubr Tribol 56:122–129
29. Rajamani GP, Shunmugam MS, Rao KP (1992) Parameter optimization and properties of friction welded quenched and tempered steel. Weld J 71:225–230 (1992)
30. Fuji A, North TH, Ameyama K, Futamata M (1992) Improving tensile strength and bend ductility of titanium/AISI 304L stainless steel friction welds. Mater Sci Technol 8:219–235
31. ASTM E8/E8 M-16a (2016) Standard test methods for tension testing of metallic materials. ASTM International, West Conshohocken, PA. www.astm.org
32. Arivazhagan N, Singh S, Prakash S, Reddy GM (2011) Investigation on AISI 304 austenitic stainless steel to AISI 4140 low alloy steel dissimilar joints by gas tungsten arc, electron beam and friction welding. Mater Des 32:3036–3050
33. Mendes R, Ribeiro JB, Loureiro A (2013) Effect of explosive characteristics on the explosive welding of stainless steel to carbon steel in cylindrical configuration. Mater Des 51:182–192

Corrosion Studies on Friction–Welded Aluminium Alloy AA6061-T6 to Copper with Nickel Interlayer

E. Ravikumar, N. Arunkumar, D. Ananthapadmanaban and V. Prabhakaran

Abstract The research studies the effect of corrosion on friction-welded AA6061 aluminium alloy to pure copper with nickel interlayer. The potentiodynamic polarization method was utilized to determine the corrosion rate in the chosen environment. All tests were performed in an aerated 0.6 M NaCl aqueous solution (6.5 pH, 30 °C) to determine characteristics of the corroded areas, specifically the welded region and the parent metals. In this method, the potential of the working electrode was varied with the corresponding current being monitored. It was observed that the parent metal-aluminium alloy (AA6061) was more corroded than the welded regions. Further, specimens welded with lower 'upset pressure' were less corroded than those welded with higher 'upset pressure'. The corrosion rate varied from 0.466 and 356.64 mA/cm^2. SEM fractography was used to determine the type and extent of corrosive action, by studying the characteristics across the cross section.

Keywords Friction welding · Dissimilar materials · Corrosion rate Optical microscopy · SEM fractography · Corrosion studies

E. Ravikumar (✉) · V. Prabhakaran
Department of Mechanical Engineering, Alpha College of Engineering,
Chennai, India
e-mail: ramakrishnar2009@gmail.com

V. Prabhakaran
e-mail: prabhameed@gmail.com

N. Arunkumar
Department of Mechanical Engineering, St. Joseph's College
of Engineering, Chennai, India
e-mail: n.arunkumar@rediffmail.com

D. Ananthapadmanaban
Department of Mechanical Engineering, SSN College of Engineering,
Chennai, India
e-mail: ananth_1out@yahoo.com

© Springer Nature Singapore Pte Ltd. 2019
A. K. Lakshminarayanan et al. (eds.), *Advances in Materials and Metallurgy*,
Lecture Notes in Mechanical Engineering,
https://doi.org/10.1007/978-981-13-1780-4_28

1 Introduction

Friction welding is a unique solid-state welding technique, particularly advantageous in joining dissimilar metals and alloys. Friction welding (FRW) is a solid-state welding process that gives out heat through friction between workpieces in relative rotation to one another. An upsetting force is used to provide a lateral push to the specimen. Friction welding is actually a forging technique and not technically a welding as no melting of constituent metals takes place. Friction welding is used with metals and thermoplastics in a wide variety of aviation and automotive applications. The friction force and the relative rotation give frictional heat. Thus, the metal reaches plastic state and by giving sufficient force to produce upset pressure a defect free weld joint is obtained [1].

In recent years, non-ferrous metals including aluminium alloys have drawn increasing attention due to their application in the marine, aerospace and automobile industries. This is because of their high strength-to-weight ratio, together with their natural ageing characteristic, which gives more strength to the Aluminium alloy [2]. The friction welding process results in minimal formation of brittle intermetallic compounds at the interface, since it is carried out at high pressure, with a short processing time and not in the molten state [3]. This is not the case with conventional welding, where greater formation of brittle intermetallic compounds with increasing aluminium content, results in reduced ductility. Garcia et al. studied the pitting corrosion resistance in chloride containing environment. The different zones of the welded joints of austenitic stainless steels [AISI-304L and AISI-316L] were studied using potentiodynamic anodic polarization and cyclic potentiodynamic polarization, a it was inferred that pitting corrosion of welded metals was higher than that of base metal [4]. Bimes et al. studied the pitting corrosion behaviour of spur martensitic weld in chloride environment, maintaining potentiostatic technique and presented the fact that HAZ was the most important zone for pitting corrosion and [5]. The AA6061 alloys contain precipitation-hardened aluminium alloy, containing magnesium and silicon as its major alloying elements, with a little copper and iron. In addition, zinc together with the magnesium or magnesium plus copper and nickel, develop various levels of strength. Those containing copper and nickel have the highest strength and have been used as a material of construction, in the food industry and in aircraft, for more than 50 years [6]. Among heat-treatable alloys, the 6xxx family exhibits moderate strength alloys that have a high level of resistance to general corrosion. Corrosion resistance approaches those of non-heat-treatable alloy [7]. Many of these Aluminium alloys are used in alkaline solutions, especially in Nuclear Industries. Therefore, it is necessary to study the corrosion behaviour of Al–Ni–Cu interlayer in alkaline condition. The polarization methods such as potentiodynamic polarization, potentiostaircase and cyclic voltammeter are commonly used for corrosion testing in laboratories. The intermetallic formation that happens when aluminium and copper are joined reduces corrosion resistance. To improve corrosion at this junction, an interlayer of nickel has to be introduced between aluminium and copper.

2   Abbreviations and Acronyms

3.5% NaCl solution	g of NaCl per 100 g of solution (= 0.6 M)
ASME	American Society of Mechanical Engineers
ASTM	American Society for the Testing of Materials
CE	Corrosion Electrode
Current density	Corrosion Current Density in [mA/cm^2]
E_{CORR}	Corrosion potential in mV
EDX	Energy Dispersive X-ray (analysis)
FRW	Friction welding
HAZ	Heat-Affected Zone
I_{CORR}	Corrosion Current in mA
mA/cm^2	Current Density in milli-Amperes per square centimetre
mm/year	Corrosion Rate in millimetres per year
MPa	Pressure in Megapascals
mV	Millivolt
pH	A measure of the hydrogen ion concentration of a solution
RE	Reference Electrode
RPM	Rotational Speed in revolutions per minute
FESEM	Field Emission Scanning Electron Microscope (fractography)
WE	Work Electrode

3   Experimental Work

3.1   Choice of Interlayer

Studies reveal that the problem of intermetallic formation while joining two dissimilar metals exists in friction welding as well. Even though the intermetallic compound formation is less when compared to fusion welding process, it still seriously affects the strength of friction-welded components during service. Therefore, methods have to be explored to reduce the formation deleterious compounds at the interface. The following are important properties of interlayer that are to be considered in joining of dissimilar metals.

3.2   Mutual Solubility

The interlayer should be preferably soluble in both the metals. This ensures a successful joint. The solubility limits can be identified from phase diagrams. Even if the interlayer is not completely soluble in the base metals, at least the intermetallic which results when such an interlayer is used should be having lesser harmful effects than the compounds that form between the base materials when joined directly. Nickel is soluble in both Aluminium and Copper.

3.3 Thermal Expansion Coefficient

The coefficient of thermal expansion of both the base materials is another important factor. Wide difference in thermal conductivities will set up internal stresses in the intermetallic zone during any temperature change of the weldment. If the intermetallic zone is brittle, the joint may fail during service. Hence, the interlayer should be such that its thermal expansion coefficient falls in between the two base metals or at least near the thermal expansion coefficient of one of the base metals to avoid abrupt change in expansion and contraction properties during thermal cycles if the joint undergoes such changes. The thermal expansion coefficients of aluminium, copper, and nickel are 22×10^{-6}, 16×10^{-6} and 13×10^{-6} m, respectively. While performing our experiments, we have chosen Nickel as interlayer because its thermal coefficient of expansion is fairly near that of Copper, it gives good corrosion resistance to the joints and also it is cheaply available. Rods of wrought aluminium alloy AA6061, pure copper and pure nickel each having diameter of 20-mm were used for the present study [8]. The chemical composition and mechanical properties of base metals (aluminium alloy and copper) and the nickel interlayer are given in Tables 1, 2 and 3. The 20-mm diameter copper rod of length 75-mm was friction-welded to a 20-mm diameter Nickel rod of length 40-mm. After the weld was complete, the 40-mm long nickel road was machined so that only 0.5-mm length of Nickel remained to be the nickel interlayer. In a similar fashion, the 20-mm diameter aluminium rod of length 75-mm was friction welded to a 20-mm diameter Nickel rod of length 40-mm. After the weld was complete, the 40-mm long nickel portion of the rod was machined so that only 0.5-mm length of Nickel remained to be the nickel interlayer.

Table 1 Chemical composition of aluminium AA6061-T6

Elements	Si	Fe	Cu	Mn	Cr	Al
wt%	0.708	0.212	0.184	0.127	0.098	Remainder

Table 2 Chemical composition pure copper

Elements	Cu	Pb	Sn	Fe	Ni	Te
wt%	99.73	0.003	0.184	0.080	0.068	0.018

Table 3 Chemical composition of nickel

Elements	Al	Mn	Si	Mo	Cr	Ni
wt%	0.033	0.003	0.002	0.004	0.002	99.5

3.4 Experiment Design

Friction-welded joints are fabricated by Taguchi L9 orthogonal array at three levels with four parameters each as shown in Table 4. Out of the samples tested, five samples were not properly welded. Hence, only four samples were investigated further.

For microstructure evaluation, inverted metallurgical optical microscopy with magnification range 100x–600x was used. In order to study corrosion, the potentiodynamic polarization test was conducted in accordance with the ASTMG3-89 standard for the weld portion of a friction-welded aluminium alloy with a dissimilar metal (in this case copper).

The samples taken for testing had dimensions 1 cm × 1 cm. In the welded samples, the weld interface was compared with parent metal. Saturated calomel electrode was used as a reference electrode and the experimental setup is presented in Fig. 1.

3.5 Electrolyte Solution and Tests

All corrosion tests were done in the weld zone. Potentiodynamic polarization tests were carried out in accordance with the ASTMG3-89 standard using software based. GILL AC Potentiostat from ACM Instruments, UK. For corrosion testing, a polarization cell was setup. This consisted of the electrolyte solution in a bath,

Table 4 Friction welding parameters as per Taguchi L$_9$ orthogonal array design

S. No.	Upset pressure (MPa)	Friction pressure (MPa)	Burn-off length (mm)	Speed of spindle (RPM)	Remarks
1	2.5	1.3	1	2000	Cu did not weld
2	2.6	1.5	2	1000	Cu did not weld
3	2.7	1.7	3	1500	Cu came out on machining
4	**2.8**	**1.8**	**2**	**1500**	–
5	**2.9**	**1.9**	**3**	**2000**	–
6	**3.0**	**1.9**	**1**	**1000**	–
7	**2.0**	**2.0**	**3**	**1000**	–
8	3.1	2.1	2	1500	Cu came out on machining
9	3.2	2.2	2	2000	Cu did not weld

Bold values indicates good tensile test results

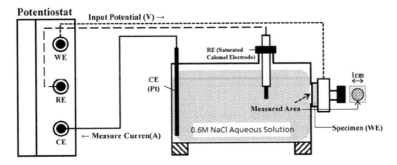

Fig. 1 Experimental test set up. CE: Counter Electrode for measuring current electrode material was platinum. RE: Reference electrode (saturated calomel electrode). WE: work electrode

a reference electrode, a counter electrode(s) and the metal test sample (i.e. the work electrode) are connected to a test sample holder. The electrodes are connected to a potentiostat. The working electrode (WE), reference electrode (RE), and counter electrodes (CE) are placed in the electrolyte solution. The solution chosen resembles the conditions of actual application of the weld joint. We have chosen a 0.6 M NaCl aqueous solution.

In the solution, an electrochemical potential (voltage) is generated between the various electrodes. The corrosion potential (E_{CORR}) is measured by the potentiostat as a potential difference between the working electrode (WE) and the reference electrode (RE). The polarization curves were determined by stepping up the potential at a scan rate of 0.5 mV s^{-1} from–250 mV to +250 mV versus open-circuit potential (E_{CORR} vs. SCE). Experimentation was done at room temperature (25 °C), with 60 min time delay to reach steady state in a freely aerated condition. During this time delay the potential versus time, and Current versus time, were also recorded. The experiments conducted were repeated twice in order to ensure reproducibility. Friction welded samples of 1 cm × 1 cm with thickness of 1 cm for SEM analysis were neatly removed from the cylindrical specimen by wire EDM. In this study a corrosive environment 0.6 mol L^{-1} NaCl aqueous solution was used. The sample was polished by various grades of emery sheets in succession and the surface impurities were removed.

3.6 Test Metals

Tables 1, 2 and 3 give the chemical compositions of the metals used in the experiments, i.e. Aluminium alloy, Copper and nickel.

3.7   Test Parameters

Nine samples were used in the experiments.
Out of these nine parameters, a few parameters—1, 2, 3, 8 and 9 are welded, but it is broken during machining. Owing to bad results, these parameters have been omitted from the remaining part of the experimental work. The remaining four parameters, which gave satisfactory results, were used to prepare specimens for corrosion testing (Table 5).

4   Results and Discussion

4.1   Corrosion Test Results

For comparison, the samples which gave minimum and maximum corrosion rates are considered (Table 6).

Table 5 Friction welding parameters considered for experimental analysis

S. No.	Sample No.	Friction pressure (MPa)	Upset pressure (MPa)	Burn-off length (mm)	Speed (RPM)
1	**4**	1.8	2.8	2	1500
2	**5**	1.9	2.9	3	2000
3	**6**	1.9	3.0	1	1000
4	**7**	2.0	2.0	3	1000

Bold values indicates good tensile test results

Table 6 Results of the corrosion tests

Sample No.	Current density (mA/cm^2)	Rest potential (electrochemical corrosion potential) (mV)	Reverse potential (mV)	Corrosion rate (penetration rates) (mm/year)	Remarks
4	**37.408**	**−722.46**	**250**	**870.12**	**Cu–Ni–Al**
5	**36.69**	**−721.96**	**250**	**412.03**	**Cu–Ni–Al**
6	**356.64**	**−728.03**	**250**	**8295.5**	**Cu–Ni–Al**
7	**0.4659457**	**−733.97**	**250**	**5.2325**	**Cu–Ni–Al**
Cu base metal	209.37	−291.99	250	4870.1	Cu
Al base metal	558.82	−734.65	250	6275.5	Al
Nickel interface metal	0.00285	−335.37	250	0.030664	Ni

Bold values indicates good tensile test results

Of all the parameters used in the friction welding of the Cu–Ni–Al, sample No. 7 shows lowest corrosion rate and a correspondingly lower corrosion current (I_{CORR}). The corrosion current is as low as 0.466 mA/cm^2. Results of corrosion tests, which are taken in the weld zone are shown in Figs. 2 and 3 for samples 7 and 6.

Figure 2 shows the highest corrosion rate (356.64 mA/cm^2) as welded in Rest potential is shown by sample No: 6 at the interface. The corresponding corrosion current namely I_{CORR} is as high as 356.64 mA/cm^2. Sample 6 has a friction welding parameter combination of high friction pressure, low burn-off length and lower

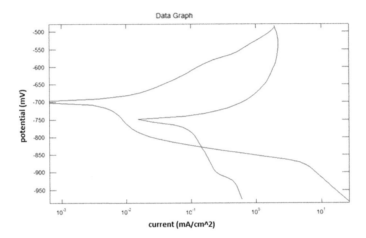

Fig. 2 Polarization curve Al–Ni–Cu (sample no. 7)

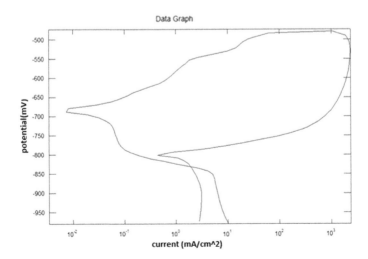

Fig. 3 Polarization curve Al–Ni–Cu (sample no. 6)

speed of rotation. It is possible that higher friction pressure led to greater sites of corrosion initiation. Lower burn-off length also leads to lesser bonding at the interface. Generally, lower speed of rotation leads to better welds. Hence, the loss of bond strength due to using lower burn-off could have been compensated by the use of low speed of rotation. The samples 4 and 5 showed relatively lower corrosion rate in the intermediate range. The corresponding corrosion currents densities were 37.408 and 36.69 mA/cm^2. The corresponding corrosion rates were 870.12 and 412.03 mm/year, respectively. Figure 3 indicates the polarization curve of Al–Ni–Cu for sample 6.

The surface of the potentiodynamic polarization subjected shows deep corrosion pits for sample No. 6. Compared to sample 6, the sample 7 friction welded surface interface showed lowest pits and corrosion affected areas. On the other hand, the samples 4 and 5 showed severe oxidized surface by the corrosion tests. The black surface appearing in the image are the deep oxidized zones and the depth of the pits is higher.

From a corrosion perspective, the combination of parameters used for sample 7 appears optimal, since it has resulted in the lowest corrosion current and rate. Hence, the parameters comprising 2 MP upset pressure, friction pressure of 2 MPa and a burn-off length of 3-mm at a rotational speed of 1000 RPM are the most ideal, relative to the other combinations for friction welding of these materials.

4.2 Microstructural Analysis

The principal high strength Cu–Ni alloys result from the fact that the addition of aluminium to a Cu–Ni binary alloy increases the strength. This is due to the formation of age-hardening precipitates. The increase in strength in these Cu–Ni–Al alloys is caused by the nanometre-scale precipitates of Ni_3Al (γ-phase), that form. It has been found that the effectiveness of this phase in increasing the mechanical properties can be improved through the addition of other elements. Cu–Ni precipitation-hardened alloys are highly suitable for marine environments due to their corrosion resistance. The main features of these materials, is that they display excellent all-round corrosion resistance, resistance to hydrogen embrittlement, high impinge, very good anti-galling characteristics, a good modulus of elasticity and have excellent machinability. In the present study, the microstructures were taken after corrosion testing to evaluate the nature and amount of corrosion products. Etchants used for Aluminium, Copper and Nickel were Keller's Reagent, Potassium Dichromate, and Aqua regia, respectively. Because of the presence of corrosion products, the microstructure appears blurred.

The critical current to obtain a passive layer is of the order of 10^{-1} as can be seen in the polarization curve given above, whereas the critical current to get a passive layer is of the order of 10^{-2} in the case of pure Aluminium. It is possible that Nickel interlayer acted as a barrier to oxidation of Aluminium and hence prevented Aluminium from oxidizing. There is evidence to show that Nickel acts as a barrier and prevents diffusion as shown by Madhusudan Reddy et al. in Steels, but the effect of Nickel interlayer on non-ferrous combinations has not been studied so far [9].

Figure 4 presents the microstructure of Ni–Cu for sample 7. Figures 5 and 6 show the Al–Cu polarization curve and microstructure of Sample 4.

Comparison of the potentiodynamic polarization values shows that Nickel base metal is least affected. The corrosion current value is least and the metal shows high resistance to the polarization. It is seen from the above corrosion tests the corrosion parameters obtained for all the interfaces in which Nickel is common the corrosion at the nickel interface is lower. sample 7, where the nickel was the interface, showed the lowest corrosion rate, as nickel resisted the corrosion by polarization well. Hence, the combination of parameters used in the above friction welding studies shows that the sample No. 7 with an upset pressure of 2 MPa, a friction pressure of 2 MPa and a burn-off length of 3-mm is optimal, to ensure least corrosion after welding.

Comparisons of the microstructures of the corrosion-tested specimens show that pure nickel corrodes very little, has very few corrosion products. Similarly, sample 7 showed considerably less corrosion affected surface than the other samples 4, 5 and 6.

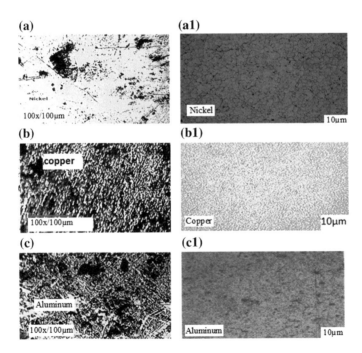

Fig. 4 **a, b, c** Microstructure after corrosion. **a1, b1, c1** Microstructure before corrosion

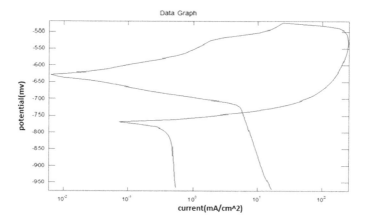

Fig. 5 Polarization curve Al–Ni–Cu (sample 4)

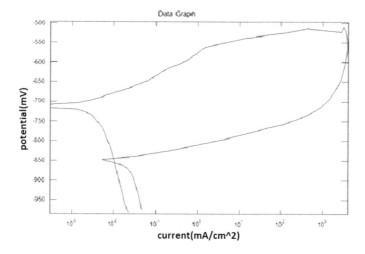

Fig. 6 Base metal—aluminium alloy—analysis of the curve

Figures 6 and 7 present the polarization curve and microstructure of base metal (aluminium alloy). Figure 7 being microstructure of base metal does not show any heat affected zone.

Talbot and Tabolt [10] studied the passive oxide film that is easily formed on the surface of Aluminium alloy. This passive layer is formed when exposure is given to air or water in the presence of chloride ions. Kenneth R et al. also reported that in the presence of chloride ions, corrosion rate is very high. Another factor which affects the corrosion rate is the heterogeneity of their microstructures [11]. Polarization resistance can be directly correlated to the general corrosion rate of metals at or near their corrosion potential, ECORR. Polarization resistance

Fig. 7 Microstructure of
base metal aluminium alloy

measurements are found to be a more accurate and rapid way to measure the general
corrosion rate. Electrochemical polarization methods are extremely useful in
understanding and evaluating the corrosion resistance of materials. Effect of
changes in the corrosive environment can also be understood. These methods could
help establish criteria for anodic or cathodic protection. They could also provide
information on susceptibility to several forms of corrosion. Venugopal et al. [12]
studied microstructural and pitting corrosion properties of friction stir welding of
AA7075 Al alloy in 0.6 Molar NaCl solution. Corrosion resistance of weld metal
was found to be better than that of TMAZ and base metals. Srinivasa Rao and
Prasad Rao [13] researched upon the mechanism of pitting corrosion of heat
treatable Al–Cu alloys and weld. In the present study, Electrochemical corrosion
test by Tafel curve extrapolation method was carried out for Samples 6 and 7 of
base alloy AA6061 and Ni and friction welded area in solution chloride of 3.5% to
determine corrosion parameters such as corrosion potential (E_{CORR}) and corrosion
current (I_{CORR}).

4.3   SEM and EDX Analysis

SEM analysis of sample 7 is presented below.
SEM of Nickel–Copper weld interface (sample 7).
Figure 8 depicts friction welded sample of aluminium and copper with nickel as the
barrier layer. The images are taken after subjecting the interface of Nickel and
copper interface zone with potentiodynamic polarization electrochemical corrosion
evaluation tests as per ASTM G 59-97(2014). The upper part of the image shows
the Nickel matrix and the lower part shows the copper matrix the diffusion region is
at the centre of the two with diffusion of nickel up to 20 μm in copper side with
dark colour. The electrolytic corrosion showed lowest corrosion rate. The image
shows almost unaffected metal matrix of Nickel and the copper diffusion zone is

Fig. 8 Interface of specimen welded observed under FESEM at 1000 rpm (on the side of copper)

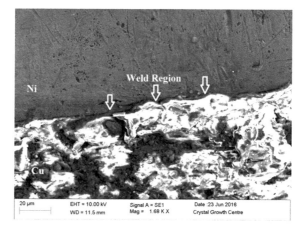

Fig. 9 Interface of specimen welded observed under FESEM at 1000 rpm (on the side of aluminium)

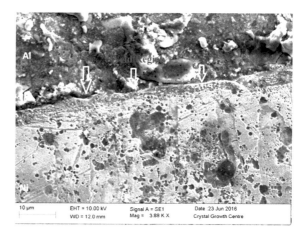

affected more on the copper phases than the nickel diffused layer. The fusion zone is not affected considerably on comparison with copper.

Observation on weld region (friction-welded Al and Ni as the barrier layer).

Friction-welded sample of aluminium and copper with nickel as the barrier layer is shown in Fig. 9. The images are taken after subjecting the interface of Nickel and aluminium interface zone with potentiodynamic polarization electrochemical corrosion evaluation tests as per ASTM G 59-97 (2014). The upper part of the image shows the aluminium matrix and the lower part shows the nickel matrix the diffusion region is at the centre of the two with diffusion of nickel up to 25 μm in aluminium side with porous appearance. The electrolytic corrosion showed lowest corrosion rate. The image shows almost affected metal matrix of Nickel by electrolytic pitting and the aluminium/nickel diffusion zone is affected more on the aluminium phases than the nickel diffused layer. The fusion zone is not affected considerably on comparison with aluminium.

Fig. 10 Micrograph of
joining compound

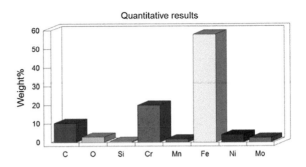

The larger diffusion of nickel (25 μm) on the aluminium side may be due to the higher thickness of aluminium.

4.4 SEM-EDS

SEM-EDS of the Nickel interface show the presence of Iron in the interlayer. It is possible that Nickel interlayer contained some iron and iron intermetallic formation has happened (Fig. 10).

5 Conclusion

This research work investigated experimentally the corrosion behaviour of Ni interlayer on aluminium-copper. From the potentiodynamic polarization test, it was found that the sample 7 shows a low corrosion rate (5.2325 mm/year) and sample 6 shows a high corrosion rate (8295.5 mm/year).

It has been established that Nickel acts as a barrier for diffusion and presence of Nickel interlayer improves the corrosion resistance. This fact is supported by SEM-EDS studies. SEM photos of Aluminium–Copper indicate that there is a brittle fracture, which seems to have changed to a slightly ductile fracture due to the presence of the Nickel interlayer. The reduction in intermetallic also seems to have played a part in the fracture becoming more ductile. Corrosion rate with Nickel interlayer has also shown a very good improvement over corrosion without interlayer.

Acknowledgements The authors would like to thank Dr. Sathikh, former Vice-Chancellor of Madras University, Met Mech lab Chennai for their help on this experimental work and I.I.T Chennai for friction welding. We also thank the Management of Sathyabama University, Alpha College of Engineering, St. Joseph's College of Engineering, and SSN College of Engineering.

References

1. Yilbas BS, Sahin AZ, Kahraman N, Al-Gami AZ (1995) Friction welding of St-Al and Al-Cu materials. J Mater Process Technol 49:431–443
2. Mahendran G, Balasubramanian V, Senthilvelan T (2010) Influences of diffusion bonding process parameters on bond characteristics of Mg-Cu dissimilar joints. Trans Nonferrous Met Soc China 20:997–1005
3. Lee WB, Bang KS, Jung SB (2005) Effects of intermetallic compound on the electrical and mechanical properties of friction welded Cu/Al bimetallic joints during annealing. J Alloy Compd 390(1–2):212–219
4. Garcia C, Martise F, De Tiedra P, Blanco Y, Lopez M (2008) Intermetallic Studies on Aluminium-Copper.Corros Sci 50:1184–1194
5. Bilmes PD, Llorente CL, Mendez CM, Garvasi CA (2009) Corrosion behaviour of Intermetallics. Corros Sci 51:876–881
6. Kaufman JG, Roy EL (2005) In: Baboian R (ed) Corrosion test and standards, applications and interpretation, 2nd edn. ASM International, Material park, OH, pp 1–8
7. Kaufman JG (2005) ASM hand book, vol 13 B. In: Cramer SD, Covino BS Jr (eds) Corrosion: materials. ASM International, Material park, OH, pp 95–124
8. Park IJ, Kims ST, Lee IS, Park YS, Moon MB (2009) Mater Trans (JIM) 50:850–854
9. Madhusudan Reddy G, Venkatramana P (2012) Role of nickel as an interlayer in dissimilar metal friction welding of maraging steel to low alloy steel. J Mater Process Technol 212(1):66–77
10. Talbot D, Talbot J (1998) Corrosion science and technology. CRC Press, LIC
11. Trethwey KR, Chamberlain J (1996) Corrosion for science and engineering, 2nd edn. Long Man Group Limited, UK
12. Venugopal T, Srinivasarao K, Prasadrao K (2004) Trans Inst Met 57:659–663
13. Srinivasa Rao K, Prasad Rao K (2004) Corrosion and welding of Aluminium to Copper. Trans Indian Inst Met 57:503–610
14. Frankel GS, Stockert L, Hunkeler F, Boehni H (1987) Metastable pitting of stainless steel. Corrosion 43:429–436
15. Stewart D, Williams E (1992) Intermetallics in Corrosion of Aluminium Copper. Corros Sci 33:457–463
16. MadhusudanReddy G, Venkatramana P (2012) Role of nickel as an interlayer in dissimilar metal friction welding of maraging steel to low alloy steel. J Mater Process Tech 212(1):66–77

Experimental Investigation of Mechanical and Tribological Properties of Al 7075—MoS$_2$/ZrO$_2$/Ni Hybrid Composite

Anish Ranjan and A. Shanmugasundaram

Abstract Self-lubricating metal matrix composites (MMCs) are replacing many conventional materials that are used in automotive, aerospace and marine applications due to superior wear resistance. The objective of this work is to investigate the effect of reinforcing solid lubricant—Molybdenum Disulfide (MoS$_2$), hard ceramic particles of Zirconia (ZrO$_2$) and metallic particles of Nickel (Ni) into base Al 7075. The hybrid Al-MMC is fabricated using stir casting process. The total weight percentage of MoS$_2$, ZrO$_2$ and Ni are varied by 10, 20 and 30%. Maximum hardness is obtained in the case of 30% reinforcement and is 137% of its base Al 7075. Dry sliding wear tests are conducted to know the effect on wear rate. The wear rate was reduced by 58.94% in the case of minimum wear condition for 30% reinforcement. SEM, EDS and XRD were done and the presence of reinforcing particles was confirmed.

Keywords Stir casting · Metal matrix composite · Microhardness
Dry sliding wear · 7075 aluminium alloy

1 Introduction

Aluminium and its alloys are having low density and a high strength to weight ratio, hence aluminium alloys are used widely where weight saving is important like in airplanes, submarines, and structural. Moreover, in automobile industries for improving fuel economy, major parts must be replaced with lighter alloys like titanium, magnesium and aluminium. Among these, aluminium is widely used because magnesium is self-explosive and titanium is expensive, also melting of

A. Ranjan · A. Shanmugasundaram (✉)
Department of Mechanical Engineering, Amrita School of Engineering,
Coimbatore, India
e-mail: a_shanmugasundaram@cb.amrita.edu

A. Ranjan · A. Shanmugasundaram
Amrita Vishwa Vidyapeetham, Coimbatore, India

© Springer Nature Singapore Pte Ltd. 2019
A. K. Lakshminarayanan et al. (eds.), *Advances in Materials and Metallurgy*,
Lecture Notes in Mechanical Engineering,
https://doi.org/10.1007/978-981-13-1780-4_29

299

titanium is difficult [1]. Among different aluminium alloys, Al 7075 has strength comparable to that of steel, very good resistance to corrosion and low relative density. These properties allow the usage of Al 7075 in engineering industries. Like all other Al alloys, Al 7075 is also having a poor tribological property which limits its usage in applications where high wear resistance is needed [2].

To improve properties of Aluminium alloy, methods like mechanical alloying and fabrication of Metal Matrix composites can be used. Mechanical alloying and sintering was done for Al–Sn alloys which result in increased wear resistance than the base material [3]. Metal matrix composites are fabricated by suspending one or more materials either in short (non-continuous) fibre, long (continuous) fibre, platelets or particle shapes into a metallic matrix [4]. Hybrid metal matrix composites are those which have more than one reinforcement being suspended in the metallic matrix [5]. Metal matrix composites are used extensively in applications such as ground transportation (auto and rail), aerospace, thermal management, industrial, military, recreational and infrastructure Industries [6]. The main reason for such a wide application of metal matrix composites is their improved properties like high strength, hardness, stiffness, wear and corrosion resistance.

Different fabrication techniques can be utilized for the production of composite materials such as stir casting, squeeze casting and powder metallurgy [4]. Using powder metallurgy route for the fabrication of MMC, a necklace distribution of B_4C was observed with minimal micro-porosity [7]. Squeeze casting was also used for the production of a composite, the author observed that parts produced by this method are application-oriented and can be manufactured as mass production [8]. Stir casting is a simple method for fabricating a composite and can be used for mass production as well. It also minimizes the cost of the product [9].

Ceramics are used commonly as the reinforcement, but these particles while increasing hardness and strength of the composite reduces the ductility significantly. SiC [10], Al_2O_3 [11], ZrO_2 [12] are some of the ceramic reinforcements used with Al 7075 alloy. Al 6063 surface was modified by the addition of fly ash and found that the hardness is improved [13]. At the same time, use of self-lubricating powders like Hexagonal BN, Graphite and Molybdenum Disulfide, also attracted the researcher's attention towards them, and the objective behind using these self-lubricating material is reducing usage of toxic lubricants and also reducing energy consumption in industrial components [14]. Researchers also used metallic particulates like Nickel and Copper, the novel advantage of using these metallic particulates is that appreciable amount of ductility is retained with an increase in yield strength of the composite. Ni was embedded in the Aluminium 1050 matrix using friction stir processing and ductility values were found to be much better than ceramic reinforced [15]. Hybrid composites were also fabricated in previous studies like Al_2O_3 and Gr [16], SiC and B_4C [17] as reinforcements and Al 7075 alloy as a matrix.

The aim of this work is to obtain superior hardness and resistance to wear for Al 7075 by fabricating a hybrid Al-MMC. Not much work has been done in reinforcing Al 7075 with metallic particles, hence Ni is chosen as reinforcement. Also, when it comes to self-lubricating capabilities, MoS_2 sustains more loads when compared with graphite [18]. Still, very few author explored MoS_2 as reinforcement. Hence, a combination of ZrO_2, MoS_2 and Ni with varying weight percentages will be used as the reinforcement. Stir casting has been chosen for fabrication of hybrid MMC as the process is simple, flexible and applicable for large volume production.

2 Experimental Work

Bottom pouring type stir casting furnace was used for fabrication of composites shown in Fig. 1. Stirrer, furnace and the inner surface of the mould were coated with a non-stick coating to avoid sticking of metal on the surface of stirrer, mould or furnace.

Reinforcement particles and die were preheated for removal of moisture. Al 7075 alloy was then melted to a temperature of 800 °C in the furnace and mechanically stirred using an impeller driven by an electrical motor. Due to stirring a vortex motion of molten metal was obtained. To increase the wettability of the ceramic and metallic particles, 1 wt% magnesium was added to the melt [19], followed by the addition of preheated reinforcement particles of MoS_2, ZrO_2 and Ni. Mechanical stirring continued for 5 min and then poured into the preheated mould. A similar procedure was adopted for fabrication of other variants of composites. Weight % of MoS_2, ZrO_2 and Ni for different variants of the composite is shown in Table 1.

Microhardness for base and composites was measured using Mitutoyo hardness testing equipment. Dry sliding wear test for base Al 7075 and three variants of

Fig. 1 Stir casting setup

Table 1 Weight % of reinforcements

Variant	MoS$_2$ (%)	ZrO$_2$ (%)	Ni (%)	Al 7075 (%)
1	2	4	4	90
2	3	8	9	80
3	4	12	14	70

composites were carried out using Pin-on-disc wear equipment. For testing, pin with a diameter of 10 mm and height of 40 mm has been prepared. The dry sliding wear test was conducted as per ASTM G99-95a standard. The EN 31 steel wear disc with a hardness of 60 HRC and a diameter of 165 mm was used to rotate against the specimens. Three different loads of 10, 20 and 30 N were applied during wear test while sliding velocity of disc varied as 1, 2 and 3 m/s. The sliding distance was fixed to be 1000 m. Wear rate of the variants was determined using weight loss method. Actual densities of variants were found using simple water emersion technique and theoretical density was found using the rule of mixture. Using actual and theoretical densities porosity was estimated. A comprehensive study of the microstructure of hybrid composite was done using Scanning Electron Microscopy (SEM). Techniques like EDS and XRD was done to know whether MoS$_2$, ZrO$_2$ and Ni are present in the matrix.

3 Results and Discussion

3.1 Chemical Composition

Table 2 shows the elemental composition of the Al 7075 obtained using 'Optical Emission Spectrometer (METAVISION–1008 I)'. The elemental composition of as received Al 7075 is in accordance with the composition prescribed in the ASM Handbook.

3.2 Microhardness Testing

Average Vicker's microhardness for the base material Al 7075 is 87 HV. The microhardness of first, second and third variants are 173, 179 and 206 HV correspondingly. The increase in hardness for first, second and third variants when

Table 2 Elemental composition of Al 7075

Elements	Si	Fe	Cu	Mn	Mg	Cr	Ni	Zn	Ti	Al
Weight%	0.156	0.268	1.357	0.108	2.402	0.222	0.061	5.227	0.06	90.05

compared to the base, was found to be 99, 106 and 137%, respectively. The prime reason for the increase in microhardness is the presence of ZrO_2, MoS_2 and Ni particles which acts as an obstacle to dislocation motion. Also, because of stirring of molten metal and particles, high angle boundaries are generated at the interface [20]. High angle boundaries are heavily disoriented, and for dislocations to cross through interface both grains must be aligned to each other. Thus, extra load is required for grain alignment so that dislocation can pass through the interface, which results in higher microhardness values.

3.3 Density and Porosity

Theoretical densities can be calculated using Eq. (1) [21]

$$\delta_t = \frac{W_c}{\frac{W_m}{\delta_m} + \frac{W_{mo}}{\delta_{mo}} + \frac{W_{ZrO_2}}{\delta_{ZrO_2}} + \frac{W_{Ni}}{\delta_{Ni}}} \tag{1}$$

where δ is density and W is the mass of corresponding material. δ_t and δ_m refers to theoretical and matrix density, respectively. Similarly, W_t and W_m refer to theoretical and matrix mass, respectively. Actual density was found out using ratio of mass to volume for each variant and denoted as δ_c. Porosity in composite can be estimated using Eq. (2) [21]

$$\text{Porosity} = \left(1 - \frac{\delta_c}{\delta_t}\right) \times 100 \tag{2}$$

Density and porosity for all variants are shown in Table 3. As the weight % of reinforcements is increasing, the density of variants is also increasing. The reason behind the increase in density of composite is the higher densities of reinforcements suspended in the matrix. Porosity was observed maximum for the first variant. The reason behind the increase in porosity can be an anti-stick coating that was applied to the surface of the mould before pouring the molten metal. Due to the high temperature of molten metal, anti-stick coating would have been vaporized and trapped in the melt resulting in higher porosity.

Table 3 Density and porosity

Variant	Theoretical density (gm/cm^3)	Actual density (gm/cm^3)	Porosity (%)
As-cast	2.81	2.69	4.27
1	2.95	2.792	5.356
2	3.15	3.021	4.1
3	3.37	3.276	2.79

3.4 EDS Analysis

EDS analysis for the 1st variant of composite has been done and presence of major alloying elements of Al 7075 and reinforcements elements, i.e. Zr, O, Mo, S, and Ni has been confirmed. It is shown in Fig. 2.

3.5 XRD Analysis

XRD test has been conducted for the first variant for confirming if there is the presence of any intermetallic compound, Al 7075 precipitates, and reinforcements. From XRD pattern, the following things can be inferred:

- Zirconia is present with the reference code 01-071-6426.
- Nickel is present with the reference code 00-045-1027.
- MoS_2 is present with the reference code 01-074-0932.
- Al 7075 precipitates, i.e. Magnesium Silicide (Mg_2Si) with reference code 01-074-5963 and Magnesium Zinc ($MgZn_2$) with reference code 01-074-7051 are present.
- No intermetallic compound has been formed, this implies that reinforcement particles are thermodynamically stable and the interface between aluminium alloy and reinforcement particles tends to be free (Fig. 3).

3.6 SEM Analysis

SEM micrograph for the first variant is shown in Fig. 4 taken at 1000X magnification. Specimen with a dimension of $10 \times 10 \times 10$ mm was prepared and

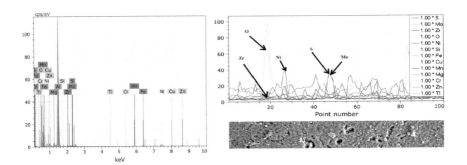

Fig. 2 EDS spectrum of first variant

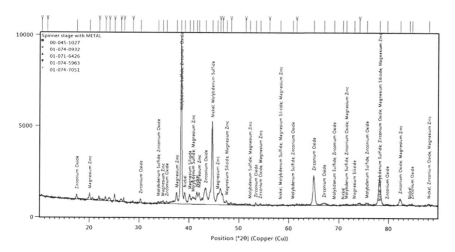

Fig. 3 XRD spectrum for first variant

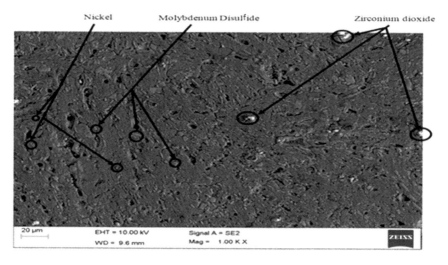

Fig. 4 Micrograph of first variant at 1000X

polished for better surface finish followed by etching using Keller's reagent. Based on the result of line EDS particles can be identified from the SEM micrograph. Globular particles are zirconium dioxide, irregular flat-shaped particles are molybdenum disulfide, and smaller particles are identified as nickel.

3.7 Dry Sliding Wear Analysis

Wear behaviour of the base metal and composite were investigated. The variation in wear rate with change in sliding velocity and load is shown in Figs. 5 and 6, respectively. From Fig. 5, it is clear that higher load results in higher wear rate. The sliding wear starts when the pin makes relative contact with the rotating counter face of EN 31 steel disc. This result in asperity-to-asperity contact, i.e. peaks and valleys of pin and disc combine together and obstruct the relative movement. The hardness of steel disc is more when compared to the composite. With the increase in load, the temperature at the interface of pin and disc increases due to friction and

Fig. 5 Variation of wear rate w.r.t. load

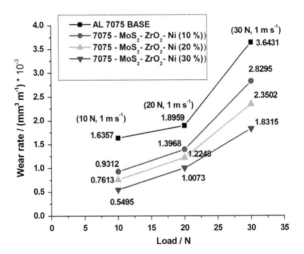

Fig. 6 Variation of wear rate w.r.t. sliding velocity

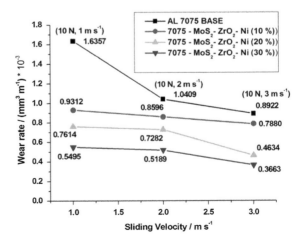

leads to softening of the specimen. Thus, the hard asperities of the counter disc penetrate into the pin. The harder asperities of the disc will break the softer asperities of pin resulting in more wear of sample. The effect is well pronounced when the further load is increased leading to higher wear rate. For all the variants at higher load, more wear rate is observed. Similar reporting has been done for the effect of load on wear rate for Al–Zn–Cu alloy [22, 23]. From the Fig. 6, it can be inferred that when sliding velocity increases wear rate reduces. At higher sliding velocity frictional force generated between the pin tip and the disc will be higher, and thus temperature at interface increases. At higher temperatures, an oxide layer is formed over the tip of the pin. Due to the oxide layer, the pin slides smoothly over the disc and hence reduces the wear [24, 25]. The difference in wear rates for maximum (load of 30 N and sliding velocity of 1 m/s) and minimum wear condition (load of 10 N and sliding velocity of 3 m/s) of base Al 7075 and the third variant is shown in Fig. 7.

Reduction in wear rate for the third variant with respect to the as-cast Al 7075 in case of maximum wear condition is 58.94% and in case of minimum wear condition is 49.73%. It can be inferred that the wear resistance of Al 7075 has been enhanced. This is because strain fields are generated around reinforcements due to the thermal discrepancy between the Al 7075 and reinforcement particles while solidifying. These strain fields offer obstruction to the generation of the cracks when pin and disc make a relative movement [26]. Apart from this, the presence of hard ZrO_2 particles resists the cutting action of counterface asperities and Molybdenum disulfide creates a protective layer between the pin and the disc [27]. These activities result in avoiding direct contact of Aluminium with counterface thus results in less metal removal.

Fig. 7 Variation in wear rate of as-cast Al 7075 and third variant

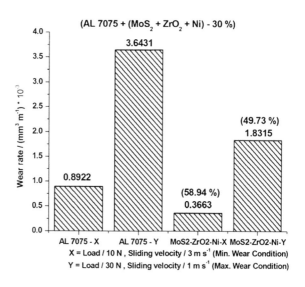

4 Conclusions

In the present work, Al 7075—MoS_2/ZrO_2/Ni hybrid composite was fabricated using stir casting. The results are as follows:

- The reinforcement particles are effective in improving the hardness and wear resistance.
- With the increase in weight % of reinforcement, resistance to dislocation motion increases resulting in high microhardness. The increase in microhardness for first, second and third variants when compared to the base, was found to be 99, 106 and 137%, respectively.
- Wear rate of composites reduces because of generation of strain fields around the reinforcements and formation of a MoS_2 layer over the pin which avoids direct contact of aluminium with the disc. The maximum reduction in wear rate in case of minimum and maximum wear condition was 58.94 and 49.73%, respectively for the third variant when compared to the base.
- SEM, EDS and XRD analysis confirm the presence of reinforcement particles in the matrix. XRD analysis of the composite revealed that reinforcement particles are thermodynamically stable because no intermetallic compound has been formed.
- The density of composite increased with increase in weight % of reinforcements because ZrO_2, MoS_2 and Ni are having a higher density than base Al 7075.
- Porosity level can be further reduced, if we preheat the mould to a higher temperature and follow squeeze casting route for the fabrication of composite.

Acknowledgements Special thanks must go to our professor Sanjivi Arul for his technical guidance and discussions. In addition, the authors would like to thank M/s. PSGTECHS COE INDUTECH, Department of Textile Technology, Neelambur, Coimbatore, Tamil Nadu, India for providing facility to carry out the material characterization work.

References

1. Sudhakar I, Madhusudhan Reddy G, Srinivasa Rao K (2016) Ballistic behavior of boron carbide reinforced AA7075 aluminium alloy using friction stir processing—an experimental study and analytical approach. Def Technol 12:25–31
2. Yu L, Cao F, Hou L, Jia Y, Shen H, Li H, Ning Z, Xing D, Sun J (2017) The study of preparation process of spray formed 7075/Al–Si bimetallic gradient composite plate. J Mater Res 32:3109–3116
3. Liu X, Zeng MQ, Ma Y, Zhu M (2008) Wear behavior of Al-Sn alloys with different distribution of Sn dispersoids manipulated by mechanical alloying and sintering. Wear 265:1857–1863
4. Kaczmar JW, Pietrzak K (2000) The production and application of metal matrix composite materials. J Mater Process Technol 106:58–67
5. Naik N, Ramahima R, Arya H, Prabhu SV, Shamarao N (2001) Impact resistance and damage tolerance characteristics of glass-carbon epoxy hybrid composite plates. Compos PartB 32B:565–574

6. Miracle DB (2005) Metal matrix composites—from science to technological significance. Compos Sci Technol 65:2526–2540

7. Jiang QC, Wang HY, Ma BX, Wang Y, Zhao F (2005) Fabrication of B4C participate reinforced magnesium matrix composite by powder metallurgy. J Alloys Compd 386:177–181

8. Vijayaram TR, Sulaiman S, Hamouda AMS, Ahmad MHM (2006) Fabrication of fiber reinforced metal matrix composites by squeeze casting technology. J Mater Process Technol 178:34–38

9. Pardeep S, Gulshan C, Neeraj S (2013) Production of AMC by stir casting—an overview. Int J Contemp Pract 2:23–46

10. Zhou W, Xu ZM (1997) Casting of SiC reinforced metal matrix composites. J Mater Process Technol 63:358–363

11. Kumar GBV, Rao CSP, Selvaraj N, Bhagyashekar MS (2010) Studies on Al6061-SiC and Al7075-Al2O3 metal matrix composites. J Miner Mater Charact Eng 9:43–55

12. Hernández-Martínez SE, Cruz-Rivera JJ, Garay-Reyes CG, Elias-Alfaro CG, Martínez-Sánchez R, Hernández-Rivera JL (2015) Application of ball milling in the synthesis of AA 7075–ZrO2 metal matrix nanocomposite. Powder Technol 284:40–46

13. Shanmugasundaram A, Arul S, Sellamuthu R (2017) Effect of fly ash on the surface hardness of AA 6063 using GTA as a heat source. Metall Res Technol 114(511):1–8

14. Omrani E, Dorri Moghadam A, Menezes PL, Rohatgi PK (2016) New emerging self-lubricating metal matrix composites for tribological applications. In: Ecotribology: research developments, pp 63–103

15. Yadav D, Bauri R (2011) Processing, microstructure and mechanical properties of nickel particles embedded aluminium matrix composite. Mater Sci Eng A 528:1326–1333

16. Baradeswaran A, Elaya Perumal A (2014) Study on mechanical and wear properties of Al 7075/Al2O3/graphite hybrid composites. Compos Part B Eng 56:464–471

17. Uvaraja VC, Natarajan N (2012) Optimization of friction and wear behaviour in hybrid metal matrix composites using Taguchi technique. J Miner Mater Charact Eng 11(8):757–768

18. In KA, Kompozitov S, Mos A, Vinoth KS, Subramanian R, Dharmalingam S, Anandavel B (2012) Mechanical and tribological characteristics of stir-cast Al-Si10 Mg and self-lubricating Al-Si10 Mg/MoS2 composites. Mater Technol 46:497–501

19. Radhika N, Subramanian R, Venkat Prasat S (2011) Tribological behaviour of aluminium/alumina/graphite hybrid metal matrix composite using Taguchi's techniques. J Miner Mater Charact Eng 10:427–443

20. Lee W-B, Lee C-Y, Yeon Y-M, Jung S-B (2005) Effects of the local microstructures on the mechanical properties in FSWed joints of a 7075-T6 Al alloy. Zeitschrift fur Met 96:940–947

21. Agarwal B, Broutman L, Chandrashekhara K (2006) Analysis and performance of fiber composites. Wiley, New York

22. Ilangovan S, Arul S, Shanmugasundaram A (2016) Effect of Zn and Cu content on microstructure, hardness and tribological properties of cast Al-Zn-Cu alloys. Int J Eng Res Africa 27:1–10

23. Shanmugasundaram A, Arul S, Sellamuthu R (2017) Investigating the effect of WC on the hardness and wear behaviour of surface modified AA 6063. Trans Indian Inst Met 71:1–9

24. Zhu HG, Ai YL, Min J, Wu Q, Wang HZ (2010) Dry sliding wear behavior of Al-based composites fabricated by exothermic dispersion reaction in an Al-ZrO2-C system. Wear 268:1465–1471

25. Ilangovan S, Shanmugasundaram A, Arul S (2016) Influence of specimen temperature on wear characteristics of AA6063 aluminium alloy. J Surf Sci Technol 32:93

26. Michael Rajan HB, Ramabalan S, Dinaharan I, Vijay SJ (2014) Effect of TiB2 content and temperature on sliding wear behavior of AA7075/TiB2 in situ aluminum cast composites. Arch Civ Mech Eng 14:72–79

27. Kumar NGS, Shankar GSS, Basavarajappa S, Kumar GSS (2015) Dry sliding wear behavior of Al 2219/Al2O3-MoS2 metal matrix hybrid composites produced by stir casting route. Int J Eng Res Adv Technol 1:28–33

Comparative Hardness Studies and Microstructural Characterization of 87Sn–7Zn–3Al–3In and 87.5Sn–6 Zn–2Al–2.5In Lead-Free Soldering Alloys

D. Arthur Jebastine Sunderraj, D. Ananthapadmanaban
and K. Arun Vasantha Geethan

Abstract Lead is an element which was being used traditionally in soldering because of its low melting point. However, lead is poisonous and increasing research is being carried out on developing lead-free soldering alloys. In this work, two lead-free alloys, namely 87Sn–7Zn–3Al–3In and 87.5Sn–6Zn–2Al–2.5In were melted by induction melting with pure argon atmosphere. Vicker's hardness test was conducted and also the pasty range of the alloy was found using Differential Scanning Colorimetry (DSC). Hardness values were higher than normal lead-based soldering alloys and the pasty zone was comparable to normal lead-based alloys. The lead-free solders were characterized with Energy Dispersive X-Ray Spectra (EDS). Microstructural characterization was also done to corroborate results.

Keywords Lead free alloys · Pasty · Microstructural characterization

1 Introduction and Literature Review

Tin is being used as an alloying element instead of lead in recent times for use in solders. This is due to the poisonous nature of lead, which is increasingly being replaced in solders. As far back as in 2001, lead and lead-containing compounds

D. Arthur Jebastine Sunderraj (✉) · K. Arun Vasantha Geethan
Department of Mechanical Engineering, St. Joseph's Institute of Technology, Chennai 600119, India
e-mail: arthurjebastine@gmail.com

K. Arun Vasantha Geethan
e-mail: kavgeeth@gmail.com

D. Ananthapadmanaban
Department of Mechanical Engineering, SSN College of Engineering, Kallavakkam, Chennai, India
e-mail: ananthapadmanaband@ssn.edu.in

© Springer Nature Singapore Pte Ltd. 2019
A. K. Lakshminarayanan et al. (eds.), *Advances in Materials and Metallurgy*,
Lecture Notes in Mechanical Engineering,
https://doi.org/10.1007/978-981-13-1780-4_30

were listed as being among the top 17 chemicals threatening human beings. This fact has been reiterated by the Environmental Protection Agency (EPA) of the US [1]. Lead has also been restricted in its use by a directive adopted by the European Union [2]. Hence, it is obvious that environmental effects of lead have forced researchers to look for alternative lead free alloys in order to reduce harmful waste from electrical and electronic industries.

There are some qualities which are desirable for lead free solders. These are low melting temperature of solder, good wetting properties, adequate strength, good thermal conductivity. Most importantly, it should not have any harmful effects on the environment. Lead-free solders such as Sn–Ag, Sn–Ag–Cu eutectic solders with the melting temperature of 221, 227 and 217 °C respectively are being used in electronics industries [3–5].

Tin has replaced lead as the major component of the alloy and Zn as the major alloying element. Sn–Zn solder is less expensive, non-hazardous and possesses nominal mechanical properties. Some Sn–Zn alloys have found use in electronics industries [4, 6]. Sn–Zn alloys have superior fatigue properties compared to Pb–Sn alloys [6, 7]. These alloys also have a fairly low melting point and good wettability [8]. However, the properties of binary Sn–Zn solders are unable to meet the stringent requirements that are needed for use in electronics industries. This being the case, ternary and quaternary Sn–Zn solders have been researched upon by researchers [9–11].

Addition of Aluminium in Sn–Zn alloys was found to improve the wettability and some mechanical properties [11]. Barbosa et al. have found improvement in tensile strength when Bismuth is added along with Al in Sn–Zn alloy [12].

Fujitsu has developed Sn–Zn–Al ternary alloys with higher productivity and good joint reliability [13].

Erwin Siahaan has proved that addition of Zn to the extent of 15% to Sn gave a melting point of 189 °C as compared to 183 °C for the Lead–Sn system [14].

Although some research has been performed on microstructural changes of cast Sn–Zn alloys with Cerium, Lanthanum and Gadolinium, Indium additions have not been attempted so far as a quaternary addition to Tin–Zinc alloys.

2 Experimental Work

2.1 Preparation of Alloy

Tin, Aluminium, Zinc and Indium (purity 99.999%) were taken and melting was performed out in an induction furnace under pure argon atmosphere.

A coin-like specimen with a diameter of approximately 50 mm was obtained as the output from the furnace.

The molten alloys were homogenized at 810 °C for 5 h in an Alumina crucible. This crucible serves twin purposes. It acts as a mould as well as a crucible to hold

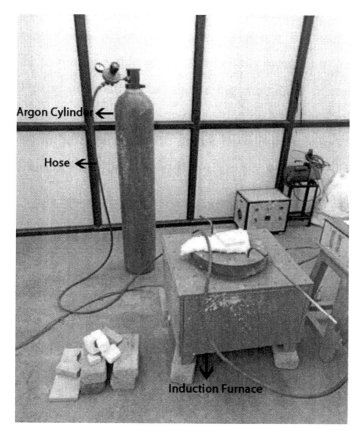

Fig. 1 Shows induction furnace used in this work

molten alloy during homogenization. A cooling rate of 6–8 °C was used while cooling the specimen. Figure 1 depicts the induction furnace used.

2.2 Chemical Composition

Overall chemical composition was determined using Energy Dispersive Spectroscopy (EDS). The microstructure was also analysed and chemical composition at the needle-like and leaf-like portions of the microstructure were separately analysed.

Figure 1 depicts the cast specimen (50 mm diameter) obtained. The specimen is then faced, turned and ground. Figures 2, 3 and 4 show the specimen at different stages of specimen preparation. The specimen is then polished before hardness testing and Differential Scanning Calorimetry. The polished specimen is depicted in Fig. 5.

Fig. 2 Shows coin-sized cast specimen

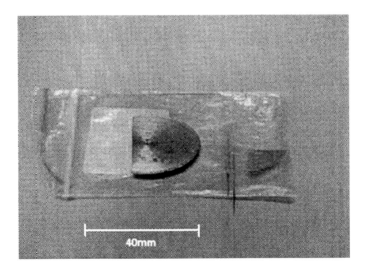

Fig. 3 Shows specimen after turning and facing

2.3 *Hardness Testing*

Vickers micro hardness testing method was used to calculate hardness. The test procedure ASTM E-384 which was used to find microhardness, specifies a range of light loads using a diamond indentor.

Fig. 4 Shows specimen after grinding

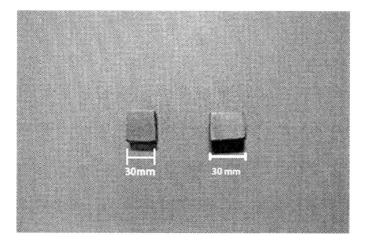

Fig. 5 Shows specimen after polishing

2.4 Differential Scanning Calorimeter

DSC is a technique used to determine temperature at which phase changes take place. The phase changes detected are melting, boiling temperatures, latent heats, glass transition temperature, curing temperature, etc. The technique uses changes in specific heat with change in temperature. In our work, DSC curves were used to determine the solidus and liquidus temperatures. Between the liquidus and the solidus temperatures, the alloy exists in a mushy state. The mushy zone can be deduced as (Liquidus temperature–Solidus temperature). Figure 7 shows the Differential Scanning Calorimeter. All DSC analyses were performed between 25 and 250 °C. Heating and cooling rates used were 10 °C. Figure 7 shows the Differential Scanning Colorimeter.

3   Results and Discussion

3.1   Chemical Composition

Overall chemical composition as determined by EDS is given in Table 1.

The chemical composition of the alloy should be so chosen that it is low enough that it will not damage any temperature sensitive parts of the electronic circuits. At the same time, it should be high enough that the joint withstands the temperatures that the circuit is being used. Generally, some work has been done on Indium % from 4 to 25 wt%. In this work, research on low wt% of Indium (2.5–3%) solder is being attempted, mainly because of the cost of Indium and also the reasoning that preliminary set of experiments are being done with lower Indium content. The Indium content is planned to be increased in future work.

3.2   DSC Results

Figures 6 and 7 show DSC results for the two new alloys that have been manufactured in this work, namely 87 Sn–7Zn–3Al–3In and 87.5–Sn–6Zn–2Al–2.5In.

Figure 6 shows the DSC curve for 87Sn–7Zn–3Al3 In onset of phase change at around 189.6 °C. There is a sudden spurt in the DSC graph which is indicative of phase change. The curve peaks at 202.5 °C, when the reaction begins to slow down and then ends at 205.1 °C. Since, this is an alloy and the only phase change expected in this region is the soldi to liquid phase change, it can be inferred that this solid to liquid phase change has happened over the range of temperature, which is indicated in Table 2. In this case, added Indium is 2.5% and there is a fairly medium range of phase transformation.

Figure 7 shows the DSC curve for 87.5Sn–6Zn–2Al2.5In onset of phase change at around 190.9 °C. There is a sudden spurt in the DSC graph which is indicative of phase change. The only phase change expected in this range of temperature is the solid to liquid phase transformation. So, it can be inferred that melting has taken

Table 1 Overall composition	Sample	Sn	Zn	Al	In
	1	87	7	3	3
	2	87.5	6	2	2.5

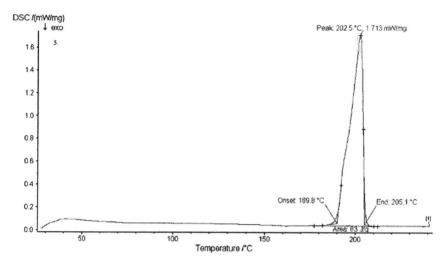

Fig. 6 Shows DSC of 87Sn–7Zn–3Al–3In

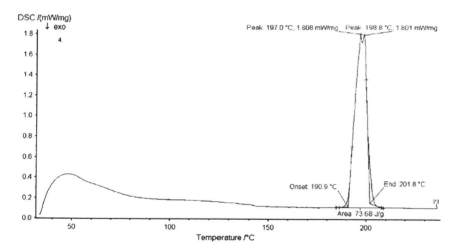

Fig. 7 Shows DSC of 87.5Sn–6Zn–2Al–2.5In

place over this range of temperature. The curve peaks at 202.5 °C, when the reaction begins to slow down and then ends at 205.1 °C. In this case, added Indium is 3% and the range of phase transformation is lesser than that shown in Fig. 6.

The range of melting as seen from the DSC curves is 10.9 °C for the 87Sn–7Zn–3Al–3In alloy (as seen from the difference between onset and end of the peak). Similarly, the range of melting is 15.3 °C for the 87.5Sn–6Zn–2Al–2.5In alloy. It appears that addition of higher Indium in the alloy reduces the range of melting.

Table 2 Shows a summary of melting point and mushy ranges of various solder alloys

Solder alloy	T_{on}	T_{off}	$T_{on}-T_{off}$	T_m	References
Sn–37Pb	179.5	191.0	11.5	183	[7]
Sn–9Zn	194.8	200.1	5.3	198.1	[8]
SAC	215.9	225.4	9.5	220.2	[9]
82Sn–15Zn–3Al	194.2	215.6	19.4	210.9	[10]
50Sn–43Zn–7Al	199.1	221.3	22.2	211.3	[10]
87Sn–7Zn–3Al–3In	190.9	201.8	10.9	198.8	Current study
87.5Sn–6Zn–2.5In	189.9	205.1	15.3	202.5	Current study

3.3 Melting Point and Mushy Zone Range

Melting point and mushy range of solder alloys change with composition of the alloy. Jian Zhou et al. have studied this variation extensively in Sn–Zn alloys and have found that addition of Bismuth reduces the melting point. However, studies by Jian Zhou et al. show that the range of mushy zone increases [15]. Studies by Chian Chun Liu et al. on Sn–Zn–Bi–In alloys showed that both solidus and liquidus temperature are lowered on addition of Indium, thereby decreasing the pasty range [16].

It is seen that in alloys like Sn–9Zn, the range of mushy zone is very low of the order of 5.3 °C, whereas in alloys like 50Sn–43Zn–7Al, the mushy zone is around 22.2 °C. In our work, for the 2 alloys that have been manufactured, the mushy zone is 10.9 and 15.3 °C, respectively, 10.9 °C mushy zone is lower than 11.5 °C as reported in Ref. [7]. Higher the mushy zone, the longer the alloy takes to solidify. If the mushy zone is too small, the alloy solidifies quickly, whereas if the mushy zone is too large, the alloy gives freedom to solder in a larger range of temperature, but there could be a danger of oxidation effects. So, a moderate mushy zone that we have obtained in our study seems to strike the right balance.

3.4 Hardness Test Results

Sn–37Pb solders normally used for soldering has a hardness of 12 H_V [17]. So, on an average both the new alloys have hardness higher than the traditionally used lead based soldering alloys (Table 3).

Table 3 Gives the results of hardness testing

Specimen	Hardness, $HV_{0.2}$					Average
Sn–7Zn–3Al–3In	18.4	19.9	19.1	18.8	19.3	19.1
Sn–6Zn–2Al–2.5In	18.6	17.7	18.1	18.3	17.9	18.1

3.5 Microstructural Analysis

Microstructures of both alloys show uniform distribution of needle shaped Zn-rich phase and leaf shaped Al rich phase, with the former being effective in improving the strength of the joint. In the 1st alloy, the Chemical composition of the needle and leaf shaped morphologies are as given here—Leaf shaped morphology-Al-88%, Zn-12% Needle shaped morphology-In-72%, Zn-27.96% and Sn-0.02%. In the second alloy, composition of the leaf and needle shaped morphologies are as given here—Leaf shaped morphology-Al-91.6%, Zn-8.4% Needle shaped morphology-In-66.36%, Zn-31% and Sn-2.64% (Figs. 8, 9, 10 and 11).

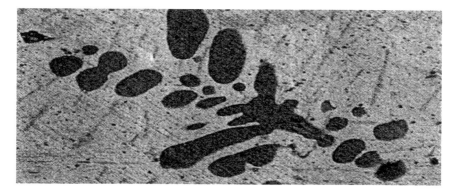

Fig. 8 Examination on leaf shape of the specimen

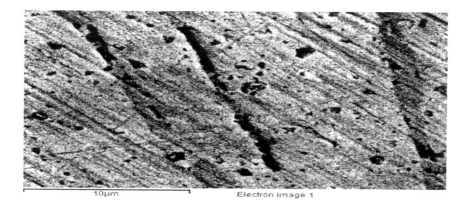

Fig. 9 Examination on needle shape of the specimen

Fig. 10 Examination on leaf shape of the specimen

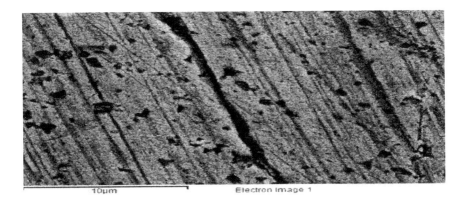

Fig. 11 Examination on needle shape of the specimen

3.6 EDAX Analysis

EDAX Analysis for both the alloys is presented in Figs. 12, 13, 14, 15. In the case of both alloys, the leaf shaped morphology showed no intermetallics, whereas the needle shaped morphology showed intermetallics.

Cerium and Lanthanum with wt% of the order of 0.1 wt% have been added to Sn–Zn eutectic alloy and rodlike Zinc rich phases have been observed. The rod like phases decreased with increase in rare earth element wt% [18]. In the current work, since the weight % of rare earth element Indium added is relatively high, it is possible that the needle like phase is rich in Indium (72 and 66.36% respectively) and Zinc percentage is lesser, agreeing with published literature on Cerium and Lanthanum addition. Gadolenium addition has been found to refine grains of Mg–5Sn–Zn–Al [19].

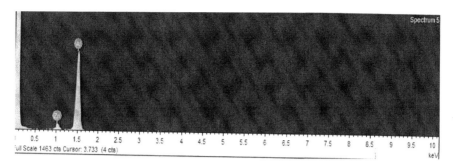

Fig. 12 Chemical composition of the leaf shape specimen

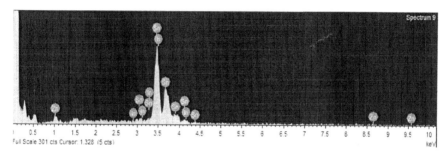

Fig. 13 Chemical composition of the needle shape specimen

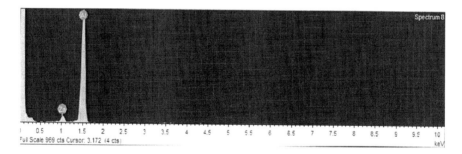

Fig. 14 Chemical composition of the leaf shape specimen

In some cases, there have been reports of intermetallic formation suppression as a result to the addition of rare earth metals. A study by Huan Lee et al. showed that addition of Praseodymium reduced the formation of intermetallics at the junctions of a Sn–Zn–Ga solder [20].

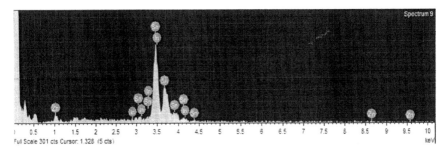

Fig. 15 Chemical composition of the needle shape specimen

4 Conclusion

Mushy zone range is in a moderate range for both the new alloys prepared, namely 10.9 and 15.3 °C respectively. Hardness values are higher for both the new alloys compared to normal Sn–Pb-based solder alloys. Hence, addition of Indium increases hardness. This may give longer life of solder. Both alloys show a Tin-rich matrix with Needle-shaped and leaf-shaped morphologies in the second phase. Addition of Indium changed the composition of both the leaf- and needle-shaped morphologies in the microstructure.

Acknowledgements The authors wish to acknowledge the help rendered by R. Arun and V. Arunkumar, Alumni, St. Josephs Institute of Technology and the encouragement given by the management, St. Josephs Institute of Technology, Semmancherry. They also thank the Management, SSN College of Engineering for providing a conductive atmosphere for doing research.

References

1. Tu KN, Zeng K (2001) Tinlead(Sn-Pb)solder reaction in flip chip technology. Mater Sci Eng R34(1):1–58
2. (2003) The European parliament and the council of the European Union. Official J Eur Union L 37:19–23
3. Fallahi H, Nurulakmal MS, Fallahi Arezodar A, Abdullah J (2012) Effect of iron and indium on IMC formation and mechanical properties of lead-free solder. Mater Sci Eng A 553:22–31
4. Hu X, Li K, Min Z (2013) Microstructure evolution and mechanical properties of Sn0.7Cu0.7Bi lead-free soldersproduced by directional solidification. J Alloy Compd 566:239–245
5. Hamada N, Useugi T, Takigawa Y, Higashi K (2012) Effects of Zn addition and aging treatment on tensileproperties of Sn–Ag–Cu alloys. J Alloy Compd 527:226–232
6. Geranmayeh AR, Nayyeri G, Mahmudi R (2012) Microstructure and impression creep behavior of lead-free Sn–5Sb solder alloy containing Bi and Ag. Mater Sci Eng A 547:110–119
7. Hu X, Li Y, Min Z (2013) Interfacial Reaction and Mechanical Properties of Sn-Bi Solder joints. J Mater Sci Mater Electron. http://dx.doi.org/10.1007/s10854-012-1052

8. Liu CY, Chen YR, Li WL, Hon MH, Wang MCJ (2007) Behavior of Sn-0.7Cu-xZn lead free solder on physicalproperties and micro structure. Elec Mater 36:11

9. Anderson IE, Foley JC, Cook BA, Harringa J, Terpstra RL, Unal O (2001) Alloying elements in near eytectic Sn-Ag-Cu solder alloys for improved microstructural stability. J Electron Mater 30(9):PP1050–1059

10. McCormack M, Jin S, Kammlott GW, Chen HS (1993) New Pb-free solder alloy with superior mechanical properties. Appl Phys Lett 63(1):15–17

11. Miric AZ, Grusd A (1998) Lead free alloys. Soldering Surf Mounting Technol 10(1):19–25

12. Soares D, Vilarinho C, Barbosa J, Silva R, Pinho M, Castro F (2004) Effect of the Bi content on the mechanical properties of a Sn-Zn-Al-Bi solder alloy. Mater Sci Forum 455–456: 307–311

13. Kitayama M, Shono T (2005) Development of Sn-Zn-Al lead free alloys. Fujitsu Sci Tech J 41(2):225–235

14. Siahaan E (2017) Behavior of Sn-0.7Cu-xZn lead free solder on physical properties and micro structure. In: IOP conference series, vol 237

15. Zhou J, Sun Y, Xue F (2005) Properties of low melting point Sn–Zn–Bi solders. J Alloy Compd 397(1–2):260–264

16. Liu CC, Whang ZH, Zhang G, Suganama K (2015) Thermal property, wettability and interfacial characterization of novel Sn–Zn–Bi–In alloys as low-temperature lead-free solders. Mater Des 84:331–339

17. Steward T, Liu S (2002) Database for solder properties with emphasis on new lead free solders. National Institute for Standards and Technology and Colorado School of Mines

18. Wu CML, Law CMT, Yu DQ, Wang L (2003) The wettability and microstructure of Sn-Zn-RE alloys. J Electron Mater 32(2):63–69

19. Fang CF, Meng LG, Wu YF, Wang LH, Zhang XG (2013) Effect of Gd on the microstructure and mechanical properties of Mg-Sn-Zn-Al alloy. Appl Mech Mater 312:411–414

20. Ye H, Xue S, Luo J, Lee Y (2013) Properties and interfacial microstructure of Sn-Zn-Ga solder joint with rare earth Pr addition. Mater Design 46:316–322

Study of Austempered Ductile Iron (ADI) on Varying Austempering Temperatures

Gurmeet Singh, Sahil Sharma and Dilkaran Singh

Abstract Austempered ductile iron (ADI) proven itself as a significant manufacturing material due to its remarkable properties like withstanding high fatigue behavior, fracture toughness, wear resistance, ductility, etc. This paper reports the experimental results obtained on subjecting ADI specimens to thermal treatment, i.e., austempering and austenitization. All austempered ductile iron ADI specimens are subjected to mechanical and metallurgical testing. Within the given range of varying austempering temperatures, tensile strength and metallurgical changes were analyzed. The tensile studies show that as the austenitization temperatures increased from 900 °C, strength values increased, but a reverse trend was observed at temperatures higher than 1100 °C. With increasing the austenitization temperatures, the carbon percentage was observed to increase, leading to decrease in elongation, and an increase in the hardness. For metallurgical analysis, SEM factographs were employed to test the heat-treated ADI specimen. The microstructure images show that high austenitization temperatures give coarse properties to the ADI. At austempering temperature 420 °C, it shows the rise of retained austenite from 35 to 46% of volume.

Keywords Austempering · Austempered ductile iron (ADI) · Mechanical testing
Scanning electrode microscopy (SEM) · Factography

1 Introduction

Austempered ductile cast iron (ADI) has appeared as a significant industrial raw material in the latest time, because of its remarkable properties, i.e., high ductility with improved strength, high fatigue, better fracture behavior, and higher wear

G. Singh (✉) · D. Singh
ME Department, TIET (Deemed University), Patiala 147004, India
e-mail: gurmeet.singh@thapar.edu

S. Sharma
ME Department, Amity University, Noida, India

© Springer Nature Singapore Pte Ltd. 2019
A. K. Lakshminarayanan et al. (eds.), *Advances in Materials and Metallurgy*,
Lecture Notes in Mechanical Engineering,
https://doi.org/10.1007/978-981-13-1780-4_31

resistance [1–4]. These engineering advantages bring the ADI as a first choice for using extensively in structural and wear resilient applications in several types of industries, i.e., automotive industry, machinery and equipment used in defense and earth moving, etc. [5–9]. ADI is being developed from high-quality ductile or nodular cast iron. ADI is acquired with the isothermal heat treatment process which is known as "austempering" [3]. Its unique microstructure is the reason for its remarkable properties which consists of ferrite with high-carbon austenite. The developed ADI is very much different from the austempered steels in which the microstructure contains ferrite with carbide. Enormous amount of silicon in ductile iron conquers the precipitation of carbide during austempering process and helps to retain the extensive amount of stable high-carbon austenite [10, 11]. A very little quantity of alloying elements such as Ni, Mo, and Cu are generally added to ADI and lead to adequate hardenability when quenched to the austempering temperature without the formation of pearlite. The purpose of adding the alloying is to conquer the pearlite formation process and directed the austenite to transform into bainite. Ni and Cu do not segregate as much as Mn and Mo, and in any case, they partition preferentially into the solid phase [12]. They do not affect the mechanical property like hardenability, but when combined with Mn or Mo, there is a useful increase in the maximum section size that can be austempered successfully [13]. Ni may be added within the range of 0.5–3.5 wt%, whereas Cu varies within the range of 0.5–1.0 wt% [14].

With reference to the previous studies on ADI, Zammit et al. [15] carried out a comparative experimental study about sliding wear behavior of ground ADI with shot peened ADI. The tribological behavior prior to shot peening and the subsequent change in wear resistance of Cu–Ni ADI have been considered. There is a subsequent increase in surface roughness, and hardness has been reported. Jacuinde et al. [16] have analyzed the wear resistance of boron micro-alloyed ADI. An increase in ductility but decrease in hardness was observed, along with increased carbon diffusion which led to a more homogeneous structure. Zhang et al. [17] studied the fatigue behavior of high-strength ADI. Zero limit of fatigue was observed in high-cycle system. A constant decrease in the S–N curves was stated in the observations. Meneghetti et al. [18] analyzed an impact of heat treatment on fatigue behavior of spheroidal graphite irons. Data confined in a single scatter band was reported. ADI is a solution to high demand of durable materials. ADI is also more cost-effective than its counterpart materials, and thus, is extensively used in automotive and defense sectors. It also presents an effective solution to the problems faced with conventional cast iron. Previous studies conducted a different kind of experiment using ADI. But its behavior with change of austempering temperature is very less explained. Some environmental conditions also play vital role. In this paper, the ADI specimens have undergone heat treatment process (Austenitization and austempering) and the prime objectives of this study are to study the mechanical and metallurgical changes. Various mechanical properties like tensile stress and compressive strength are studied. Microstructural observations are performed by scanning electron microscopy (SEM). Hardness testing, factography, and X-ray diffraction (XRD) studies are also carried out. The experimental results and conclusions thus obtained are promising in the field of ADI.

2 Experimental Procedures

The heat treatment procedure of specimens takes place in three steps [19, 20]:

- The specimens are heated to the conventional austenitization temperature, i.e., between 840 and 900 °C, and soaked in salt bath, for the time depended on the size and chemical composition of the casting (normally 1 h/25 mm of section). Soaking is done to saturate the austenite with carbon and alloys, which leads to the high hardenability.
- Further, casting is quenched in salt, oil, or lead bath maintained at 250–425 °C. It held in bath for this temperature sufficiently long to permit the austenite to transform completely to bainite. The holding time is very significant because the final structure is dependent on it. It depends on the temperature of the quenching bath, section size, and the chemical composition of the iron.
- Air cooling is done from bath temperature to room temperature. No tempering is done after austempering.

The composition of specimens undergone heat treatment process are as follows [21] (Table 1).

2.1 Specimen Preparation

For tensile testing, specimen has been prepared according to ASTM standard [21]. Figure 1 shows the specimen specifications.

The 12 ADI samples have been undergone for different heat treatment conditions as described in Table 2.

These heat-treated samples are quenched in salt bath with controlled temperature of ±5 °C and composition of salt bath is as shown in Table 3.

Table 1 Chemical distribution of specimens

Elements	C	Ni	S	Si	Mn	Mg	Cu	Mo	P	Cr
Percentage contribution	3.2–3.6	0.14	0.013	2–2.5	<0.23	0.04	0.51	0.001	0.03	0.02

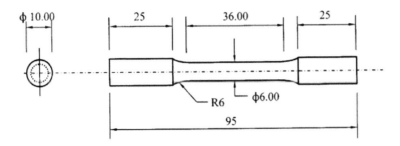

Fig. 1 Tensile specimen specifications

Table 2 Austenitization and austempering temperature with process time

Specimen	Austenitization temp. (°C)	Austenitization time (t_γ) (min)	Austempering temp. (°C)	Austempering time (t_a) (min)
1	900	60	420	60
2	900	60	420	120
3	900	120	420	60
4	900	120	420	120
5	1000	60	420	60
6	1000	60	420	120
7	1000	120	420	60
8	1000	120	420	120
9	1100	60	420	60
10	1100	60	420	120
11	1100	120	420	60
12	1100	120	420	120

Table 3 Composition of salt bath for austempering

Constituents	KNO_3 (M = 101.11 g/mol)	$NaNO_3$ (M = 84.99 g/mol)
wt%	50	50

3   Results and Discussion

3.1   Tensile Strength

All specimens undergone for tensile testing and each sample show that there are unique tensile properties due to the different heat treatment conditions. Tables 4, 5, and 6 show the results of mechanical testing of ADI specimens.

Table 4 Effect of austenitizing at 900 °C on tensile strength of ductile iron

Austenitizing temp. (°C)	Time (T1) (in min)	Austempering temp. (°C)	Time (T2) (in min)	Tensile strength (MPa)
900	60	420	60	963
		420	120	944
	120	420	60	955
		420	120	935
Without treatment				711

Table 5 Effect of austenitizing at 1000 °C on tensile strength of ductile iron

Austenitizing temp. (°C)	Time (T1) (in min)	Austempering temp. (°C)	Time (T2) (in min)	Tensile strength (MPa)
1000	60	420	60	1014
		420	120	1016
	120	420	60	1152
		420	120	947

Table 6 Effect of austenitizing at 1100 °C on tensile strength of ductile iron

Austenitizing temp. (°C)	Time (T1) (in min)	Austempering temp. (°C)	Time (T2) (in min)	Tensile strength (MPa)
1100	60	420	60	989
		420	120	1037
	120	420	60	824
		420	120	994

3.2 Effect of Austenitizing Temperature on Tensile Strength

The samples were treated at 900 °C and further, austempered for 1 h and 2 h at 420 °C. The tensile strength was 963 and 944 MPa, respectively. The austenitizing temperature raised to 1000 °C, and tensile strength increased to 1014 and 1016 MPa for 1 hour and 2 hours austempering. On further increasing the austenitizing temperature to 1100 °C, the tensile strength decreased to 989 MPa when 1 hour austempered and increased to 1037 MPa when austempered for 2 hours (Fig. 2).

The samples were heated at 900 °C and further, austempered for 1 h and 2 h at 420 °C. The tensile strengths were 955 and 935 MPa, respectively. The austenitizing temperature raised to 1000 °C, and tensile strength increased to 1152 and 947 MPa for 1 hour and 2 hours austempering. And on further increasing the austenitizing temperature to 1100 °C, the tensile strength decreased to 824 MPa when 1 hour austempered and increased to 994 MPa when austempered for 2 hours.

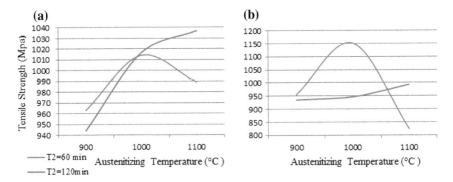

Fig. 2 Impact of austenitizing temperature on tensile strength when austenitized for **a** 60 min, **b** 120 min

3.3 Influence of t_a on Strength

See Fig. 3.

3.4 Hardness

Hardness test performed on austempered ductile iron samples and results can be described as in the graph (Fig. 4). At different austenitization temperatures for two hours and austempered at 420 °C for 1 h, hardness values obtained 309, 365, and 404 VHN, respectively. It can be concluded from hardness values that when austenitizing temperature raised, the hardness value also increased due to the presence of high-carbon austenite.

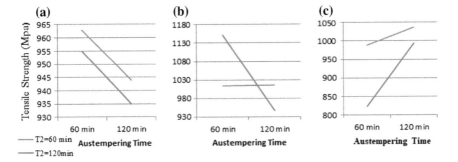

Fig. 3 Influence of ta on strength at different austenitizing temperatures

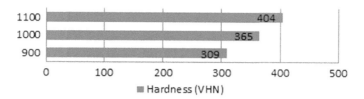

Fig. 4 Hardness variation with austenitizing temperature

3.5 Microstructural Analysis

ADI microstructure which typically exists ferrite with austenite forms a unique ausferretic structure which completely diverse to microstructure of austempered steel. With a rise in austenitising temperature, certain changes can be observed in the structure, i.e., carbon capacity of austenite boost and also rise in grain size of austenite. The details can be observed as follows (Figs. 5, 6, and 7).

At high austempering temperature, less rate of nucleation results in rare needles of ferrite, spreading to larger in size. Thus, coarser bainite can be seen at 420 °C. In high austenitizing temperature of 1000 °C, it was observed that microstructure became coarser and also an increase in volume fraction of austenite.

The superior mechanical properties are mainly due to the existence of feather-like upper bainitic structure that is composed of high dislocation density acicular ferrite and high retained austenite. Normally, volume fraction of austenite is about 20–40%.

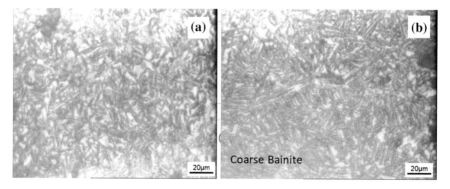

Fig. 5 Images of samples treated at 900 °C. **a** T1 = 2 h and T2 = 1 h, and **b** T1 = 2 h and T2 = 2 h

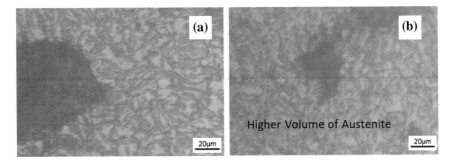

Fig. 6 Images of samples treated at 1000 °C. **a** T1 = 2 h and T2 = 1 h, and **b** T1 = 2 h and T2 = 2 h

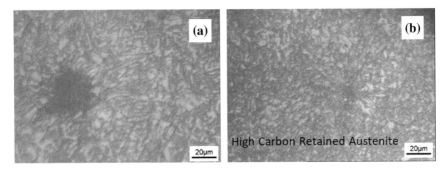

Fig. 7 Images of samples treated at 1100 °C. **a** T1 = 2 h and T2 = 1 h, and **b** T1 = 2 h and T2 = 2 h

3.6 Fracture Study

The rupture surface, for austempered specimens at 420 °C which clearly flash sharp dimples, also shows the existence of "riverlike" structure in few areas for samples which austempered at 420 °C due to intercellular region of the blocky austenite. The rupture surface of ADI shows large quantity of dimples spread homogeneously over the entire surface but, some of the fractures occurred by quasi-cleavage. The

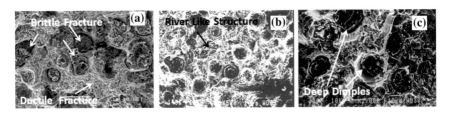

Fig. 8 Fractographs of samples treated at **a** 900 °C, **b** 1000 °C, **c** 1100 °C

observed dimples in this factograph indicate the locations of the existing retained austenite (Fig. 8).

4   Conclusions

1. The microstructure of the ADI specimens is coarsened by raising the austenitization temperature.
2. Tensile tests concluded that as austenitization temperature increased from 900 °C, strength increased but at higher temperature of 1100 °C strength values decreased.
3. Carbon quantity of retained austenite also rises with a rise of austenitizing temperature.
4. At 420 °C of temperature, it has been observed that quantity of retained austenite rises from 35% at 900 °C to 46% of volume at 1100 °C.
5. The total carbon present also makes a rise with a rise of austenitising temperature.
6. It has been observed that with a gain in austenitizing temperature, hardness of specimens gets increased due to the existence of high-carbon austenite.

References

1. Blackmore PA, Harding RA (1984) The effects of metallurgical process variables on the properties of austempered ductile irons. J Heat Treat 3(4):310–325
2. Anowak JF, Norton PA (1985) A guide to mechanical properties possible by austempering, 1.5% Ni, 0.3% Mo iron. AFS Trans 88:123 (1985)
3. Dorazil E (1991) High strength austempered ductile cast iron. Prentice Hall, USA
4. Lerner YS, Kingsbury GR (1998) Wear resistance properties of austempered ductile iron. J Mater Eng Perform 7(1):48–52
5. Tanaka Y, Hidehiko K (1992) Development and application of austempered spheroidal graphite cast iron. Mater Trans 33(6):543–557
6. Hayrynen KL, Brandenberg KR, Keough JR (2002) Applications of austempered cast irons. Trans Am Foundrymen's Soc 2:929–939
7. Seah KHW, Sharma SC (1995) Machinability of alloyed austempered ductile iron. Int J Mach Tools Manuf 35(10):1475–1479
8. Zimba J, Simbi DJ, Navara E (2003) Austempered ductile iron: an alternative material for earth moving components. Cement Concr Compos 25(6):643–649
9. Kobayashi T, Hironobu Y (1988) Development of high toughness in austempered type ductile cast iron and evaluation of its properties. Metall Trans A 19(2):319–327
10. Rouns TN, Rundman KB, Moore DM (1984) On the structure and properties of austempered ductile cast iron (retroactive coverage). Trans Am Foundrymen's Soc 92:815–840 (1984)
11. Ahmadabadi M, Nili H, Ghasemi M, Osia M (1999) Effects of successive austempering on the tribological behavior of ductile cast iron. Wear 231(2):239–300
12. Bayati H, Elliott R (1995) Austempering process in high manganese alloyed ductile cast iron. Mater Sci Technol 11(2):118–129

13. Darwish N, Elliott R (1993) Austempering of low manganese ductile irons. Mater Sci Technol 9(7):572–585
14. Kim YJ, Hocheol S, Hyounsoo P, Jong DL (2008) Investigation into mechanical properties of austempered ductile cast iron (ADI) in accordance with austempering temperature. Mater Lett 62(3):357–360
15. Zammit A, Abela S, Wagner L, Mhaede M, Grech M (2013) Tribological behaviour of shot peened Cu–Ni austempered ductile iron. Wear 302(1):829–836
16. Bedolla-Jacuinde A, Guerra FV, Rainforth M, Mejía I, Maldonado C (2015) Sliding wear behavior of austempered ductile iron micro alloyed with boron. Wear 330:23–31
17. Zhang J, Qingpeng S, Ning Z et al (2015) Very high cycle fatigue property of high-strength austempered ductile iron at conventional and ultrasonic frequency loading. Int J Fatigue 70:235–240
18. Meneghetti G, Ricotta M, Atzori B (2014) A synthesis of the low-and medium-cycle fatigue behaviour of as-cast and austempered ductile irons based on the plastic strain energy. Procedia Mater Sci 3:1173–1178
19. Trudel A, Gagne M (1997) Effect of composition and heat treatment parameters on the characteristics of austempered ductile irons. Can Metall Q 36(5):289–298
20. Moore DJ, Rouns TN, Rundman KB (1985) The effect of heat treatment, mechanical deformation, and alloying element additions on the rate of bainite formation in austempered ductile irons. J Heat Treat 4(1):7–24
21. Sahoo SS (2012) A study on the effect of austempering temperature, time and copper addition on the mechanical properties of austempered ductile iron. M. Tech thesis. National Institute of Technology, Rourkela

Experimental Analysis of the Influence of Combined Heat and Cryogenic Treatment on Mechanical Properties of Steel

Chandan Nashine, S. Anbarasu, Pratik S. Bhansali and Gunamani sahoo

Abstract The main emphasis of the research is to analyse the variation in the behaviour of properties of steel after the combined heat and cryogenic treatment. Preheat treatment was done, where the steels were kept at 600 °C for 60 min and then was quenched using saline bath, followed by the cryogenic treatment at 77 K. For the proper comparison of the variation of properties, the cryogenic treatment of steel was carried out at varying soaking time steps of 2, 4, 6 and 8 h for appropriate analysis of the material. Using optical microscopy, it was observed that as compared to the untreated steels, the impacts of combined heat and cryogenically treated samples were having an excellent microstructure and surface texture. The presence of saturated austenite was initially higher for the untreated material but was found to be lowering with the increase in cryogenic soaking time of material. SEM examination visibly specifies the formation of fine and coarse grains on treated and untreated steels, respectively. Microhardness of cryogenically treated materials was amplified by around 29%, when subjected to combined heat and cryogenic treatment. XRD analysis was done to analyse the variation in the friction coefficient for both the sets of specimens, and it was found that 8 h cryogenically treated specimens were having significantly reduced coefficient of friction as compared to untreated steels.

Keywords Cryogenic treatment · Soaking time · Microhardness
Friction coefficient

C. Nashine (✉) · P. S. Bhansali · G. sahoo
Department of Mechanical Engineering, NIT Rourkela, Rourkela, India
e-mail: chandan.nashine@gmail.com

P. S. Bhansali
e-mail: bhansalipratiks@gmail.com

G. sahoo
e-mail: guna.nit.rkl@gmail.com

S. Anbarasu
NIT Rourkela, Rourkela, India
e-mail: anbarasus@nitrkl.ac.in

© Springer Nature Singapore Pte Ltd. 2019   335
A. K. Lakshminarayanan et al. (eds.), *Advances in Materials and Metallurgy*,
Lecture Notes in Mechanical Engineering,
https://doi.org/10.1007/978-981-13-1780-4_32

1 Introduction

Steels are extensively utilized engineering material in the modern cryogenic application like low-temperature turbo-pump components, cold flow pipes, low-temperature probes, etc. They mostly contain 0.03–2.2% of carbon which varies upon the requirements of various sectors they are chosen. Since steels have the edge over other engineering material due to their luxury of availability and economy, they are preferred for multiple complex designs. There are distinctive sorts of material treatment forms which are utilized for enhancing its different mechanical properties by changing the microstructure of steel [1, 2]. Properties of heat-treated steels mainly rely upon three distinct stages: heating temperature, soaking duration and the different cooling rates [3]. In recent decades, broad intrigue has appeared in the impact of low-temperature treatment on the execution of hardware steels [4–6]. Low-temperature treatment is, for the most part, delegated as either 'cold treatment' at lower temperatures down to around 193 K or 'Deep Cryogenic treatment' at around 78 K. Cryogenic treatment is not an alternative for the heat treatment. Cryogenic treatment is an ideal technique for lessening the percent of present austenite [5]. Cryogenic treatment comprises cooling the material in extinguish condition and after that promptly placing it in below zero centigrade degree and afterwards tempering for stress relieving. Expanding protection from wear, lowering of internal stresses, consistency of measurements and statement of miniaturized scale carbides in the field can be viewed as the most vital benefits of utilizing cryogenic heat treatment. The less the temperature of the cryogenic condition, change in properties is performed with greater quickness [7]. With deep cryogenic treatment connected sharply in the completion of quenching, remaining austenite is diminished, and spots for the nucleation of carbides made during hardening lead to martensite formation. Cryogenic treatments may create the changes in austenite phase, as well as can deliver metallurgical changes inside the martensite. This offers many advantages where flexibility and wear protection are interesting in solidified steels [4]. Although there has been a lot of research in the field of improving the properties of steel, there is very limited research in the era of combined heat and cryogenic treatment and their varying effect with respect to soaking time. In this research work, an experimental analysis is conceded to compare the effect of combined heat and cryogenic treatment of steel with varying soaking time.

2 Experimental Work

The assessment sample material was mild steel round bars. The cylindrical specimens with 30 mm outer diameter and 80 mm span were prepared from the round steel bars. First, the solid cylindrical specimens were given exact dimension using wire electro-discharge machining (EDM) tool for the higher accuracy of the analysis. The machining speed of Wire EDM was chosen as 2 mm per minute established on the preliminary experimental experience. 'To minimize the risk of distortion and cracking specimens were preheated at 600 °C for 60 min in a heating furnace'. Then, the steels were soaked for 30 min in a saline immersion, 'to reassure the homogeneity of temperature all through the complete area of steel', and also, to avoid the decarburization of the specimen surface [1]. Further, the cryogenic conduct of other sets of steels which were already heat-treated was done. The samples remained soaked in liquid nitrogen for different time periods. Soaking times of 2, 4, 6 and 8 h were applied individually for four different specimens.

2.1 Microstructure Test

Microstructure tests of both untreated and cryogenically treated samples are analysed using optical microscopy. Both cryogenically treated and untreated samples were first cleaned using kerosene and then polished with SiC emery of 800 grades and then finishing was done using diamond paste. Further etching of material was done using acetone. These micrograph images were taken at the magnification of 1000X for clear visibility of residuals.

2.2 Vickers Hardness Test

Microhardness tests for both the treated and untreated sets of steels were done. For this test, the steels were given the required dimension required as per the standards. During the test, the Vickers implies a 136° pyramidal diamond indentor which makes a right-angled depression [2]. For the ease of understanding, the mean of microhardness was considered to form the obtained values in the range. Calculation of the microhardness of the test steels was obtained by the ratio of applied force to the area of right-angled depression.

3 Results and Discussion

3.1 Microstructure

SEM inspection was carried out to analyse the variations for both groups of the sample that directly impacts the strength of the material when these are subjected to combined heat treatment and cryogenic treatment. The SEM images were taken at 2000X magnification factor for proper visibility of grains. No polishing or etching was done for SEM analysis.

From Fig. 1, it can be observed that there has been a significant change in the microstructure of the cryogenically treated samples with the change in soaking time. The SEM result shows that the grains of a cryogenically treated sample were finer than the untreated sample. This change in microstructure and grains may be due to the sudden soaking of specimens after heat treatment and simultaneous cryogenic treatment.

The white narrow spots detected in the micrograph (Fig. 2) show the existence of saturated austenite. It was detected that as the soaking time increases the white patches were reducing drastically. The black patches in the figure show the existence of carbide, which were very low in the untreated sample but are present in higher percentage in the cryogenically treated samples. The black patches or the existence of carbide in the micrograph increases with the increase in soaking time. The 8 h cryogenically treated specimen contains the maximum percentage of carbide and minimum percentage of saturated austenite, which clearly indicates the transformation of austenite phase into martensite (Table 1).

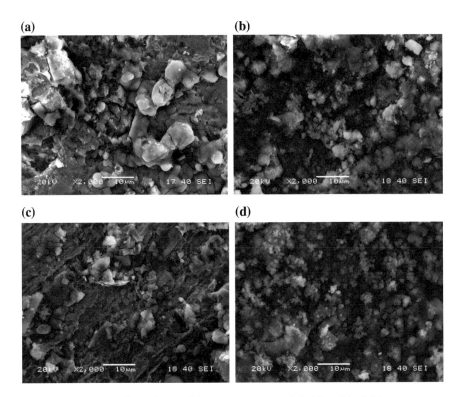

Fig. 1 Microstructures for various soaking times: **a** untreated, **b** 4 h, **c** 6 h, **d** 8 h

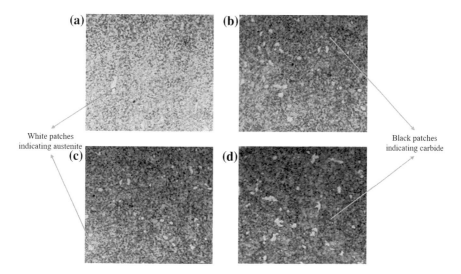

White patches
indicating austenite

Black patches
indicating carbide

Fig. 2 Micrographs for various soaking times using optical microscopy: **a** untreated, **b** 4 h, **c** 6 h, **d** 8 h

Table 1 Elemental composition of steel

Element	Atomic no.	Series	Norm. C (wt%)
Fe	26	K-series	94.37
C	6	K-series	0.80
O	8	K-series	4.38
Na	11	K-series	0.45
		Total	100

3.2 Microhardness

Microhardness results obtained from the Vickers hardness test are mentioned in Fig. 3; it is being observed that the microhardness of the treated steels was greater than the untreated sets of steels. The 8-h-treated sample showed the microhardness in the varying range of 256–260 which was considerably higher than the untreated samples. This increase in the microhardness is due to the increase in carbide content after the treatment of steel (Fig. 4; Table 2).

The increase in the carbide content w.r.t soaking time as shown in the above micrographs seems to be the reason for continuous increase in the microhardness of the steels.

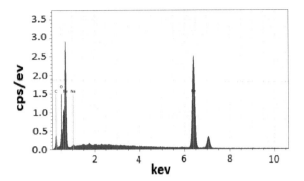

Fig. 3 Energy-dispersive X-ray spectrography (EDS) of the specimen material

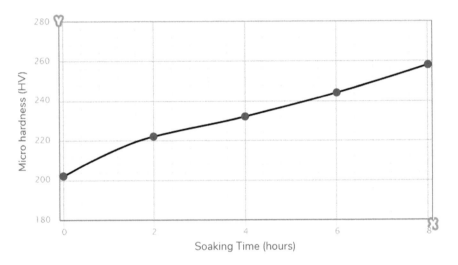

Fig. 4 Microhardness of treated steels at different soaking times

Table 2 Microhardness range of the test specimens

Specimen	Microhardness value HV (Vickers hardness test)
Untreated sample	200–204
2 h cryogenically treated sample	220–224
4 h cryogenically treated sample	230–234
6 h cryogenically treated sample	242–246
8 h cryogenically treated sample	256–260

Fig. 5 Comparision of friction coefficient of untreated and treated test steels

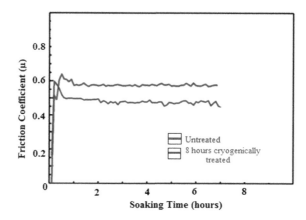

3.3 Friction Coefficient

XRD analysis confirms the reduction in friction coefficient of treated sets of steel as compared to the untreated ones. The friction coefficient of untreated samples was around 0.6 which was much higher than the heat-treated and 8-h-treated steels. Figure 5 shows that the friction coefficient of 8-h soaked specimen was around 0.45, and as the soaking time increases the friction coefficient decreases.

4 Conclusions

- Microstructure analysis from micrograph shows that the austenite content was decreasing with the increase in soaking time, whereas the carbide content for 8-h-treated steels was found to be maximum.
- SEM analysis gives a bright idea regarding the variation in the microstructure of treated steels with different soaking times. The microstructure and grains of the treated material were finer than the untreated materials.
- Microhardness of the steel was found to be affected by the simultaneous heat and cryogenic treatment. With increase in the soaking time, there has been a continuous increment in the microhardness of the steels.
- Microhardness of the 8-h-treated samples of steels was 29% higher than the untreated sets of steel.
- XRD analysis shows that the friction coefficient of 8 h cryogenically treated set of steels was 15% lower than the untreated set steels.
- Simultaneous heat and cryogenic treatment of steels were found to be safe and a useful tool for the enhancement in the mechanical property of steels.

Acknowledgements Authors are obliged to the XRD laboratory and SEM laboratory of NIT Rourkela for their kind corporation and internal support during July 2017.

References

1. Singh M, Singh H (2014) Influence of deep-cryogenic treatment on the wear behavior and mechanical properties of mild steel. Int J Res Eng Technol 4:169–173
2. Molinari A et al (2001) Effect of deep cryogenic treatment on the mechanical properties of tool steels. J Mater Process Technol 118(1):350–355
3. Tobolski EL, Fee A (2000) Macroindentation hardness testing. ASM handbook, volume 8: mechanical testing and evaluation. ASM International, Ohio, pp 203–211. ISBN 0-87170-389-0
4. Li X et al (2017) Influences of thermomechanical treatments on the cryogenic treatability of a slightly unstable austenitic stainless steel. Mater Manuf Process 1–9
5. Bhawar V et al (2017) Influence of Deep Cryogenic Treatment (DCT) on thermo mechanical performance of AISI H13 tool steel. J Mater Sci Chem Eng 5(01), 91
6. Amini K, Akhbarizadeh A, Javadpour S (2017) Cryogenic heat treatment—a review of the current state. Metall Mater Eng 23(1):1–10
7. Zhirafar S, Rezaeian A, Pugh M (2007) Effect of cryogenic treatment on the mechanical properties of 4340 steel. J Mater Process Technol 186(1):298–303

Mechanical Properties' Evaluation of Hemp Fibre-Reinforced Polymer Composites

R. Bhoopathi and M. Ramesh

Abstract Self-curing epoxy thermoset resins have been focused considerable attention by industrialist because of their superior mechanical strength, good resistance to moisture and chemical process, high bonding strength and low shrinkage during curing. In the present improbable environment, the hemp plant fibres/epoxy incorporated composites are serving better than traditional materials in order to low price, high strength, low density, sustainability, low abrasive wear and corrosion resistance. The important objectives of this experimental work are to develop the chemically treated and untreated hemp/glass fibre-reinforced composite materials and investigate the mechanical strengths like tensile strength, flexural strength and impact strengths of prepared samples as per ASTM standard. It has been observed that the chemically treated hemp/glass fibre-reinforced samples exhibited maximum flexural strengths, and these materials are recommended for automobile sectors as alternative material by replacing synthetic fibre incorporated composite materials. The interfacial behaviour, internal constituents, fibre failure mode, fibre pullout, voids and delamination of fractured area are studied by conducting morphological studies with aid of scanning electron microscopy.

Keywords Hemp/glass fibre composites · Chemical treatment · Mechanical properties · Sustainability · SEM analysis

R. Bhoopathi
Department of Mechanical Engineering, Sri Sai Ram Engineering College,
Chennai 600044, India
e-mail: bhoopathir.mech@gmail.com

M. Ramesh (✉)
Department of Mechanical Engineering, KIT-Kalaignar karunanidhi
Institute of Technology, Coimbatore 641402, India
e-mail: mramesh97@gmail.com

© Springer Nature Singapore Pte Ltd. 2019
A. K. Lakshminarayanan et al. (eds.), *Advances in Materials and Metallurgy*,
Lecture Notes in Mechanical Engineering,
https://doi.org/10.1007/978-981-13-1780-4_33

1 Introduction

The role and advantages of plant fibre-reinforced composite material products and components are recognised by many automotive, aviation, textile and construction industries than traditional materials like metals. Due to the severe environmental problem, most of interior and exterior parts of the automobile components are expected as eco-friendly biomaterials by the several manufacturers. The plant fibres such as flax, hemp and sisal fibre incorporated composite components and products have been great potential in various industries. The plant fibre incorporated composite products hold the superior role in the fields like automotive, textile, aerospace, paper production, thermal insulating materials and building construction industries because of strength-to-weight ratio [1, 2]. The plant fibre incorporated composite materials serve as replacing materials to the man-made fibre composites. Plant fibre composite materials hold the several merits like good specific modulus, less density, low cost, eco-friendly, easy recyclable, lightweight, better insulator, good acoustic characteristics and superior mechanical strengths [2–8]. Because of low density and high value of porosity, the hemp fibre and shives composite material holds the superior thermal insulating performances. The different composite specimens were fabricated and studied the physical and mechanical strengths by incorporating hemp plant fibre with different quantities of biodegradable resin materials and hydrophobizers [1].

Shahzad conducted the experimental studies on treated hemp plant fibre-reinforced composites with biodegradable, thermoplastic and thermoset resin matrices and compared the mechanical strengths of different composite samples. The chemical treatment was carried out to improve the bonding characteristics between outer surface of each fibre and matrix constituent. The experimental readings of the tested composite samples were showed the improved mechanical strengths. The physical–mechanical strength of the plant fibres mainly depends on the main constituents like cellulose, hemicellulose and lignin. These strengths of plant fibres are varied due to the factors like chemical content, source of origin, age, retting and extracting techniques, and climate conditions of the fibre. The variation in the basic chemical elements of the raw hemp natural fibre across fibre length is the biggest drawbacks, which leads to different mechanical properties on composite materials [7]. Zhang et al. [9] prepared the activated carbon by heating the hemp stem at 500 °C for 1 hour and KOH activation process. The structural and thermal characteristics on reaction activities of activated carbon were observed and analysed using FTIR spectroscopy, nitrogen adsorption–desorption isotherm and thermo-gravimetric–mass spectrometry.

Xia et al. [10] were fabricated the hybrid composite using hemp natural fibre mat, aluminium foil sheet and thermoset resin for effective electromagnetic field interference prevention by resin transfer moulding methods. From the experimental results, they found that the excellent mechanical properties and effective electromagnetic field interference prevention performances suggested as good engineering materials in the field of electromagnetic interference. The effects of treated short hemp fibres volume content on the solidification behaviour, mechanical strength and thermal characteristics of the tested composites were analysed and studied using DMA, DSC, FTIR, TG analysis and SEM analysis. The experimental study shows the improved mechanical strengths by adding the THF within the resin [11].

Sullins et al. [12] studied the mechanical properties of alkaline-treated hemp fibre incorporated polypropylene composites with different weight percentages. The experimental results of 5 wt% polypropylene with treated hemp fibre composites hold the good mechanical properties. Sepe et al. [13] fabricated and studied the mechanical strengths of alkaline- and silane-treated hemp/epoxy fibre and untreated hemp/epoxy incorporated composite laminates by vacuum infusion process at ambient temperature. The silane-treated hemp/epoxy composites are performed better in tensile and flexural properties than the NaOH-treated composites. Nowadays, most of the researchers focusing their research work on plant fibre incorporated composite components to replace the man-made fibre/synthetic fibre incorporated composite components due to their environmental consequences during the recycling process. Similarly, the load carrying capacity of chemically modified plant fibre incorporated composite components are excellent than the man-made fibre/synthetic fibre incorporated composite components.

2   Materials and its Processing

The raw hemp fibres are purchased from M/s. Chandra Prakash & Company Ltd., Jaipur, India. The epoxy resin, hardener (HY911), silicone gel, thinner solution and sodium hydroxide (NaOH) powders are purchased from M/s. Sakthi Fibre Glass Ltd., Chennai, India. The chemical constituents, physical and mechanical strengths of raw hemp fibres are shown in Tables 1 and 2.

The different sets of composite plates are fabricated by reinforcing the hemp fibre with matrix materials using hand layup method [3]. The alkaline solution has been used to release the chemical constituents like cellulose, hemicellulose, and lignin and pectin from fibre in order to improve the mechanical and physical strengths of the composites [7]. The orientation between the raw hemp fibre and

Table 1 Chemical constituents of hemp fibres [2, 7, 11, 14–21]

Components	Range
Cellulose (wt%)	55–90
Hemicellulose (wt%)	15–22.4
Lignin (wt%)	4–13
Pectin (wt%)	0.8–1.6
Waxes (wt%)	0.8
Moisture (wt%)	9–12
Biomass (Mg DM/ha/y)	7–34.0
Ash (%)	0.8

Table 2 Physical and mechanical strengths of raw hemp fibres [3, 7, 15–17, 22–25]

Property	Value
Density (g/cm^3)	68–81
Specific apparent density	1500
Tensile property (MPa)	310–1235
Specific tensile strength (MPa)	210–510
Elastic modulus (GPa)	20–70
Specific Young's modulus (GPa)	20–41
Failure strain (%)	0.9–4.2
Specific modulus (GPa-cm^3/g)	0.8
Diameter [μm]	17–24
Length (mm)	8.3–14
Aspect ratio (l/d)	549
Micro-fibril angle (Θ)	6.2

glass fibre is maintained at 0° and 90° to withstand the uniaxial and biaxial loads. Initially, the NaOH solution was prepared using pellets, and then the chemical solution was prepared with water by mixing the 5% of NaOH solution. The chemical treatment was carried out on the outside part of raw hemp fibres by immersing into the chemical bath for 1 day under undisturbed environment. Then, the fibres are taken away from the NaOH solution and cleaned by distilled water. The washed and treated fibres are kept under atmospheric conditions for 2 days to dry the water content on the fibres [12, 13]. The epoxy thermoset resin and catalyst are mixed together with the proportion of 10:1 and stirred continuously for 30 min. Using thinner solution, the uppermost layer of the base plate was cleaned and silicone gel was coated over entire uppermost layer of the base plate in order to prevent the sticking of composite with the base plate. Finally, the composite plates are fabricated by hand-lay process by reinforcing the treated fibre by maintaining the 0° and 90° orientation between adjacent layers of fibre [13]. Then, the fabricated laminates are kept under constant load on hydraulic press for 1 day to get the uniform cross section.

3 Mechanical Testing

3.1 Tensile Strength

The chemically processed and unprocessed hemp fibre incorporated epoxy composites were produced by hand-lay process and from each laminate, three sets of specimen have been prepared for the test as per ASTM D638 standards [2, 12]. Each test specimen has been tested using computerised UTM machine by applying the tensile load and the corresponding stress against the loads are noted for compression of results.

3.2 Flexural Strength

Three test specimens from each laminate of chemically processed and unprocessed hemp/epoxy composites are produced based on the ASTM D790 standards [2, 12, 13]. The flexural strength of each laminate was tested by applying three-point flexural loads on the test specimen using computerised UTM machine. The flexural strength readings and its corresponding displacement readings of all tested specimen have been noted for comparison of results.

3.3 Impact Strength

The three sets of impact test specimen have been prepared from fabricated hemp/epoxy composite laminates according to the ASTM A370 standards [3]. The V notches are produced on edge of the test specimen to define the failure path after that the edges of test specimen are clearly finished using abrasive sheet. Finally, the maximum energy absorbed during the failure of each test specimen were noted by applying the impact load with aid of computerised charpy impact tester for result comparison.

4 Experimental Result Analyses

4.1 Tensile Strength Analysis

The tensile strengths of chemically processed and unprocessed hemp/epoxy composite samples were tested by applying pull loads on ends of each sample with the help of computerised UTM machine, and the tensile values of each sample were presented and analysed in the Fig. 1. The chemically processed hemp/epoxy

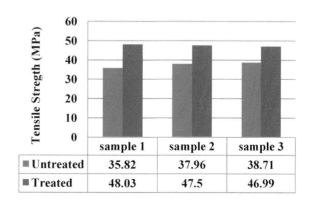

Fig. 1 Tensile strength comparison of hemp/epoxy samples

	sample 1	sample 2	sample 3
■ Untreated	35.82	37.96	38.71
■ Treated	48.03	47.5	46.99

composite samples exhibit the superior tensile strength than the chemically unprocessed samples and hold the maximum value of 48.03 MPa.

4.2 Flexural Strength Analysis

The flexural strengths of chemically processed and unprocessed hemp/epoxy composite samples were tested by applying three-point bending loads on top and bottom surfaces of each sample with the help of computerised flexural testing machine, and the flexural readings of each sample were presented and analysed in Fig. 2. From Fig. 2, it was observed that the chemically processed hemp/epoxy composite samples exhibit the superior flexural strength than chemically unprocessed samples and noted the maximum flexural reading of 0.52 MPa.

4.3 Impact Strength Analysis

The V notches were cut on mid-portion of longer dimension of the each chemically processed and unprocessed hemp/epoxy composite sample. The notched samples were tested by applying sudden impact loads with the help of computerised charpy impact testing machine, and the corresponding impact energy stored on each sample was presented and analysed in Fig. 3. From Fig. 3, it was observed that the chemically processed hemp/epoxy composite samples exhibit the superior impact strength than chemically unprocessed samples and the maximum impact energy of 8.62 J/mm^2 was noted.

Fig. 2 Flexural strength comparison of hemp/epoxy samples

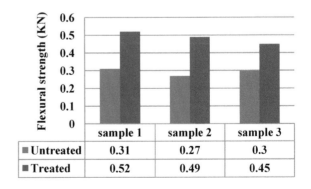

Fig. 3 Impact strength comparison of hemp/epoxy samples

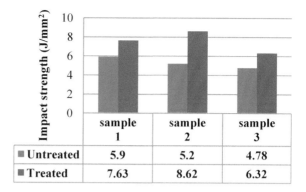

	sample 1	sample 2	sample 3
■ Untreated	5.9	5.2	4.78
■ Treated	7.63	8.62	6.32

5 Morphological Analyses

The scanning electron microscopy was used to analyse the ruptured surface of the chemically processed and unprocessed hemp/epoxy composite tested samples. The morphological characteristics of tested samples under the tensile loading, flexural loading and impact loading are shown in Figs. 4 and 5. From the SEM micrographs, the interfacial behaviours of tested samples like fibre dislocation, matrix

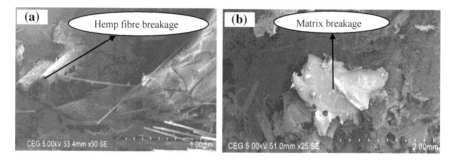

Fig. 4 SEM micrographs of untreated hemp/epoxy composite samples

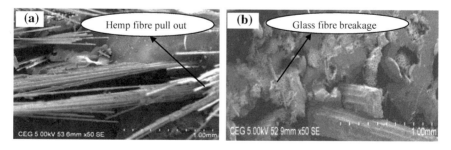

Fig. 5 SEM micrographs of treated hemp/epoxy composite samples

voids, fibre pull out, matrix crack, fractured edges of the chemically processed and unprocessed hemp/epoxy fibre composites were clearly analysed [2, 3, 12].

6 Conclusions

The chemically processed and unprocessed hemp fibre and glass fibre incorporated composite laminates were fabricated, and the mechanical strengths of the test samples like tensile property, flexural property and impact property of the specimen have been noted using computerised UTM machine and impact testing machine. The following conclusion has been derived based on the test results.

- The chemically processed hemp/epoxy fibre composites hold the high value of tensile strength when compared to chemically unprocessed composites. The tensile value of the chemically processed composite test specimen varies from 46.99 to 48.03 MPa. Similarly, the untreated sample holds the strength ranging from 35.82 to 38.71 MPa.
- The maximum flexural strength readings have been noted on the chemically processed hemp/epoxy fibre composite test specimen, and the test results vary from 0.45 to 0.52 KN and for unprocessed, it is 027–0.31 KN.
- Almost similar range of test readings of chemically processed and unprocessed hemp/epoxy fibre composites has been noted from the noted impact test readings. It varies from 6.32 to 7.63 J/mm^2.
- The morphological characteristics like voids, fibre pullout, ruptured surfaces and fibre dislocation of the fibre and matrix failure are clearly noted.
- From these experimental investigations, the chemically processed hemp/epoxy composite components can be suggested for replacing the man-made or synthetic fibre incorporated composite components in automotive and construction applications.

References

1. Vaitkus S, Wo DLU (2017) Investigations on physical-mechanical properties of effective thermal insulation materials from fibrous hemp. Procedia Eng 172:586–594
2. Maslinda AB, Abdul Majid MS, Ridzuan MJM, Afendi M, Gibson AG (2017) Effect of water absorption on the mechanical properties of hybrid interwoven cellulosic-cellulosic fibre reinforced epoxy composites. Compos Struct 167:227–237
3. Lau K, Hung P, Zhu M, Hui D (2018) Properties of natural fibre composites for structural engineering applications. Compos B 136(136):222–233
4. Bhoopathi R, Ramesh M, Deepa C (2014) Fabrication and property evaluation of banana-hemp-glass fiber reinforced composites. Procedia Eng 97:2032–2041
5. Bhoopathi R, Ramesh M, Rajaprasanna R, Sasikala G, Deepa C (2017) Physical properties of glass-hemp-banana hybrid fiber reinforced polymer composites. Indian J Sci Technol 10 (7):1–7

6. Bhoopathi R, Deepa C, Sasikala G, Ramesh M (2015) Experimental investigation on mechanical properties of hemp-banana-glass fiber reinforced composites. Appl Mech Mater 766–767:167–172
7. Shahzad A (2012) Hemp fiber and its composites-a review. J Compos Mater 46(8):973–986
8. Ramesh M, Palanikumar K, Reddy KH (2017) Plant fibre based bio-composites: sustainable and renewable green materials. Renew Sustain Energy Rev 79:558–584
9. Zhang J, Gao J, Chen Y, Hao X, Jin X (2017) Results in physics characterization, preparation, and reaction mechanism of hemp stem based activated carbon. Results Phys 7:1628–1633
10. Xia C, Yu J, Shi SQ, Qiu Y, Cai L, Wu HF (2017) Natural fiber and aluminum sheet hybrid composites for high electromagnetic interference shielding performance. Compos B 114:121–127
11. Qadeer Dayo A, Chang Gao B, Wang J, Bin Liu W, Derradj M, Hussain Shah A, Ahmed Babar A (2017) Natural hemp fiber reinforced polybenzoxazine composites: curing behavior, mechanical and thermal properties. Compos Sci Technol 144:114–124
12. Sullins T, Pillay S, Komus A, Ning H (2017) Hemp fiber reinforced polypropylene composites: the effects of material treatments. Compos Part B 114:15–22
13. Sepe R, Bollino F, Boccarusso L, Caputo F (2017) Influence of chemical treatments on mechanical properties of hemp fiber reinforced composites. Compos B. https://doi.org/10.1016/j.compositesb.2017.09.030
14. Schluttenhofer C, Yuan L (2017) Challenges towards Revitalizing hemp: a multifaceted crop. Trends Plant Sci 22:917–929
15. Faruk O, Bledzki AK, Fink HP, Sain M (2012) Biocomposites reinforced with natural fibers: 2000–2010. Prog Polym Sci 37:1552–1596
16. Yan L, Kasal B, Huang L (2016) A review of recent research on the use of cellulosic fibres, their fibre fabric reinforced cementitious, geo-polymer and polymer composites in civil engineering. Compos B 92:94–132
17. Binoj JS, Edwin Raj R, Sreenivasan VS, Rexin Thusnavis G (2016) Morphological, physical, mechanical, chemical and thermal characterization of sustainable indian areca fruit husk fibers (Areca Catechu L.) as potential alternate for hazardous synthetic fibers. J Bionic Eng 13:156–165
18. Vandenhove H, Hees M (2005) Fibre crops as alternative land use for radioactively contaminated arable land. J Environ Radioact 81(2–3):131–141
19. Zimniewska M, Kozlowski R, Rawluk M (2004) Natural vs. man-made fibres—physiological viewpoint. J Nat Fibers 1(2):69–81
20. Kumar R, Obrai S, Sharma A (2011) Chemical modifications of natural fiber for composite material. Pelagia Res Lib Der Chem Sin 2:219–228
21. Yan L, Chouw N, Huang L, Kasal B (2016) Effect of alkali treatment on microstructure and mechanical properties of coir fibres, coir fibre reinforced-polymer composites and reinforced-cementitious composites. Constr Build Mater 112:168–182
22. Romhány G, Czigány T, Karger-Kocsis J (2017) Failure assessment and evaluation of damage development and crack growth in polymer composites via localization of acoustic emission events: a review. Polym Rev 57(3):397–439
23. Kabir MM, Wang H, Aravinthan T, Lau KT (2011) Effects of natural fibre surface on composite properties: a review. In: Proceedings of the 1st international post- graduate conference on engineering, designing and developing the built environment for sustainable wellbeing, pp 94–99
24. Celino A, Freour S, Jacquemin F, Casari P (2013) The hygroscopic behavior of plant fibers: a review. Front Chem 1(43):1–12
25. Deoray N, Kandasubramanian B (2017) A review on three-dimensionally emulated fiber embedded lactic acid polymer composites: opportunities in engineering sector. Polym Plast Technol Eng 2559:1–15

Influence of Homogenization Temperature on Mechanical Properties from Outer to Inner Zone of Al–Cu–Si Alloy Castings

E. Naveen, S. Ilangovan and Sanjivi Arul

Abstract Impact of variation in mechanical properties from the outer to inner zone of aluminium–copper–silicon (Al–1.5Cu–0.5Si) sand cast alloy was studied by applying three different homogenization thermodynamic states (425, 500 and 575 °C respectively) at a constant soaking time of 10 h. The effect of hardness, strength, and percentage deformation on specimens were measured at outer, middle and inner zones of the sectioned rods in both as-cast and homogenized conditions. The hardness and tensile strength of the specimens were decreased from outer zone to inner zone in both as-cast and all homogenized conditions; whereas its individual value with respect to various zones decreased with increased homogenization temperatures. It was also observed that the percentage elongation of the alloy varied inversely with the hardness and tensile strength values, and also that the hardness values varied proportional to the tensile strength.

Keywords Homogenization treatment · Al–Cu–Si cast alloy · Microstructure Hardness and tensile strength

1 Introduction

Among the most commercially used engineering materials, aluminium and its alloys occupy the third place [1]. In the cast and wrought forms, along with or without heat treatment processes, it is widely used. For over five decades, aluminium and its alloys are considered next to ferrous materials in the metal market. Being the most versatile of all engineering and construction materials, its demand has rapidly increased. The properties of aluminium can be significantly improved and enhanced by alloy additions and heat treatment processes [2–4]. The number of cast aluminium alloys developed considering its outstanding properties of low

E. Naveen (✉) · S. Ilangovan · S. Arul
Department of Mechanical Engineering, Amrita School of Engineering, Amrita Vishwa Vidyapeetham, Coimbatore, India
e-mail: enaveen2018@gmail.com

© Springer Nature Singapore Pte Ltd. 2019
A. K. Lakshminarayanan et al. (eds.), *Advances in Materials and Metallurgy*,
Lecture Notes in Mechanical Engineering,
https://doi.org/10.1007/978-981-13-1780-4_34

density and excellent castability have made it the widely employed among materials in weight sensitive areas of automotive, aeronautics and space flight applications [5]. Further, several factors that affect the properties of aluminium alloys are heat treatment processes, alloying elements and their amount, etc. [6]. Homogenization is one of the heat treatment processes, which also change the mechanical properties of aluminium alloys. The heat treatment cycle steps consist of heating, soaking and cooling [7]. The workability behaviour of a material is found to improve by regulating the uniform phase distribution using various heat processing methods. Workability improvement means, to increase the capability of a material by deformation process [8, 9]. Workability can be defined as the amount of material that undergoes deformation without failure [10, 11]. It is well-known that the as-cast component microstructure consists of columnar and equi-axed grains distributed from outer to inner zones of the ingots. To modify the microstructure of cast ingots, homogenization heat treatment is carried out at elevated temperatures. The homogenized ingots get more easily deformed than that of non-homogenized ingot due to uniform distribution of alloying elements and similarity in microstructure.

The present paper investigates the impact of homogenization temperature on Al–1.5Cu–0.5Si alloy, particularly selected for its suitability for sand and permanent mould castings and relatively good machining properties; when compared to difficult and easy machinable other aluminium alloy groups. This particular alloy castings show improved machinability after heat treatment in most of the general engineering applications, where moderate mechanical properties are required. Further, the casting characteristics of Al–1.5Cu–0.5Si alloy permit it to be used for the production of reasonably thin sections and for pressure tight castings. Moreover, in the heat-treated condition, the casted alloy maintains a relatively high static loading performance. Based on the above advantages and the Al–Cu and Al–Si phase diagrams, the Al–1.5Cu–0.5Si alloy is taken into consideration and carefully analyzed for sand casting and homogenization for various temperatures of 425, 500, and 575 °C respectively. The mechanical properties were then studied at outer, middle and inner zones.

2 Material Fabrication and Research Methodology

The required Al–1.5Cu–0.5Si cast rods of size diameter 110 mm and length 120 mm were fabricated by melting alloying elements LM0-aluminium (pure aluminium) ingots, silicon and electrolytic copper wire in an electric arc furnace and sand casting it. The Al–Cu–Si alloy rod after fabrication was tested and guaranteed for chemical composition as per standards using the Arc spectroscopic analysis and the results reported in Table 1. The cast rods were then pre-machined to Ø 100 mm diameter and length 100 mm as shown in Fig. 1. The homogenization treatment was then

Table 1 Chemical constituents of the Al–1.5Cu–0.5Si alloy (wt%)

Al	Si	Fe	Cu	Mn	Mg	Cr	Ni	Zn	Pb
97.89	0.5	0.18	1.55	0.01	0.03	0.002	0.019	0.023	0.02

Fig. 1 Machined cast rods

performed at 425, 500 and 575 °C for 10 h under inert Nitrogen atmosphere and cooled in the furnace itself. These temperatures were selected on the basis of Al–Cu and Al–Si phase diagrams between the liquidus and solidus curves. Electric resistance type muffle furnace was used to homogenize the blocks. The homogenized blocks were then marked and made to sections of size 20 mm × 20 mm × 100 mm respectively, representing different zones such as inner, middle and outer (Fig. 2). The blocks were cut using wire cut Electric Discharge Machining (EDM) process as shown in Fig. 3. From the above EDM wire cut standard size blocks; specimens were prepared to study the hardness and tensile strength (as shown in Fig. 4) from outer zone to the inner zone in both homogenized and as-cast conditions. The Specimens were prepared and tested as per ASTM E 8 M-04 for tensile test and ASTM E 92-82 for Vickers microhardness test.

3 Results and Discussion

3.1 Microstructure

The microscopic examination of 500 °C homogenized Al–1.5Cu–0.5Si alloy was performed at outer, middle, and inner zones of the cast and machined ingots. Figure 5a–c show the microstructural images captured by an optical microscope with magnification intensity (200X).

A 0.5% Hydrogen Fluoride (HF) as etchant was applied to the specimen, preparing it for standard metallographic technique examination. In all the three cases, fine eutectics were found present in the matrix of Aluminium. From the microstructure analysis, based on available literature, it could be inferred that the

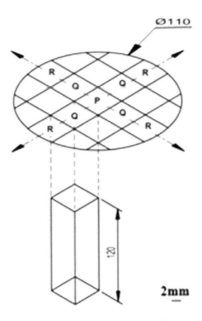

Fig. 2 Shows the inner, middle and outer (P, Q, R) zones of the specimen

Fig. 3 Photographic image of the specimen blocks

black phase is AlFeMgSi, the wide spread area with light brown phase, namely CuAl$_2$ and; the bright brown phase traces of Al–Mg and Al–Si intermetallic [12]. Figures 6, 7 and 8 show XRD and EDX plots of outer, middle, and inner zones taken from specimen homogenized at 575 °C and tested under ASTM standard test conditions. The LINE EDX clearly points out the presence of various alloying elements and the XRD shows the intermetallic formation of the compounds of these alloying elements as constituents. The SEM microstructural images of the outer, middle, and inner zones of the Al–1.5Cu–0.5Si alloy specimen, homogenized at 575 °C for a homogenizing time of 8 h is shown in Fig. 9a–c. The SEM samples

Fig. 4 Photographic image of the tensile test specimen

(a) **(b)** **(c)**

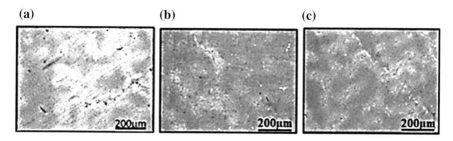

Fig. 5 **a, b, c** Microstructure images of outer, middle and inner zones taken by optical microscope at 500 °C

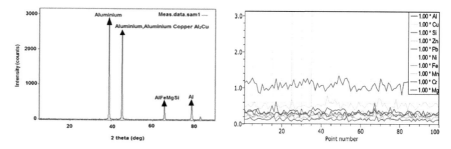

Fig. 6 XRD and EDX plots of outer zone homogenized at 575 °C

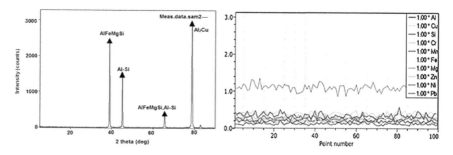

Fig. 7 XRD and EDX plots of middle zone homogenized at 575 °C

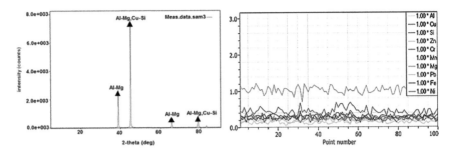

Fig. 8 XRD and EDX plots of inner zone homogenized at 575 °C

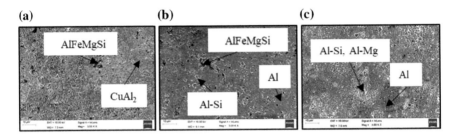

Fig. 9 **a, b, c** SEM micrographs of outer, middle and inner zones homogenized at 575 °C

are prepared as per standards and the analysis is done on the same BRUKER secondary microscope mounted CARL ZEISS GEMINI ∑IGMA Scanning Electron microscope (SEM) machine using a primary lens of magnification intensity 3000X. The SEM images show the intermetallic compounds precipitated out as given by the XRD results.

3.2 Microhardness

Figure 10 shows the microhardness of the as-cast and homogenized specimens at three different zones such as outer, middle, and inner respectively. The Vicker's microhardness indentation test is performed on a "MITUTOYO MVK—Hll" Hardness testing machine for 10 observations per sample per zone for inner, middle and outer zones for an applied load of 100 gf for time 15 s both for as-cast and homogenized conditions. It is clearly seen that the rate of cooling of the rods have effect on the hardness of the alloy before and after homogenization treatments. In all the cases, the hardness values decreased from outer to inner of the specimens. During homogenization treatment the hardness increased above the as-cast condition hardness values for temperatures 425 and 500 °C and decreased for

Fig. 10 Hardness versus homogenization temperature and position

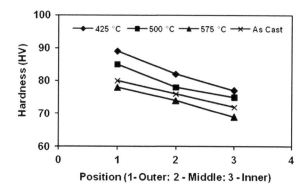

temperature 575 °C respectively. The maximum hardness is observed in the specimen subjected to 425 °C and the values decreased as the homogenization temperature increased.

Further investigation of Table 2 observed the decrease of hardness values from outer (80 HV) to inner (72 HV) for the as-cast alloy. This is due to the higher cooling rate of the outer zone as compared to the middle and inner zones resulting in varying microstructure such as columnar grains at the outer zone and equi-axed grains in the middle and inner zones. The variation in microstructure and hardness can also be due to the decrease in temperature or solidifying rates between the outer and inner zones producing thermal stresses leading to an increased accumulation of stress that vary the as-cast rods behaviour from outer to inner zones [12–14]. The specimen zones also showed the decrease of hardness values from outer to inner zones for all temperatures and; as the homogenization temperature increased, the hardness value decreased improving the workability.

It can be further discussed as follows:

1. For the as-cast condition, the variation in hardness between outer and inner zones is 8 HV. This variation is due to the variation in the cooling rates and microstructural changes between outer and inner zones.
2. In homogenized condition, the variation in hardness between outer and inner zones is 16, 12, and 9 HV for homogenization temperatures of 425, 500, and

Table 2 Hardness value of the specimens subjected to various conditions

Condition	Hardness (HV)			Diff. in hardness (outer–inner) (HV)
	Outer	Middle	Inner	
As-cast	80	76	72	8
Homogenized at 425 °C	89	82	73	16
Homogenized at 500 °C	87	78	75	12
Homogenized at 575 °C	78	74	69	9
Difference in hardness between 425 and 575 °C	11	8	4	

575 °C respectively. Thus, the variation in hardness between zones decreased with increasing homogenization temperatures and was found to be least for the as-cast condition (for all temperatures) and minimum for 575 °C (for homogenized conditions). The reduction in hardness variation between zones is due to (a) increased ductility of the alloy (b) increased grain size (c) more homogenization of alloying elements (d) change of the as-cast microstructures from columnar grains and equi-axed grains to grains of varying sizes and; (e) difference in cooling rates.

3. Moreover, Table 2 also shows the variation in hardness within zones (outer, middle and inner), between homogenization temperature limits. The hardness variation within zones between homogenization temperature limits was least for the inner zone and it increased from inner (4 HV) to outer (11 HV) as shown in the Table 2. This can be due to variation in temperature gradient between homogenization temperatures (425, 500 and 575 °C) and room temperature (≈30 °C) causing cooling rate variations from outer to inner zone (i.e., the outer zone cools faster than the inner).

3.3 Tensile Strength

The tensile testing is done at a speed of 1 mm/min on a "Tinius Olson Universal testing machine" for the three zones for all the homogenized and as-cast conditions for standard specimen size. The variation in tensile strength of the alloy specimen with respect to various zones of the ingot specimen is given below (Fig. 11). The alloy strength under tensile load is observed to decrease from outer to inner zones for all temperatures including both as-cast and homogenized conditions. This is quite similar to Fig. 9 (Hardness vs. Zones), where the hardness values are found to reduce for all temperatures from outer to inner zones. Thus, hardness and tensile strength follow the same trend of decreasing values from outer to inner zones respectively [13, 14]. The yield strength of the material is determined as 62 MPa.

Fig. 11 Tensile strength versus homogenization temperature and position

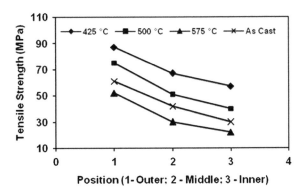

Fig. 12 Percentage (%) Elongation versus homogenization temperature and position

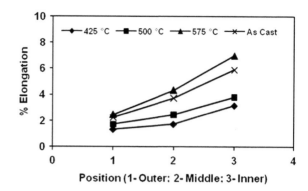

The variation in percentage (%) elongation with respect to homogenization temperatures for the outer, middle and inner zones is shown in Fig. 12. Figure clearly shows the increase in % elongation across all zones with increase in homogenization temperatures. This is also found true for the as-cast. It can be due to increased tensile strength obtained in the inner zone for both as-cast and all homogenized conditions than that of middle and outer zones because of lower cooling rate.

Table 3 shows the variation of tensile strength values between the three zones and the difference in tensile strength values between the outer and inner zones for all temperatures (homogenized and as-cast conditions). It can be observed that tensile strength decreases from outer to inner zones for all conditions.

Also, it is observed that the difference in tensile strength between outer and inner zone value decreases for homogenized ingots with increasing homogenization temperatures. The difference in tensile strength value between outer and inner zones is least for ingot homogenized at 575 °C amongst all ingots (both as-cast and homogenized). Similarly, the difference in tensile strength within a zone is

Table 3 Tensile strength and percentage elongation values of the specimens subjected to various conditions

Condition	Tensile strength (MPa)			% elongation			(values)$_{outer}$–(values)$_{inner}$	
	Outer	Middle	Inner	Outer	Middle	Inner	Tensile strength (MPa)	% elongation
As-cast	61	42	30	2.25	3.75	5.92	31	3.67
Homogenized at 425 °C	87	67	57	1.34	1.73	3.17	37	1.83
Homogenized at 500 °C	75	51	40	1.73	2.48	3.83	35	2.1
Homogenized at 575 °C	52	30	22	2.48	4.35	7	30	4.52
(values)$_{425\ °C}$–(values)$_{575\ °C}$	35	37	35	1.14	2.62	3.83		

maximum for middle zone among all temperatures. Table 3 gives the variation of % elongation values between the three zones and the difference in % elongation values between the outer and inner zones for all temperatures (homogenized and as-cast conditions).

It can be observed that % elongation increases from outer to inner zones for all conditions. Also, it is observed that the difference in % elongation value between outer and inner zones increases with increasing homogenization temperatures. The difference in % elongation value between outer and inner zones is least for ingot homogenized at 425 °C amongst all the ingots (both as-cast and homogenized). Similarly, the difference in % elongation values within a zone with respect to all temperatures is maximum for the inner zone.

4 Conclusions

The microhardness value of the as-cast specimens measured at three different zones decreases from outer (80 HV) to inner (72 HV). It is due to the higher cooling rate of the outer zone as compared to the middle and inner zones. Within the specimen, the hardness value decreases from outer to inner zones for all temperatures. As the homogenization temperature increases, the hardness value decreases. Similarly, the tensile strength of the alloy also decreases from outer to inner zones for all temperatures and conditions (both as-cast and homogenized). As the homogenization temperature increases, correspondingly the tensile strength also decreases in all the three zones. Therefore, it can be concluded that both hardness and tensile strength are functions of each other. The % elongation increases from outer to inner zone with all temperatures both for as-cast and homogenized conditions. The percentage elongation is found to be maximum for the inner zone sample homogenized at 575 °C. The variation in percentage elongation within a zone is least for outer zone, and between zones is least for the ingot homogenized at 425 °C.

References

1. Dwivedi DK (2010) Adhesive wear behavior of cast aluminium-silicon alloys: overview. Mater Des 31(5):2517–2531
2. Isadare AD, Aremo B, Adeoye MO, Olawale OJ, Shittu MD (2013) Effect of heat treatment on some mechanical properties of 7075 Aluminium Alloy. Mater Res 16(1):190–194
3. Radhika N, Subramanian R (2014) Effect of aging time on mechanical properties and tribological behaviour of aluminium hybrid composite. Int J Mater Res 105(9):875–882
4. Arunagiri S, Radhika N (2016) Studies on adhesive wear characteristics of heat treated aluminium LM25/AlB$_2$composites. Tribol Ind 38(3):277–285
5. Zhang LY, Jiang YH, Ma Z, Shan SF, Jia Y, Fan C, Wang W (2008) Effect of cooling rate on solidified microstructure and mechanical properties of aluminium-A356 alloy. J Mater Process Technol 207(1):107–111

6. Samuel FH, Samuel AM, Liu H (1995) Effect of magnesium content on the aging behavior of water-chilled Al–Si–Cu–Mg–Fe–Mn (380) alloys castings. J Mater Sci 30(10):2531–2540

7. Birol Y (2012) Effect of cooling rate on precipitation during homogenization cooling in an excess silicon AlMgSi alloy. Mater Charact 73:37–42

8. Rao KP, Doraivelu SM, Gopinathan V (1982) Flow curves and deformation of materials at different temperatures and strain rates. J Mech Working Technol 6(1):63–88

9. Rao KP, Prasad M (1984) Deformation processing of aluminium alloy 2618, developments of processing map. Aluminium 60:259–265

10. Semiatin SL, Jonas JJ (1984) Formability and workability of metals: plastic instability and flow localization. In: Asm Series in Metal Processing, 2. American Society for Metals, Metals Park, Ohio, pp 43–116 (1984)

11. Dieter GE (1984) Workability testing techniques. American Society for Metals, Metals Park, Ohio, pp 1–7 (1984)

12. Totik Y, Gavgali M (2003) The effect of homogenization treatment on the hot workability between the surface and center of AA2014 ingots. Mater Charact 49:261–268

13. Ilangovan S, Sai Krishna V, Gopath NK (2014) Study of effect of cooling rate on mechanical and tribological properties of cast Al-6.5Cu aluminium alloy. Int J Res Eng Technol 3(5):62–66

14. Ilangovan S (2014) Effects of solidification time on mechanical properties and wear behaviour of sand cast aluminium alloy. Int J Res Eng Technol 3(2):71–75

Mechanical Characterization and Robustness of Self-compacting Concrete with Quarry Dust Waste and Class-F Fly Ash as Fillers

B. Mahalingam, P. Sreehari, Srinath Rajagopalan, S. Ramana Gopal and K. Mohammed Haneefa

Abstract Self-Compacting Concrete (SCC) is a special type of concrete which does not require any form of external forces to get compacted. However, it behaves similar or better to conventionally vibrated concrete when it gets hardened. The present study focuses on developing SCC with a constant powder content of 600 kg/m^3 with 450 kg/m^3 of cement. The remaining portion of class-F fly ash (150 kg/m^3) was replaced step by step with a waste material from granite crushing industries called as quarry dust waste (QDW); which is available in abundance at crushed sand factories as a waste material resulted from the washing of crushed granite to remove very fine particles. The effects of replacement were studied at fresh and hardened states of SCC. Apart from the mechanical properties such as compressive, flexural, and split tensile strengths, the ultrasonic pulse velocity assessment was performed to ensure the integrity of test specimens. Robustness, which is the ability of SCC to perform similar way in the case of any small fluctuations in material design or properties is also studied in the present paper. The study revealed that the quarry dust waste can be incorporated in making SCC with reliable fresh and hardened properties. Additionally, the robustness of SCC with quarry dust waste is good and within acceptable limit. Moreover, the incorporation of quarry dust waste makes the concrete more sustainable.

B. Mahalingam (✉) · P. Sreehari · S. Rajagopalan · S. Ramana Gopal
Department of Civil Engineering, SSN College of Engineering, Chennai, India
e-mail: mahalingamb@ssn.edu.in

P. Sreehari
e-mail: sreeharip@ssn.edu.in

S. Rajagopalan
e-mail: srinathr@ssn.edu.in

S. Ramana Gopal
e-mail: ramanagopals@ssn.edu.in

K. Mohammed Haneefa
Department of Civil Engineering, IIT Madras, Chennai, India
e-mail: mhkolakkadan@gmail.com

© Springer Nature Singapore Pte Ltd. 2019
A. K. Lakshminarayanan et al. (eds.), *Advances in Materials and Metallurgy*,
Lecture Notes in Mechanical Engineering,
https://doi.org/10.1007/978-981-13-1780-4_35

Keywords Self-compacting concrete · Fly ash · Ultrasonic pulse velocity
Robustness · Sustainability

1 Introduction

Concrete is the most widely used material in world, next to water. Since 1950,
accounting for a population hike of threefolds; use of concrete has increased by 34
times [1]. Rapid infrastructure developments and affordable housing programmes in
developing countries necessitate the use of economic and durable concretes. Use of
supplementary cementitious materials and energy efficient materials make concrete
more sustainable and durable. Durability of concrete is directly proportional to the
efficiency of compaction to consolidate the concrete. Presence of voids in concrete
due to improper compaction will result in significant reduction in compressive
strength and subsequent premature distresses in concrete. Moreover, issues will be
more defenceless in highly congested reinforced structures, where the passing and
filling of fresh concrete become more tedious or sometime impossible.
Self-Compacting Concrete (SCC) is a special type of concrete which does not
require any form of external forces to get compacted. However, it behaves similar
or better than conventionally vibrated concrete when it gets hardened. The concept
of SCC was first introduced by Okamura and Ouchi [2] and was taken up by many
researchers. Based on state-of-the-art information available on SCC, properties of
hardened SCC is similar to normal concrete [3, 4]. Nanthagopalan and Santhanam
[5] reported SCCs with crushed stone as fine aggregate and concluded that more
volume of paste is required to achieve similar flow behaviour compared to normal
concrete with conventional river sand as fine aggregates. Presently, SCC is
extensively used in heavily reinforced civil structures like nuclear power plants,
where the flow of normal concrete is limited due to high contents of reinforcement
bars. SCC material design is typified by high volume of paste content to ensure
required passing, filling, and segregation resistance during its fresh stage. In general, a range of 0.35–0.42 is used as volume of paste with a minimum powder
content of 500 kg/m^3. Apart from cement, fly ash or any other fillers are used along
with compatible super plasticizer and viscosity modifying agents.

2 Experimental Works

A 53 grade ordinary Portland cement (OPC) was used in the study. River sand and
crushed granite were used as fine and coarse aggregates respectively. The present
study focuses on developing SCCs with a constant powder content of 600 kg/m^3
with 450 kg/m^3 of cement. The remaining portion of class-F fly ash (150 kg/m^3)
was replaced step by step with a waste material from granite crushing industries
called as Quarry Dust Waste (QDW) which is available in abundance at crushed

Table 1 Mix proportions of SCC mixes used

Mix	Cement (kg/m³)	Fly ash (kg/m³)	Quarry dust waste (kg/m³)	Fine aggregate (kg/m³)	Coarse aggregate (kg/m³)	Water (litres/m³)	% Superplasticizer
Mix 1	450	0	150	1080	720	190	0.60
Mix 2	450	50	100	1080	720	190	0.65
Mix 3	450	100	50	1080	720	190	0.70
Mix 4	450	150	0	1080	720	190	0.70

sand factories as a waste material resulted from the washing of crushed granite to remove very fine particles. The grain size of QDW is less than 300 μm. The effects of replacement were studied at fresh and hardened states of SCC. Apart from the mechanical properties such as compressive, flexural, and split tensile strengths; ultrasonic pulse velocity assessment was performed to ensure the integrity of test specimens. A polycarboxylic ether-based superplasticizer (Master Glenium Sky 8233) was used for ensuring flow properties. Robustness, which is the ability of SCC to perform in a similar way in the case of any small fluctuations in material design or properties is also studied in the current study. Table 1 provides mix proportions of SCC mixes used in the study. The fresh properties of SCCs were assessed using slump flow test, T_{500}, J-ring, L-box, U-box and V-funnel tests conforming to European guidelines [6]. The dosage of superplasticizer (SP) for different mixes was determined using marsh cone test which vary from 0.6 to 0.70% by weight of cement. The marsh cone test results are plotted as log of flow time verses %SP. The optimum dosage is the corresponding SP% at which the curve becomes flattened (Fig. 1).

Fig. 1 Marsh cone test for SCC

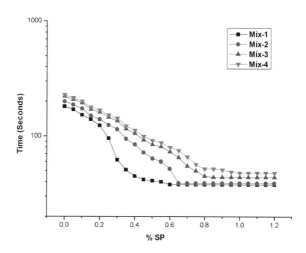

3   Results and Discussion

3.1   Fresh Properties

Table 2 presents fresh properties of SCCs developed in the study. Incorporation of QDW affects the flow properties adversely. The flow was getting reduced with the increased proportions of QDW. This effect may be due to absorption of water by fine stone particles. However, the slump flows were within the acceptable limits to qualify as SCC by EFNARC, the European guidelines for self-compacting concrete [6]. The slump flow values ranged from 680 to 640 mm for different replacement of QDW. All other fresh properties such as T_{500}, J-ring, L-box, U-box and V-funnel were within the limits specified by EFNARC. Segregation resistance was assessed by repeating a V-funnel test after 5 min. All the mixes used in the study passed the segregation resistance test. Even though the fresh properties of QDW incorporation affect the fresh properties; this study revealed that the adverse effects can be controlled or diminished by use of proper water contents and superplasticizer dosages (Fig. 2).

Table 2 Fresh properties of SCC mixes

Mix	Spread flow (mm)	T_{500} (s)	J-ring test (h1–h2) (mm)	L-box (h1/h2)	U-box (h1–h2) (mm)	V-funnel (s)	V-funnel (Sec) after 5 min.
Mix 1	640	8	9	0.87	22	13.5	15
Mix 2	645	3.98	14	0.89	15	13.3	14
Mix 3	650	3.97	8	0.91	18	9	12
Mix 4	680	2.5	5	0.90	15	11	11

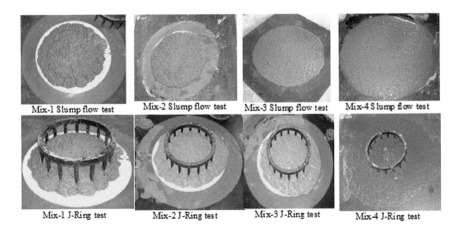

Mix-1 Slump flow test   Mix-2 Slump flow test   Mix-3 Slump flow test   Mix-4 Slump flow test

Mix-1 J-Ring test   Mix-2 J-Ring test   Mix-3 J-Ring test   Mix-4 J-Ring test

Fig. 2 Snap shots of flow spread and J-ring test of SCC mixes

3.2 Mechanical Properties

Figure 3 represents 7 and 28 days compressive strengths of mixes used in the current study. 7 days compressive strengths ranged from 19.04 to 35.47 MPa, whereas the 28 days ranged from 28.88 to 44.58 MPa. Compressive strengths were reducing as the replacement QDW increased. This may be due to the inert nature of quarry dust which does not contribute to the hydration process of cement for imparting strength to concrete. On contrary, fly ash is a pozzolanic material which reacts with calcium hydroxide in the presence of water resulting in the formation strength impacting extra calcium silicate hydrate (CSH). Figure 4 gives flexural strengths and split tensile strengths of SCC mixes developed in the study. Similar to the compressive strength observations, both the flexural and split tensile strengths exhibited the same downward trend with increased QDW replacement. Split tensile strengths of SCC mixes used ranged from 2.12 to 3.65 MPa, whereas as the flexural strengths vary from 2.54 to 3.91 MPa. Ultrasonic pulse velocity (UPV) test as per IS: 13311 (Part 1)—1992 was conducted on cube and beam specimens to ensure the compaction of SCC [7]. The UPV values were 4.01, 4.04, 4.23 and 4.45 km/sec for Mix1 to Mix 4 respectively. The UPV test results on hardened SCC results confirmed the self compaction of the mixes in the finishing stage.

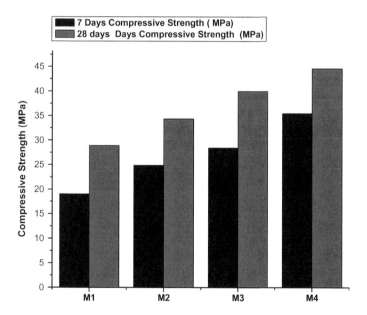

Fig. 3 Compressive strengths of SCC mixes

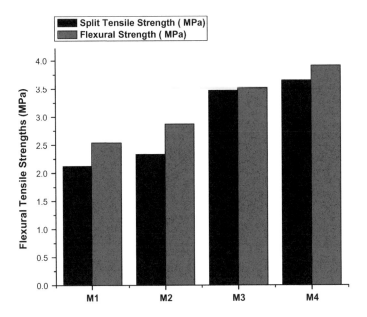

Fig. 4 Flexural tensile strengths of SCC mixes

3.3 Robustness of SCC

Robustness of SCC refers to the to the ability of SCC mixes to retain their fresh properties even with small variations in material proportioning or properties [8]. In this study water content was altered ± 10 kg/m^3 to study the robustness. This is to accommodate the possible variations during the mixing procedures/instrumental/ minor errors in material quantities/water contents in ready mix concrete plants (RMCs). Water contents of 180 and 200 kg/m^3 were used to do trials on SCC with same mix design or all mixes used in the current study. Slump flow, T$_{500}$ and compressive strengths were investigated. Figures 5, 6 and 7 pictorially depict the results of robustness studies.

At a water content of 180 kg/m^3; for a mix of complete replacement of flyash (150 kg/m^3) exhibited less robustness in terms of fresh properties. Slump flow reduced drastically to 515 from 640 mm. Other changes were from 645 to 590, 650 to 600 and 680 to 620 mm for 100, 50 and 0 kg/m^3 of QDW respectively (Fig. 5). However, the fact should be noted that these flow values are enough to qualify the material as SCC. Similarly, for T$_{500}$ showed an upward trend as expected as shown in the Fig. 6 (i.e., the less water content reduces the fluidity and increase the time for 500 mm horizontal flow). All mixes with 180 kg/m^3 water content exhibited higher strengths compared to the mixed with 190 kg/m^3. The compressive strengths were varied from 32.01 to 51.17 MPa for the mixes with 180 kg/m^3 (Fig. 7). At a water content of 190 kg/m^3 the range was from 28.88 to 44.58 MPa. The improved strengths may be due to the reduced water to binder ratio.

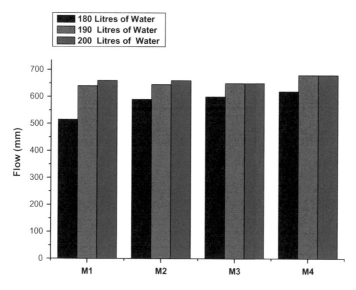

Fig. 5 Robustness study for flow of SCC mixes

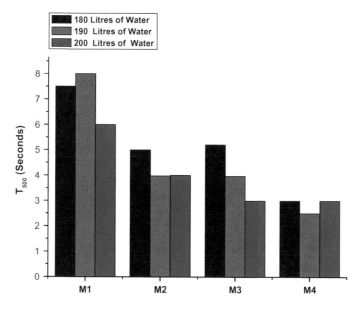

Fig. 6 Robustness study for T_{500} of SCC mixes

At a water content of 200 kg/m^3, slump flow values were almost same for all (640–680 mm). Visual index was good as there were no any signs for bleeding or segregation. All other fresh properties exhibited improved values. However, the

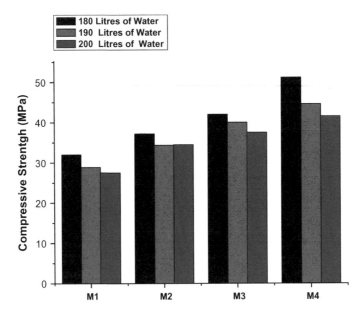

Fig. 7 Robustness study for compressive strengths of SCC mixes

compressive strengths were reduced compared to 180 and 190 kg/m^3 water contents. The compressive strengths varied from 27.5 to 41.57 MPa. The reduction
compressive strengths may be due to the increase in water to binder ratios. The
maximum reduction in strength was 7% and can be considered as reliable in
concrete mix design.

4 Conclusions

From the current study following conclusions are drawn

(a) Quarry dust waste (QDW) can be effectively used as filler material to develop
    SCC. Fresh and hardened properties are within the acceptable limits.
(b) The incorporation of QDW affects the fresh and hardened properties. However,
    the SCC made out of QDW well qualifies the passing, filling and segregation
    tests.
(c) Robustness of SCC with QDW is good and variations are within the tolerable
    limits.
(d) Incorporation of QDW in SCC may solve the issues related to its disposal and
    hence may prove to be sustainable approach.

Acknowledgements The authors are grateful to SSN Trust for their financial support through faculty research funding. The authors are thankful to M/S BASF India for providing Super Plasticizers.

References

1. Scrivener KL, John VM, Gartner EM (2016) Eco-efficient cements: potential economically viable solutions for a low-CO_2 cement-based materials industry. United Nations Environment Programme, Paris
2. Okamura H, Ouchi M (2003) Self compacting concrete (invited paper). J Adv Concr Technol I (1):5–15, April 2003
3. Santhanam M, Subramanyam S (2004) Current developments in self compacting concrete. Indian Concr J, 11–22
4. Persson B (2001) A comparison between mechanical properties of self compacting concrete and the corresponding properties of normal concrete. Cem Concr Res 31:193–198
5. Nanthagopalan P, Santhanam M (2011) Fresh and hardened properties of self-compacting concrete produced with manufactured sand. Cement Concr Compos 33(3):353–358
6. The European guidelines for self compacting concrete (2005) Specification, production and use. In: EFNARC, pp. 1–63
7. IS 13311-Part 1 (1992) Method of Non-destructive testing of concrete, Part 1: Ultrasonic pulse velocity. Bureau of Indian Standards, New Delhi
8. Bonen D, Deshpande Y, Olek J, Shen l, Struble L, Lange D, Khayat K (2007) Robustness of self-consolidating concrete. In: 5th RILEM symposium on self-compacting concrete, pp. 33–42

An Experimental Investigation of Al–Zn–Cu Alloy on Hardness, Microstructure and Wear Parameter Optimization Using Design of Experiments

J. Sreejith and S. Ilangovan

Abstract Al–25Zn–3Cu ternary alloy was prepared using stir casting. The hardness and microstructure of as-cast and homogenized specimen were studied. From the microstructure examination results, it is clear that the as-cast dendritic structure was eliminated after homogenization. Adhesive wear test was carried out using pin-on-disc wear tester as per L_9 orthogonal array. Load, velocity and sliding distance were the control parameters taken for the experiments. Then, signal-to-noise (S/N) ratio analysis was done by considering "smaller-the-better" as the goal of experiment. Using Analysis of Variance (ANOVA), the critical parameters, which affect the response were obtained and results were comparable with the S/N analysis. Also, confirmation experiments were carried out to validate the developed linear regression equation. Finally, Scanning Electron Microscope (SEM) and Energy Dispersive X-ray (EDX) analysis were used to detect the wear mechanisms in the worn out samples.

Keywords Al–Zn–Cu alloy · Micro-hardness · Specific wear rate
ANOVA · S/N ratio · Regression analysis

1 Introduction

Zinc–aluminium alloys were used as bearing materials in the past because of its characteristic mechanical properties like strength and hardness. Also, studies revealed that zinc–aluminium alloys have higher wear resistant property when compared to SAE 660 bronze under both dynamic and static conditions [1]. Copper or silicon can be added to the monotectoid Zn–Al alloys to improve its wear resistance and seizure resistance. But, monotectoid Zn–Al containing silicon has

J. Sreejith · S. Ilangovan (✉)
Department of Mechanical Engineering, Amrita School of Engineering,
Amrita Vishwa Vidyapeetham, Coimbatore, India
e-mail: s_ilangovan@cb.amrita.edu

© Springer Nature Singapore Pte Ltd. 2019
A. K. Lakshminarayanan et al. (eds.), *Advances in Materials and Metallurgy*,
Lecture Notes in Mechanical Engineering,
https://doi.org/10.1007/978-981-13-1780-4_36

superior wear behaviour to the traditional alloys like phosphor bronze or cast iron [2]. Nevertheless, the advantages of zinc based alloy, has some problems like low ductility and dimensional instability [3, 4]. Brittleness of zinc and copper rich phases are the reason for the said low ductility problems. Metastable ($CuZn_4$) to stable (Al_4Cu_3Zn) transformation is the main cause for the dimensional instability issue [5]. Hence, the necessity to find an alternate alloy was inevitable. Hence, increasing the aluminium content in Zn–Al alloy which satisfy all the properties of bearing materials and also eliminating low ductility and dimensional instability problems were made. Al–40Zn–3Cu and Al–40Zn–3Cu–2Si alloys were developed as an alternative for zinc based alloys and a stable θ ($CuAl_2$) phase was formed which eliminated the dimensional instability problems [6]. Further, Al–40Zn–3Cu alloys possess higher mechanical and wear properties than SAE 65 bronze [7]. Al–25Zn possess both highest tensile strength and hardness in the binary Al–Zn alloys. Moreover, after the addition of copper in the Al–25Zn base alloy, the mechanical properties were superior in Al–25Zn–3Cu [8]. Higher solidification time has a negative impact on the mechanical properties of Al–6,5Cu alloy [9]. Hence, the present research work is more focussed on the effect of wear control parameters on tribological behaviour of Al–25Zn–3Cu using Taguchi's design of experiments (DOE) statistically. Furthermore, the regression analysis using ANOVA and confirmatory experiments to validate the regression equation is also investigated.

2 Experimental Work

The required ternary Al–25Zn–3Cu alloy (wt%) was fabricated by bottom pouring type stir casting process. Aluminium, zinc and copper in pure form were poured in the furnace and the temperature of the furnace was kept at 780 °C (obtained from phase diagram). The stirring speed and stirring time was set at 350 rpm and 10 min respectively. The die was preheated to 250 °C and the size of the die selected was Ø 30 mm × 250 mm. Argon gas was used during the melting process to avoid the reaction of atmospheric gases with molten metals. After pouring of the molten metal into the die, the casting was air cooled to room temperature. Homogenization of the prepared cast rods was done at 350 °C for 8 hours. After holding the castings for 8 h at 350 °C, they were furnace cooled to 50 °C. Then the casting rods were air cooled to room temperature again. Figure 1(a, b) represents the microstructure of as-cast and homogenized specimens respectively. From the figures, it is clear that the uniform distribution takes place after homogenization for all the specimens. Also, some dendritic structure is visible in the as-cast condition and it reduces the properties of the alloy. Hence, after homogenization the properties can be enhanced due to the reduction of dendritic structures and uniform distribution of elements.

Hardness was investigated for both as-cast and homogenized specimen. Emery sheets of grade 1/0 and 2/0 were used to polish the specimens prior to hardness examination. The calculated micro-hardness values of the as-cast and homogenized Al–25Zn–3Cu alloy are 96.1 and 134.4 HV. From the results, it was observed that

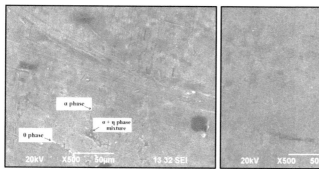

(a) As-cast specimen   (b) Homogenized specimen

Fig. 1 Microstructure of the specimens

the homogenized specimen's hardness is higher than the as-cast specimen. Hence, this implies that the homogenization of alloys has greater significance in the hardness because homogenization provides uniform distribution of metals in the alloy. To be more specific, the coarse interdendritic formation of the compounds at the time of solidification of Al–Zn–Cu alloy is eliminated or dissolved [10]. Design of experiments using Taguchi's technique is more suitable for conducting the experiments because this method helps to minimize the number and duration of experiment. The control parameters selected for conducting the experiments were load, velocity and sliding distance.

Three levels are taken for each control parameter and it is shown in Table 1. L_9 orthogonal array was selected for designing the experiments which implies nine experiments should be executed to study the response. L_9 orthogonal array is used to understand the effect of three or four independent parameters with each parameter having three set values. Specific wear rate is the response to be investigated in the present work and the objective is to minimize the response. The specimens for dry sliding wear test (Ø 12 × 40 mm) were prepared from the casted alloy using machining process. Figure 2 shows the pin-on-disc wear tester which is used to test the specimen. The density of the required Al–Zn–Cu alloy is 3270 kg/m^3 and it was calculated by water displacement method. MINITAB software is used to obtain S/N ratios and means of the Al–Zn–Cu alloy. The track diameter for the experiments is set to be constant as 150 mm. The initial and final weights were noted after each experiment for the calculation of SWR in mm^3/Nm.

Sl. No.	Load (N)	Velocity (m/s)	Sliding distance (m)
1	10	1.5	750
2	15	2.25	1250
3	20	3.0	1750

Table 1 Wear parameters and their levels

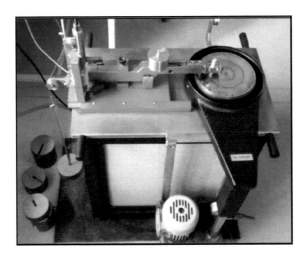

Fig. 2 Pin-on-disc wear tester

Table 2 Experimental results obtained from L_9 orthogonal array

Exp. No.	1	2	3	4	5	6	7	8	9
Load (N)	10	10	10	15	15	15	20	20	20
Velocity (m/s)	1.5	2.25	3	1.5	2.25	3	1.5	2.25	3
Sliding distance (m)	750	1250	1750	1250	1750	750	1750	750	1250
SWR (mm³/Nm)	0.061	0.053	0.035	0.075	0.050	0.052	0.062	0.085	0.058
S/N (dB)	24.20	25.43	29.05	22.46	25.96	25.54	24.07	21.31	24.68

3   Results and Discussion

This section deals with the discussions about the effect of control parameters on SWR, S/N ratio analysis, ANOVA and regression analysis.

The specific wear rate and S/N ratio of the alloy acquired for different combinations of control parameters are shown in Table 2. From the S/N ratio plot shown in Fig. 3b, the optimum level of each parameter is identified (load = 10 N, velocity = 3 m/s and sliding distance = 1750 m). Graphical representation of the mean plot for SWR is shown in Fig. 3a and it depicts the trend of responses for the chosen control parameters.

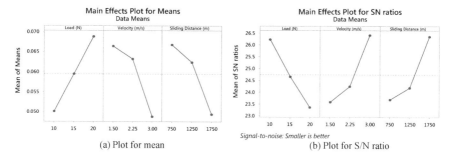

Fig. 3 Main effect plots—SWR

3.1 Impact of Load on SWR

From Fig. 3b, it can be concluded that the optimum load for the minimum SWR value is 10 N. Also, it can be inferred that the specific wear rate increases with the applied load [11, 12]. The contact pressure between the counter face and the specimen increases due to the increase in the load. Also, temperature at the contact interface increases due to high load [12]. As a result, the alloy strength decreases at high loads and material loss increases. At higher loads, the formation of debris occurs which finally results in deformation [11].

3.2 Impact of Sliding Velocity on SWR

In the present research work, the optimum value for SWR is at a sliding velocity of 3 m/s. From Fig. 3b, it is evident that the SWR is inversely proportional to the sliding velocity [13]. This phenomenon is due to the higher contact time between the counter face and specimen at low speeds since time is inversely proportional to speed [11]. Also, at higher velocities, the tendency to the formation of oxide layer is high due to the increase in temperature at the interface. As a result, the lower SWR occurs at higher sliding velocities.

3.3 Impact of Sliding Distance on SWR

From Fig. 3b, it can be noted that the wear rate is inversely proportional to the sliding distance. Hence, a minimum wear rate is obtained at a sliding distance of 1750 m. During the initial sliding, the fine surface of the alloy contacts with the counter face of the disc. As a result, the removal of the material will be more at initial sliding of the specimen. As sliding of the alloy increases, the wear rate decreases because of the formation of the hard layer at the surface [11]. Also, as the sliding distance increases, the material from the disc will adhere into the specimen. As a result, SWR is minimal at higher sliding distance.

3.4 Signal-to-Noise Ratio Analysis

Table 3 shows the response table for the S/N ratio of the selected alloy. Variation of mean within the levels is indicated by the delta (Δ) value. Delta value corresponds to the difference between the maximum and minimum response values in the response table. Higher the delta value implies the critical parameter. From Table 3, it is evident that the load is the most important parameter which influence the SWR, followed by velocity and sliding distance.

3.5 Analysis of Variance (ANOVA) for SWR

Significance of each parameter on the SWR response is analysed by using ANOVA for the significance and confidence level of 5 and 95% respectively. ANOVA for the SWR response is shown in Table 4. The confidence of replicating the experimental results is indicated by the significance level. The significance of the parameter is indicated by its P-value and most significant parameter possesses a P-value less than 0.05.

Table 3 SWR response table—"smaller-the-better"

Level	1	2	3	Delta (Δ)	Rank
Load (N)	24.63	24.66	23.36	2.87	1
Velocity (m/s)	23.58	24.24	26.43	2.85	2
Sliding distance (m)	23.69	24.19	26.37	2.68	3

Table 4 ANOVA for SWR

Source	DF	Seq SS	Adj SS	Adj MS	F	P	Pct (%)
Regression	3	0.001454	0.001454	0.000485	9.23	0.018	
Load (N)	1	0.000530	0.000530	0.000530	10.10	0.025	30.86
Velocity (m/s)	1	0.000469	0.000469	0.000469	8.94	0.030	27.31
Sliding distance (m)	1	0.000455	0.000455	0.000455	8.67	0.032	26.49
Error	5	0.000262	0.000262	0.000052			15.25
Total	8	0.001717	0.001717				

3.6 Regression Analysis

Based on the significance parameters obtained from ANOVA, a linear regression equation is developed. This developed regression equation relates the SWR response with the control parameters like load, sliding velocity and sliding distance. Equation (1) gives the developed regression equation.

$$SWR = 0.0796 + 0.001880L - 0.01179V - 0.000017S, \tag{1}$$

Where, L—load in (N), V—velocity in (m/s) and S—sliding distance in (m).

The above mentioned equation implies that the load has a positive impact on SWR but both sliding velocity and sliding distance have negative influence on the SWR response. Also, the R-square value obtained from the ANOVA analysis is 84.71% and the P-values for all the control parameters are below 0.05. Hence the model is significant from the above obtained results. Figure 4 represents the comparison between experimental SWR and regression SWR for all the nine experiments conducted using L_9 orthogonal array.

3.7 Confirmation Tests

Validation of the regression model was done using the confirmation tests for the alloy. Table 5 gives the comparison between the experimental and regression results after the confirmation tests. From these results, the percentage of error is calculated and the calculated error is within the acceptable limits. Hence, the model is also acceptable.

Fig. 4 Experimental SWR versus regression SWR

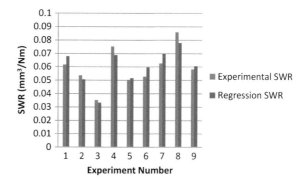

Table 5 Confirmation results

Sl. No.	Load (N)	Velocity (m/s)	Sliding Distance (m)	Experimental SWR (mm³/Nm)	Regression SWR (mm³/Nm)	Error (%)
1	13	1.75	1000	0.0652	0.0664	1.8
2	16	2	1300	0.0611	0.0640	4.5
3	19	2.5	1600	0.0603	0.0586	2.8

3.8 SEM Analysis

Deep and continuous scratches were observed in Fig. 5a. Hence the maximum SWR is depicted in Fig. 5a. Figure 5b corresponds to the minimum SWR conditions due to the minimal smearing action. Hence the adhesion and the material removal will be less and it leads to the minimum SWR of the alloy. The SEM image shown in Fig. 5b corresponds to the optimum condition ($L = 10$ N, $V = 3$ m/s, $S = 1750$ m).

3.9 EDX Analysis

EDX analysis of the alloy before and after wear test is shown in Fig. 6a, b respectively. It is clear that the presence of iron, carbon, and oxygen in the specimen after wear test is higher when compared to the specimen before the wear. And the presence of iron and carbon from the steel disc leads to the minimum wear rate condition of the alloy due to the adhesion property.

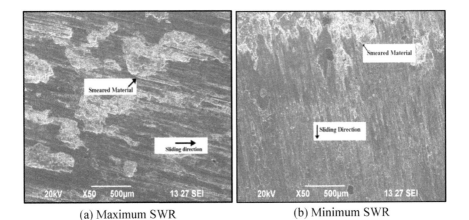

(a) Maximum SWR   (b) Minimum SWR

Fig. 5 SEM images of worn out surfaces

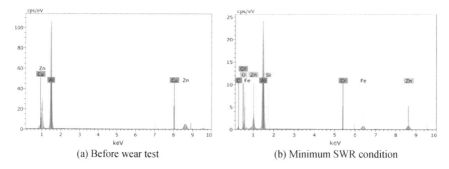

(a) Before wear test   (b) Minimum SWR condition

Fig. 6 EDX analysis of the specimens

4   Conclusions

- Hardness of the alloy was enhanced from 96.1 to 134.4 HV after homogenization treatment.
- The optimum condition was obtained from S/N analysis ($L = 10$ N, $V = 3$ m/s and $S = 1750$ m).
- From ANOVA, load is the critical parameter followed by velocity and sliding distance.
- SEM and EDX analysis justifies the occurrence of minimum SWR at optimum conditions.

References

1. Savaskan T, Purcek G, Murphy S (2002) Sliding wear of cast zinc-based alloy bearings under static and dynamic loading conditions. Wear 252:693–703
2. Murphy S, Savaskan T (1984) Comparative wear behaviour of Zn–Al based alloys in an automotive engine application. Wear 98:151–161
3. Pearl P, Savaskan T, Laufer E (1987) Wear resistance and microstructure of Zn–Al–Si and Zn–Al–Cu Alloys. Wear 117:79–89
4. Savaskan T, Murphy S (1987) Mechanical properties Zn–25Al-based alloys. Wear 116:211–224
5. Zhu YH (2004) General rule of phase decomposition in Zn–Al based alloys (II)—on effects of external stresses on phase transformation. Mater Trans 45:3083–3097
6. Savaskan T, Alemdag Y (2008) Effects of pressure and sliding speed on the friction and wear properties of Al–40Zn–3Cu–2Si alloy: a comparative study with SAE 65 bronze. Mater. Sci. Eng. A. 496:517–523
7. Alemdag Y, Savaskan T (2009) Mechanical and tribological properties of Al–40Zn–Cu alloys. Tribol Int 42:176–182
8. Bican O, Savaskan T, Alemdag Y (2009) Developing aluminium–zinc-based a new alloy for tribological applications. J Mater Sci 44:1969–1976

9. Ilangovan S, Viswanathan S, Niranthar GK (2014) Study of effect of cooling rate on mechanical and tribological properties of cast Al–6.5 Cu aluminium alloy. Int J Res Technol 3:62–66
10. Shao-hua W, Ling-gang M, Shou-jie Y (2011) Microstructure of Al–Zn–Mg–Cu–Zr–0.5Er alloy under as-cast and homogenization conditions. Trans Nonferrous Met Soc China 21.1449–1454
11. Radhika N, Raghu R (2017) Investigation on mechanical properties and analysis of dry sliding wear behavior of Al LM13/AIN metal matrix composite based on Taguchi's design. J Tribol 139:1–10
12. Arunagiri KS, Radhika N (2016) Studies on adhesive wear characteristics of heat treated aluminium LM25/AlB$_2$ composites. Tribol Ind 38:277–285
13. Ilangovan S, Jiten S (2014) Dry sliding wear behaviour of sand cast Cu–11Ni–6Sn alloy. Int J Res Technol 3:28–32

Investigation of Mechanical/Tribological Properties of Composite Ersatz Articular Cartilage with Nano Fillers

K. N. D. Malleswara Rao, R. Praveen Kumar, T. Venkateswararao, B. Sudheer Kumar and G. Babu Rao

Abstract It is observed that articular cartilage damage is an increasing problem across the globe. Poly 2-hydroxyethyl methacrylate (pHEMA) hydrogels are promising implants for ersatz articular cartilage due to their similar tissue properties. However the major obstacle to their use as replacing articular cartilage is their poor mechanical properties. By the development of composite hydrogels with hydroxylapatite (HA) nano fillers the lack properties may overcome. The purpose of this study is to describe mechanical properties of composite hydrogels for replacement of cartilage. Therefore with different percentages of HA nano fillers the hydrogels (pHEMA/HA) were prepared by cast drying method. Mechanical properties of composite hydrogels are studied with reciprocating sliding tribo-tester and the porous nature is investigated by SEM. By the results it is observed that mechanical properties of composite hydrogels are improved by adding hydroxylapatite nano fillers.

Keywords Friction articular cartilage · Composite hydrogels · Friction and wear · Hydroxylapatite (HA) · pHEMA gel

K. N. D. Malleswara Rao (✉) · R. P. Kumar · T. Venkateswararao · B. S. Kumar
Department of Mechanical Engineering, Lakireddy Bali Reddy College
of Engineering (A), Mylavaram, AP, India
e-mail: kndmrao@lbrce.ac.in

R. P. Kumar
e-mail: praveenkmr435@lbrce.ac.in

T. Venkateswararao
e-mail: tvrao@lbrce.ac.in

B. S. Kumar
e-mail: sudheerbattula1984@gmail.com

G. Babu Rao
Department of Mechanical Engineering, KKR & KSR Institute of Technology
and Sciences, Guntur, AP, India
e-mail: gbr.gudavalli@gmail.com

© Springer Nature Singapore Pte Ltd. 2019
A. K. Lakshminarayanan et al. (eds.), *Advances in Materials and Metallurgy*,
Lecture Notes in Mechanical Engineering,
https://doi.org/10.1007/978-981-13-1780-4_37

1 Introduction

In tribology bio-compatible polymers have excellent performances for human body, for example from early 70 years biomaterials (artificial articular cartilage materials) can be used as articular implants which are having properties like plenty of water, lower wear and friction. However, accidental and other injuries of cartilage are unable to be avoided so it is essential to introduce biomaterials in to medicine. Researchers have experimented this facet since long time and also reported some development. Regardless of the developments, articular damages are again wide spread with increase of traffic accidents and influence the aspect of man life gravely. Additionally, higher number of immature people are suffering from articular defects, therefore it is essential to lengthen, the long life of estarz cartilage composites. It is very imperative to try to develop high mechanical properties composites to acclimate the growing demand of articular cartilage.

Hydrogel [1] is a polymer of 3D network which is formed by ion bond, the replacing bio compatible materials for natural articular cartilage are poly (2-hydroxyethyl methacrylate) and poly vinyl alcohol because of their similar tissue properties.

Polymeric composites reinforced [2] with nano fillers are most common in today's engineering materials world. HA particles [3] as a crucial bio compatibility inorganic material have been extensively studied in inorganic/organic composite field by many researchers.

With bio compatible hydrogels [4] and HA particles as the reinforcement number of samples were prepared. Then the mechanical properties were investigated on the ensuing process. Distributed fillers over the bio matrix increase mechanical properties. The main purpose of this investigation is to study the mechanical properties of pHEMA/HA composites with HA particles as nano fillers under lubrication conditions. In the future vivo studies of ersatz articular cartilage development these results may be used.

2 Materials and Methods

2.1 Sample Preparation

The pHEMA hydrogels [5] are hydrophilic polymers with high water content. Aqueous composite hydrogels were prepared by stirring with magnetic stirrer of pHEMA (10wt%) in distilled water at 80 °C. Later the completion of mixing process HA nano fillers are added to the polymer such as 0%HA, 1%HA and 2% HA (0, 1 and 2wt%). Later the solution underwent the ultrasonic bath to take off all bubbles. Polymers which are formed were allowed to reach room temperature and then frozen for 1 h (−18 °C), whereas for the polymer crosslinkage the mixture was placed at room temperature.

2.2 Characterization of Composite Hydrogels by SEM (Scanning Electron Microscopy)

The method used for high-resolution imaging of surfaces is SEM and with this morphologic characterization [6] of polymers is done. For the characterization of polymer, Jeol JSM 6460 scanning electron microscope (SEM) used which is having a resolution of 3.5 nm.

2.3 Compression Test

As per ASTM standards the compression tests [7] were conducted on the composite hydrogels at room temperature in an ASTM-D575 Instron 4411 UTM. For the test samples were prepared in a circular shape with the dimensions of 30 mm diameter and 15 mm height and placed in UTM with cross head speed of 5 mm/min. From the obtained plots the stress and strain values were determined.

2.4 Friction and Wear Tests

The tribological tests [8, 9] were conducted on the composite hydrogels in the reciprocating sliding tribo-tester [10] (Fig. 1) with a speed of 80r/min at room temperature about 20 min for each test. Form the obtained values the graphs were plotted and the coefficient of friction values are calculated with the sample loads by the friction force.

Fig. 1 Reciprocating sliding
tribo-tester

a:Femoral head
b:Load
c:Lubricant
d:Liquid bath
e:Acrylic plate
f:Composite hydrogel

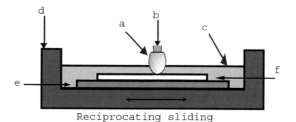

Reciprocating sliding

3 Results and Discussion

3.1 Composite Hydrogel

The samples of composite hydrogels [11] are presented in Fig. 2 by the observation it can be noted that with the addition of HA nano fillers, pHEMA/HA hydrogel became hazy (1%HA) and thick white (2%HA) where as the pHEMA hydrogel is transparent. It is pointed out that polymer with 2%HA nano fillers were macroscopically not homogeneous may be due to the polymers were sodden for such concentration of HA.

3.2 Morphological Characterization

The porous nature (Fig. 3) of composite hydrogels can be observed in the SEM images and the aggregates of HA are noticeable in 2%HA hydrogel which in result shows, for 1%HA less agglomerations and for 2%HA more agglomerations.

3.3 Compression Test

The determined stress–strain values from the acquired compression test [12] curves of the composite hydrogel are shown in Fig. 4. In the test due to the low permeability there is limited time for escape of fluid and with that small difference in the polymer volume takes place. When load is applied the hydrogel polymer is reoriented, in this time interstitial liquid starts to drain out of the polymer.

Fig. 2 Composite hydrogels samples: **a** Circular 0%HA, **b** circular 1%HA, **c** circular 2%HA, **d**, rectangle 0%HA, **e** rectangle 1%HA, **f** rectangle 2%HA

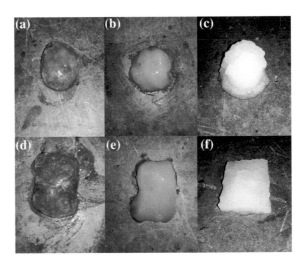

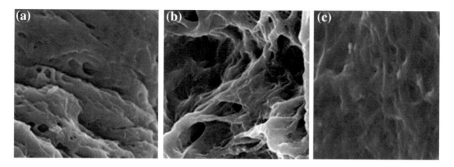

Fig. 3 Composite hydrogel samples SEM images: **a** 0%HA, **b** 1%HA, **c** 2%HA

Fig. 4 Compression test results

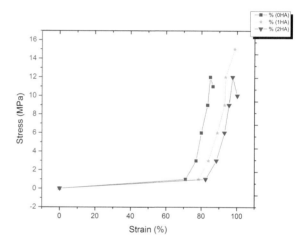

At this point for a substantial deformation it needs comparatively less load to be applied. With the applied load on the polymer the strings tends to uniform orientation and the flow of interstitial fluid causes friction which leads to material hardening.

The determined compression test values from the stress–strain curves are summarized in Table 1. It is noted that with the incorporation of HA nano fillers to the pHEMA several improvements were observed in 1%HA polymer. Which displays the increased mechanical performance and confirmed that in the SEM micrographs as HA content increased high agglomeration of nano-particles were takes place.

Table 1 Compression test results

Sample	σ (MPa)	ε (%)
0%HA	11.09 ± 0.45	93.36 ± 5.67
1%HA	17.05 ± 0.18	90.07 ± 1.065
2%HA	14.89 ± 0.29	93.58 ± 1.087

The addition of nano-fillers of the composite hydrogel exhibits advancements in the mechanical properties due to its high mechanical strength.

3.4 Tribological Characterisation of Composite Hydrogels

3.4.1 Friction Test

The rubbing properties [13] of the composite hydrogels are driven by the mechanical and also with water content presented in it. From the previous explanations it is expected that under uniform load with the drained water the surface friction increases, this creates the sliding friction [14] surface becomes larger, which leads to decrement of friction coefficient.

Tables 2 and 3 includes the friction coefficients of pHEMA/HA composite hydrogels with respect to pressure and time. The more coefficient of friction [15] was attained for the 1%HA (Figs. 5 and 6).

Table 2 Friction coefficient-pressure of composite hydrogels

Pressure/MPa	Friction coefficient		
	Sample 1 (0%HA)	Sample 2 (1%HA)	Sample 3 (2%HA)
0.25	0.078	0.08	0.079
0.75	0.062	0.07	0.079
1.25	0.063	0.072	0.079
1.75	0.082	0.103	0.121

Table 3 Friction coefficient-time of composite hydrogels

Time	Friction coefficient		
	Sample 1 (0%HA)	Sample 2 (1%HA)	Sample 3 (2%HA)
0.5	0.05	0.05	0.05
1.5	0.06	0.07	0.08
2.5	0.06	0.069	0.076
5.5	0.059	0.073	0.078
10.5	0.06	0.074	0.078
15.5	0.06	0.073	0.078
20.5	0.06	0.072	0.08

Fig. 5 Friction coefficient
versus pressure

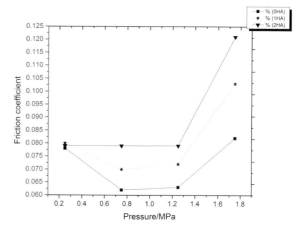

Fig. 6 Friction coefficient
versus time

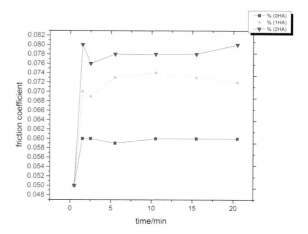

3.4.2   Wear Test

Like cartilage, all the cases of developed polymers have two phases sliding
mechanisms. The results of wear tests (Fig. 7) for pHEMA [16] and HA samples
were shown in Table 4. It is noticed that by comparing polymer values with the
previous reported articular cartilage values; the polymer values are within the
expected range and the hydrogel with 1%HA nano fillers exhibits good wear
improvements [17–21].

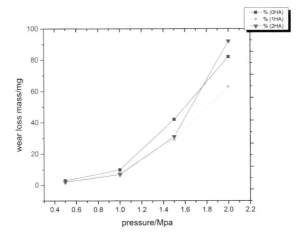

Fig. 7 Wear loss versus pressure

Table 4 Wear loss-pressure of composite hydrogels

Pressure/MPa	Wear loss		
	Sample 1 (0%HA)	Sample 2 (1%HA)	Sample 3 (2%HA)
0.5	3	2	2
1	10	8	7
1.5	42	29	31
2	82	63	92

4   Conclusion

In this study we evaluated the mechanical properties of ersatz articular cartilage material and the composite hydrogels with different nano filler were successfully produced and characterized. For the increased HA contents, gel fraction remained constant because of the agglomeration of phosphates which indicates that interaction between both material (pHEMA and 2%HA) is not good. By the observations of SEM micrographs advancements in the mechanical properties were observed when HA nanofillers were mixed to the pHEMA polymer and the best results are recorded for pHEMA/1%HA gel.

Acknowledgements I want to especially thank Mr. Konidena Venkata Ravi Sankar (MEDICAID LABS LLP) for critically reading the manuscript and helpful discussions.

References

1. Nakashima K, Murakami T, Sawae Y (2004) Evaluation of wear property of PVA hydrogel as artificial cartilage and effect of protein film on wear-resistant property. Trans JSMESer C70 (697):2780–2787
2. Susmitha B, Durga K, Malleswara rao KND et al (2018) Three dimensional finite element analysis of thin hybrid FRP skew laminates for thermo elastic behaviour of different materials. Mater Today Proc 5:1194–1200
3. DiSilvestro MR, Suh JKF (2001) Across validation of the biphasic poro viscoelastic model of articular cartilage in unconfined compression, indentation, and confined compression. J Biomech 34:519–525
4. Malleswara Rao K, Niranjan Kumar I, Praveen Kumar R (2018) Improved lubrication properties and characterization of GO nanoparticles as lubricant additives on hypereutectic Al-17Si/AISI25100 steel sliding pair. SAE Int J Fuels Lubr 11(2):201–215, https://doi.org/10. 4271/04-11-02-0010
5. Otsuka E et al (2009) A simple method to obtain a swollen PVA gel cross linked by hydrogen bonds. J Appl Polym Sci 114(1):10–16
6. Gupta S, Lin J, Ashby P, Pruitt L (2009) A fiber reinforced poro elastic model of nano indentation of porcinecostal cartilage: a combined experiment a land finite element approach. J Mech Behav Biomed Mater 2:326–338
7. Kanjickal D et al (2008) Effects of sterilization on poly (ethylene glycol) hydrogels. J Biomed Mater Res Part A 87A(3):608–617
8. Bavaresco VP, Zavaglia CAC, Reis MC, Gomes JR (2008) Study on the tribological properties of pHEMA hydrogels for use in artificial articular cartilage. Wear 265(3–4):269–277
9. Ratnam CH, Kumar BS, Malleswara Rao KND (2018) Effect of twin-pin profile tool on the microstructure and mechanical properties of friction stir welded dissimilar Aa2024 and Aa6061 aluminium alloys. Int J Mech Eng Technol 9(3):946–953
10. Mohammed R, Malleswara Rao KND, Khadeeruddin M (2013) Modeling and analysis of drive shaft assembly using FEA. Int J Eng Res Dev 8(2):62–66
11. Peerunaik M, Rao TB, Malleswara Rao KND (2013) Static and modal analysis of leaf spring using FEA. Int J Comput Eng Res 3:107–110
12. Cao L, Youn I, Guilak F, Setton LA (2006) Compressive properties of mouse articular cartilage determined in an ovel micro-indentation test method and biphasic finite element model. J Biomech Eng 128:766–771
13. Sasada T, Takahashi M, Watakabe M, Mabuchi K, Tsukamoto Y, Nanbu M (1985) Frictional behavior of a total hip prosthesis containing artificial articular cartilage. J Jpn Soc Biomater 3 (3):151–157 (in Japanese)
14. Sasada T (1988) Bio mechanics and biomaterials—friction behavior of an artificial articular cartilage. In: Proceedings of the 3rd world biomaterials congress, 2, 6
15. Wong M, Carter DR (2003) Articular cartilage functional histomorphology and mechanobiology: a research perspective. Bone 33:1–13
16. Wang M, Peng Z, Price J, Ketheesan N (2013) Study of the nano mechanical properties of human knee cartilage in different wear conditions. Wear 301:188–191
17. Bostan L, Trunfio-Sfarghiu A, Verestiuc L, Popa M, Munteanu F, Rieu J-P, Berthier Y (2012) Mechanical and tribological properties of poly(hydroxyethyl methacrylate) hydrogels as articular cartilage substitutes. Tribol Int 46:215–224
18. Somasekhar K, Malleswara Rao KND et al (2018) A CFD investigation of heat transfer enhancement of shell and tube heat exchanger using Al_2O_3-water nanofluid. Mater Today Proc 5:1057–1062

19. Baykal D, Day J, Jaekel D, Katta J, Mansmann K, Kurtz S (2012) Tribological evaluation of hydrogel articulations for joint arthroplasty applications. J Mech Behav Biomed Mater 14: 39–47
20. Ma R, Xiong D, Miao F, Zhang J, Peng Y (2010) Friction properties of novel PVP/ PVA blend hydrogels as artificial cartilage. J Bio Med Mater Res Part A 93:1016–1019
21. Bavaresco VP (2004) Tribological study of polymeric hydrogels for use as artificial articular cartilage. Ph.D. thesis, State University of Campinas, SP, Brazil

A Study on Sliding Wear Behaviour of Polyamide Six Nano-Composites with MWCNT and Copper Nano-Particles

T. Anand and T. Senthilvelan

Abstract Polyamide 6 matrix material was filled with multi -walled carbon nano tubes(MWCNT) and copper nano-particles reinforcements. The composition used for the investigation was PA6 and 0.25 wt% MWCNT with 0.2, 0.4 and 0.6 wt% copper nano-particles. Sliding wear behaviour was studied in PA6 composites. A co-rotating twin screw extruder was used to blend the materials and the specimens were prepared using injection moulding. The process parameters were Load (N), Sliding speed (m/s) and Reinforcement (%). The experiment was conducted in a Pin on Disc apparatus. Taguchi and Analysis of Variance (ANOVA) were used to analyse the influence of process parameters on wear rate. Genetic Algorithm (GA) was used to optimise the wear rate. Scanning Electron Microscope (SEM) images on worn surfaces of PA6 composites were analysed and the type of wear was observed.

Keywords Polyamide 6 · MWCNT · Copper nano particles · Taguchi technique · Genetic algorithm (GA) · Scanning electron microscope (SEM)

1 Introduction

Polyamide composite materials play a major role in the development of engineering scenario because of lightweight, reduced noise level and enhanced mechanical properties [1]. These combinations of properties cannot be found in traditional materials. The wide usage of polymeric composites made the analysis of tribological behaviour of the materials essential in view of commercial aspects. At the same time, the wear resistance of polyamide 6 materials is not completely satisfied for all engineering applications. The properties of polymeric materials can be further improved by adding reinforcements. These added reinforcements and fillers

T. Anand (✉) · T. Senthilvelan
Department of Mechanical Engineering, Pondicherry Engineering College,
Puducherry, India
e-mail: anand.tvel@gmail.com

© Springer Nature Singapore Pte Ltd. 2019
A. K. Lakshminarayanan et al. (eds.), *Advances in Materials and Metallurgy*,
Lecture Notes in Mechanical Engineering,
https://doi.org/10.1007/978-981-13-1780-4_38

improved the tribological properties like wear resistance and coefficient of friction, etc. and also higher strength [2]. The distribution of reinforcements affects such properties of a composite material.

Metal powders as reinforcements play a vital role and affect the wear rate and coefficient of friction [3]. Polyamide 6 is commercially used in various engineering applications because of its versatile performances and other properties. Studies show that the particle size very much influences the properties of the composites [4]. Nano particles reinforcement show better mechanical and other properties compared to larger sized particles. The development of nano particle filled polymer composites are growing in the recent years.

Therefore to enhance the wear properties, multi-walled carbon nano tube and copper nano-particles were added as reinforcements with polyamide 6 matrix material [5]. Copper nano particles are used as a lubricant additive to reduce the friction and wear. A co-rotating twin screw extruder was used to blend the reinforcements with matrix material. The blended PA6 composites were injection moulded and the specimens were prepared. The addition of MWCNT and copper nano particles enhanced the wear rate of PA6 composites. Pin on disc experiment was used to analyse the wear characteristics. In this work the wear properties of polyamide 6 nano composites reinforced with MWCNT and nano Cu particles were studied. The influence of Load, Sliding speed and the % of reinforcement were investigated and the optimised value is obtained using Taguchi, Analysis of Variance (ANOVA) and Genetic Algorithm (GA). The improvement of wear resistance because of the addition of nano reinforcement was discussed. Wear surfaces were analysed with the help of Scanning Electron Microscopic images.

2 Experimental Work

2.1 Material Selection

In this investigation, Polyamide 6 (PA6) matrix material reinforced with multi walled carbon nano tubes and copper nano particles in different proportions were selected for the analysis of wear rate. Table 1 shows the various proportions PA6 nano composites.

Table 1 Proportions of PA6 nano composites

Sl. No	Proportions of PA6 nano composites
1	PA6, 0.25 wt% MWCNT, 0.2 wt% nano Cu
2	PA6, 0.25 wt% MWCNT, 0.4 wt% nano Cu
3	PA6, 0.25 wt% MWCNT, 0.6 wt% nano Cu

2.2 Pin on Disc Wear Experiment

A pin on disc experimental setup with a data acquisition system has been used for the wear analysis on PA6 and its composites [6]. Figure 1 shows the pin on disc arrangement. The PA6 nano composite pins were made with 8 mm diameter and 32 mm long with an end radius of 4R. The wear behaviour of the pin against an EN-31 hardened and ground steel disc with 65 HRC and surface roughness 0.5 μm was evaluated at room temperature.

The pin was kept stationary and the disc used to be rotated with the help of a DC motor thus creating the sliding wear on the pin. The EN 31 disc has the track diameter ranging between 25 and 50 mm. A string arrangement attached with the pin holder is used to apply the load on the specimens (pins). A maximum load of 250 N can be applied on the experimental setup. Sliding distance for the entire test was 1000 m. The pin was polished and the disc was refreshed using a fine grid abrasive paper before each test. The following Eq. (1) was used to calculate the wear rate.

$$\text{Wear rate} = Q/L, \tag{1}$$

where Q = volume loss (mm^3) and L = sliding distance (m).

Though volume loss and wear rate are widely used to study the wear charactestics, wear coefficient found to be a better parameter because it is considering hardness of the pin along with the applied load and wear rate. Archard's Equation (2) is used to determine the wear coefficient [7].

$$V = (K_w WL)/3H, \tag{2}$$

where V = wear volume (mm^3), K_w = wear coefficient (mm^3/Nm), W = load (N), L = slinding distance (m) and H = hardness of the wear pin.

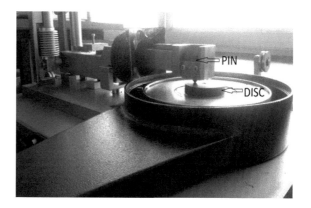

Fig. 1 Pin on disc wear experiment set up

2.3   Optimization Technique

2.3.1   Taguchi Method

Taguchi technique is a statistical and single point objective optimization approach adopted to optimise the process parameters to improve the wear rate of PA6 nano composites [8]. Taguchi method is used to determine the number of experiments and L 9 orthogonal array was selected. In this work, Load (N), sliding speed (m/s) and % of reinforcement were taken as the process variables. Table 2 shows process variables and the number of levels.

The response variable is wear rate (mm^3/m). The experimental outcomes (response variable) was analysed using signal to noise ratio and the influences of process variables on the response variables were analysed.

In this study "smaller the better" criteria was used to determine the optimum value of wear rate (3)

$$\left(\frac{S}{N}\right) = -10\log\frac{1}{n}\left(\sum_{i=1}^{n} Y_i^2\right),\tag{3}$$

where, n is the number of experiments conducted for each trial and y is the value obtained from the experiments.

2.3.2   Analysis of Variance (ANOVA)

Analysis of variance is the method widely used to identify the significant process parameters in obtaining the optimal condition [9]. Therefore ANOVA method was adopted to determine the influence of process parameters Load (N), Speed (m/s) and percentage of reinforcement on wear rate of PA6 nano composites. MINITAB software was used to measure the level of significance of each process parameters.

2.3.3   Genetic Algorithm

Genetic Algorithm (GA) is on of the important method for optimization which gives the exact solution [10]. In this study Genetic Algorithm (GA) was also used to

Table 2 Process parameters and level

Parameters	Level		
	1	2	3
Load, L (N)	5	10	15
Sliding speed, S (m/s)	2	3	4
Reinforcement, R (%)	0.2	0.4	0.6

obtain feasible optimal solution for wear rate. The advantages of Genetic Algorithm (GA) are: (1) Coded versions of the process parameters are manipulated by GA. (2) GA works on a whole population of points. (3) GA is adopting better probability technique in easy and simple way. In this study GA was adopted to obtain the best and mean values of wear rate.

3 Results and Discussion

The Polyamide 6 nano-composites reinforced with multiwalled carbon nano tube (MWCNT) and various proportions of copper nano-particles were tested for wear behaviour.

3.1 Analysis of Signal to Noise Ratio

The purposes of the experiments are to find out the maximum possible S/N ratio for the process variables used in this study. A higher signal to noise ratio indicates that the signal is higher than the noise factors. The signal to noise ratio for all the 9 trials was calculated using Eq. (3). The experimental values of wear rate and the corresponding signal to noise ratio (S/N) for a given response are mentioned in Table 3. All the computations were performed and analysed using statistical analysis software MINITAB.

The mean S/N ratio for wear rate for each level of process variables are shown in the response tables [11]. The response table for wear rate shows the delta values and rank of the process variables. Rank 1, 2 and 3 is given in descending order of delta values. It is evident that, from Table 4, load has the highest delta value and assigned

Table 3 Experimental results and S/N ratios of wear rate

Exp. No.	Load (N)	Speed (m/s)	Reinforcement (%)	Wear rate (mm^3/m)	S/N ratio for wear rate	Wear coefficient
1	5	2	0.2	0.003995	47.970	0.1701
2	5	3	0.4	0.005954	44.504	0.2572
3	5	4	0.6	0.004942	45.521	0.2165
4	10	2	0.4	0.004981	46.054	0.1076
5	10	3	0.6	0.002991	50.484	0.0655
6	10	4	0.2	0.004995	46.030	0.1064
7	15	2	0.6	0.000984	60.140	0.0144
8	15	3	0.2	0.003995	47.970	0.0567
9	15	4	0.4	0.001649	55.655	0.0237

Table 4 Response table for wear rate

Level	Load (N)	Speed (m/s)	Reinforcement (%)
1	45.66	51.39	47.32
2	47.52	47.65	48.74
3	54.59	48.74	51.72
Delta	8.92	3.74	4.39
Rank	1	3	2

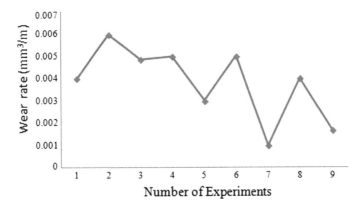

Fig. 2 Experimental runs with wear rate

rank 1. The Reinforcement % has the next highest delta value and assigned rank 2. Therefore load and % of reinforcement have higher impact on wear rate and speed has lower impact on the wear rate.

Figure 2 shows the measured wear rate for multi-walled carbon nanotube and copper nanoparticles reinfocred polyamaide 6 nanocomposites by orthogonal array. The seventh experiment provided the least wear rate 0.000984 mm³/m.

The influence of individual process parameters affecting the wear rate are shown in Fig. 3. The slope of each curve shown in the figure indicates the influence of process parameter on the output [12]. The noise level is low for better wear rate. Therefore, the signal to noise ratio is high with reduced wear rate. Figure 3 clearly shows that the *SN* ratio is increased with increased reinforcement % and thus the wear rate is reduced.

3.2 ANOVA

Table 5 shows the results of analysis of variance for wear rate. ANOVA was carried out on *SN* ratios to predict the significance of process parameters. From the table reinforcement % has the most significant value and it plays major role on wear rate.

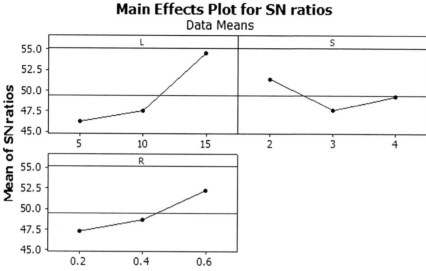

Fig. 3 Main effect plots for *SN* ratios

Table 5 Results of the analysis of variance

Source	DF	Seq SS	Adj SS	Adj MS	F	P
Load, *L* (N)	1	0.0000143	0.0000005	0.0000005	1.44	0.354
Speed, *S* (m/s)	1	0.0000011	0.0000008	0.0000008	2.52	0.254
Reinforcement, *R* (%)	1	0.0000016	0.0000020	0.0000020	6.02	0.134
L * *S*	1	0.0000032	0.0000000	0.0000000	0.01	0.928
L * *R*	1	0.0000030	0.0000042	0.0000042	12.88	0.070
S * *R*	1	0.0000013	0.0000013	0.0000013	3.94	0.186
Error	2	0.0000007	0.0000007	0.0000003		
Total	8	0.0000252				

3.3 Optimization Using Genetic Algorithm

In this study, genetic algorithm tool available in MATLAB was used to determine optimum condition for wear rate. The regression equation obtained (4) was used to form the fitness function (5). In the fitness function $x(1)$, $x(2)$ and $x(3)$ are indicating the process parameters [13].

$$\text{Regression equation: Wear} = 0.00674 - 0.000309 \text{ Load} + 0.000438 \text{ Speed}$$
$$- 0.00256 \text{ Reinforecement\%} \qquad (4)$$

Table 6 Parmeters for optimization in GA

Population type	Double vector	Stopping criteria: generation	300
Population size	50	Lower bound	5, 2, 0.2
Scaling function	Rank	Upper bound	15, 4, 0.6

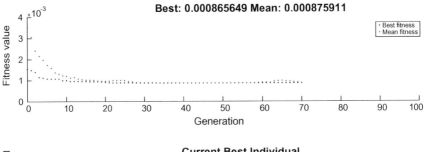

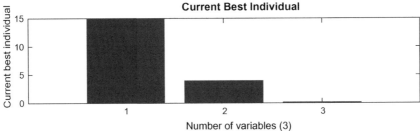

Fig. 4 Best and mean values of wear rate

Fitness function: $y = (0.00674 - 0.000309 * x(1) + 0.000348 * x(2) - 0.000256 * x(3))$

$$(5)$$

The following were the GA parameters considered for the optimization shown in Table 6.

From the results observed from genetic algorithm the optimum conditions for minimal wear rate were load 15 N, speed 4 m/s and reinforcement 0.6%. The objective function value calculated was 0.000865 mm³/m. The best and mean values of wear rate were shown in Fig. 4. The fitness value gets decreased through generations.

3.4 Scanning Electron Microscope

Scanning Electron Microscope is used to study the microstructure and wear mechanism experienced during dry sliding wear testing of MWCNT and copper

(a) **(b)** **(c)**

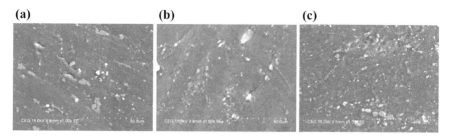

Fig. 5 PA6, 0.25 wt% MWCNT with **a** 0.2 wt% nano Cu, **b** 0.4 wt% nano Cu, **c** 0.6 wt% nano Cu

nano powder reinforced Polyamide composites. The surface of the composite pin has completely in contact with the steel disc during sliding and machine marks are also observed. The SEM images are shown in Fig 5. The worn surface of the samples shows that there are longitudinal grooves mainly present and small irregular pits are also observed. This shows that the PA6 nano composites were undergone certain abrasive wear and very small adhesive wear.

3.5 Validation Test

The validation test was conducted to confirm the optimal value obtained from Taguchi and genetic algorithm methods. The optimal conditions obtained from Taguchi method were load 15 N, speed 2 m/s and reinforcement % as 0.6 and from genetic algorithm method were load 15 N, speed 4 m/s and reinforcement 0.6%. The validation test was conducted for both the conditions and observed the results of 0.000932 and 0.000901 mm^3/m.

4 Conclusions

In this investigation the wear analysis using pin on disc experiment setup was carried out on Polyamide 6 and its composites. The wear analysis was performed on PA6 and its composites for 5, 10, and 15 N loads. The volume loss and wear rate were the parameters obtained from the experiment and the results were compared.

- Wear rate was minimised with increase in reinforcement %.
- Taguchi and Genetic algorithm were used for the optimization of PA6 composites and optimal conditions were validated.
- Worn surfaces were analysed using scanning electron microscopic images and abrasive wear was found on the composites.

- The optimal conditions for minimal wear rate were load 15 N, speed 4 m/s and reinforcement 0.6% obtained in GA and load 15 N, speed 2 m/s and reinforcement % as 0.6 obtained in Taguchi method.

Acknowledgements The authors are grateful to Central Institute of Plastics Engineering and Technology, Chennai and Anna University, Chennai for providing the facility to prepare and test the PA6 composites.

References

1. Palabiyik M, Bahadur S (2002) Tribological studies of Polyamide 6 and high density polyethylene blends filled with PTFE and copper oxide and reinforced with glass fibres. Wear 253:369–376
2. Wang S, Ge S, Zhang D (2009) Comparison of tribological behavior of nylon composites filled with zinc oxide particles and whiskers. Wear 266:248–254
3. Zhang Z (1997) Friction and wear properties of metal powder filled PTFE composites under oil lubricated conditions. Wear 210:151–156
4. Li J, Liang ZN (2010) Sliding wear performance of TiO_2/short carbon fiber/Polyamide 66 hybrid composites. Polym Plast Technol Eng 49:848–852
5. Bahadur S, Polineni VK (1996) Tribological studies of glass fabric-reinforced polyamide composites filled with CuO and PTFE. Wear 200:95–104
6. Bortoleto EM (2013) Experimental and numerical analysis of dry contact in the pin on disc. Wear
7. Kauzlarich JJ, Williams JA (2001) Archard wear and component geometry, J04100, IMechE (2001)
8. Palanikumar K, Karunamoorthy L (2006) Multiple performance optimization of machining parameters on the machining of GFRP composites using carbide (k10) tool. Taylor & Francis, Materials and Manufacturing Process (2006)
9. Siddiquee AN (2014) Optimization of deep drilling process parameters of AISI 321 steel using Taguchi method. Procedia Mater Sci
10. Sardinas RQ (2006) Genetic algorithm based multi objective optimization of cutting parameters in turning process. Eng Appl Artif Intell
11. Sivapragash M, Kumardhas P (2016) Taguchi based genetic approach for optimizing the PVD process parameter for coating ZrN on AZ91D magnesium alloy. Mater Design
12. Mathan kumar N, Senthil kumaran S (2016) Wear behaviour of Al 2618 alloy reinforced with Si_3N_4 and ZrB_2 in situ composites at elevated temperatures. Alexandria Eng J, Elsevier
13. Suresh PVS (2002) A genetic algorithm approach for optimization of surface roughness prediction model. Int J Mach Tools Manuf

Study of Mechanical and Wear Behaviour of Monotectoid Based Zinc–Aluminium Alloy

V. Sathya Prabu and S. Ilangovan

Abstract In this work, a monotectoid based Zinc–Aluminium (Zn–Al) alloy was prepared using stir casting process. The Micro-Hardness of as-cast and homogenized conditions was tested and the result showed that the homogenized alloy has a slight improvement in hardness than that of the as-cast alloy. The Microstructure of the Zn–Al revealed the dendritic structure in as-cast and non-dendritic in homogenized conditions which were analysed through a scanning electron microscope (SEM). Adhesive wear is carried out by Pin-on-Disc wear tester. Wear rate of the alloy rises with applied load conducted at different speeds. The specific wear rate and the friction coefficient vary with load. The diffusion of disc material into the specimen material changes the alloy composition. It was observed through energy dispersive X-ray spectroscopy (EDX) analysis. The worn-out surfaces were studied by using SEM analysis.

Keywords Zn–Al alloy · Homogenization · Wear rate · Microstructure and wear morphology

1 Introduction

Zn-Al alloys have been used as an alternative material for bronze, economically, in various engineering applications and for lower energy requirement for shaping and better strength. The properties are increased by interchanging its copper content with silicon and also by strontium [1]. Wear property of Zn–Al based alloys were compared by adding either copper or silicon with cast iron or phosphor bronze and identified that silicon addition was superior to other alloys [2]. It was noted that in aluminium alloy (Al–38Zn–2Cu) addition of copper increases the mechanical properties [3]. In Zn alloy addition of silicon it's up to 2% increases the mechanical

V. Sathya Prabu · S. Ilangovan (✉)
Department of Mechanical Engineering, Amrita School of Engineering,
Amrita Vishwa Vidyapeetham, Coimbatore, India
e-mail: s_ilangovan@cb.amrita.edu

© Springer Nature Singapore Pte Ltd. 2019   405
A. K. Lakshminarayanan et al. (eds.), *Advances in Materials and Metallurgy*,
Lecture Notes in Mechanical Engineering,
https://doi.org/10.1007/978-981-13-1780-4_39

properties [4]. The mechanical property of the alloy also has been improved by the surface refining method by addition of silicon content up to 16%, but in this case wear behaviour has been reduced [5]. Dimensional stability, microstructure and wear behaviour were analysed by addition of silicon and copper content in Zn–Al alloy. From this, Zn–Al–Si was best under both dimensional stability and wear property in high speed and load conditions [6]. The strength of Zn–Al–Si alloy was reduced after ageing. An alloy of reduced silicon and added copper has good mechanical properties, but lower density [7]. Zn alloys have been also tested under the static and dynamic condition and it was identified that the wear resistance was superior to SAE 660 bronze. Among the tested alloys monotectoid alloy was better than eutectoid alloys, but the copper content was affecting these monotectoid alloys [8]. Alloys have also been analysed by reinforcing with 10% of SiC particles of Zn alloys with varying sliding speed and applied pressure. Wear rate was found increasing with the pressure, but bronze shows better wear rate [9]. A newly developed leaded-tin bronze and standard zinc–aluminium alloy were compared with bronze. From this, it was understood that bronze had a higher wear rate than these alloys [10]. Zn alloy properties are also varied by replacing copper content by nickel, silicon and nickel plus silicon [11]. Variation of aluminium content was studied by swaging operation and identified that the properties have been improved [12]. Zn based alloy were studied under some modified conditions and it has been identified that the modified Zn-based alloys produced higher wear due to the frictional heating [13]. In this work, the effect of non-addition of copper and silicon on the mechanical, wear and the surface behaviour of the monotectoid based Zn–Al alloy is explored.

2 Experimental Work

2.1 Preparation of the Alloy

Zn–Al alloy was manufactured using bottom pouring type stir casting furnace. Stir casting was used for uniform distribution of material, reduction in porosity and cost reduction. The Zn–Al was added to the furnace of pure form as shown in Table 1. The furnace was kept at a temperature of 700 °C. The aluminium was added to the furnace first for melting and after sometime the zinc was added to melt as the melting temperature of Zn is lesser than aluminium. The stirring was done at 350 RPM with a mean time of 10 min. The die size of Ø 30 mm × 250 mm was preheated to a temperature of 250 °C. Argon gas was used to reduce the contamination of atmospheric gases. The molten metal is poured into the permanent metal mold and kept for cooling.

Table 1 Chemical composition of the alloy

Alloy no.	Alloy	Zinc (wt%)	Aluminium (wt%)
1	Zn–22.3Al	77.7	22.3

2.2 Homogenization Treatment

The casting is homogenized to obtain a uniform crystal structure. The furnace was preheated to a temperature of 400 °C and the cast rods are kept at a standing temperature and time of 350 °C and 8 h respectively. The process was kept in a nitrogen atmosphere and left for furnace cooling up to 50 °C.

2.3 Mechanical Testing

2.3.1 Micro-Hardness

Micro-Hardness testing is a method of determining a material's hardness when test samples of small regions are to be measured. Both as-cast and homogenization samples are checked. The pieces are polished with emery grade sheets 1/0 and 2/0. The Mitutoyo MVK—H11 Micro-Hardness tester was used for measuring Vickers micro-hardness. Five readings were taken at different locations with a load of 100 gf for time 15 s and the average hardness value was calculated.

2.3.2 Ultimate Tensile Strength and Percentage Elongation

The UTS of the specimen is measured by slowly extending it until it fractures. The % elongation of the gauge section is also noted for the applied force. The ultimate tensile strength (UTS) of the alloys is measured using Tinius Olson universal testing machine. The percentage elongation of each specimen is directly taken from the machine-generated plot. The sample dimensions are in gauge length 30 mm and diameter 6 mm.

2.4 Friction and Wear Testing

Adhesive wear experiments for specimens were analysed by using a pin on disc wear tester of DUCOM make (Fig. 1). The machine wear track was made of a steel disc (EN 31, HRC 60). The load was applied by dead weights in the rear of the machine. The maximum track diameter was 100 mm. The sliding velocity is adjusted by adjusting the RPM of the disc with track diameter constant. The disc can rotate at 2000 RPM at full speed. The specimen is prepared for the dimensions of 12 mm diameter and length of 15 mm. The parameters considered for the wear experiment are the velocity of 1, 1.5, 2 m/s and a load of 10, 15, 20 N respectively with a sliding distance of 600, 900, 1200 m. The time and tracking diameter are kept constant at 10 min and 100 mm respectively. The density was calculated using

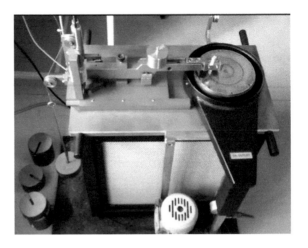

Fig. 1 Pin-on-disc wear tester

Archimedes method. From this parameter, the various experiments were performed and weight loss was calculated by measuring the initial and final weight of the wear specimens. The wear rate is defined as a volume by sliding distance and the specific wear rate is defined as wear rate by load. The COF indicates the friction with contact surfaces which was calculated using Winducom software.

3 Results and Discussion

3.1 Microstructure Analysis

The Microstructure was obtained by Scanning Electron Microscope. Prior to this, the piece used was polished and etched with Keller's solution. The microstructure of as-cast consists of the dendritic structure of aluminium (α) and zinc (β) which was not uniformly distributed as revealed in Fig. 2a. The Zn–22.3Al was kept under homogenization treatment for uniform microstructure as revealed in Fig. 2b.

3.2 Mechanical Testing

The mechanical analysis like Micro-Hardness, Ultimate Tensile Strength (UTS) and Percentage Elongation (%) was conducted for monotectoid based alloy Zn–22.3Al. The as-cast alloy exhibits lowest Micro-Hardness, UTS and highest % Elongation. After the homogenization treatment from Table 2 the Micro-Hardness, UTS of Zn–22.3Al was improved. In previous research work, the mechanical properties have been improved either by adding copper or silicon content. In this work, we have slightly improved the mechanical property of this monotectoid alloy by

(a)   (b)

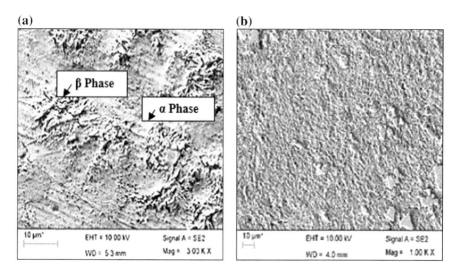

Fig. 2 Microstructure of **a** as-cast, **b** after homogenization treatment

Table 2 Density and mechanical properties of the alloy

Alloy	Micro-hardness (HV)	UTS (MPa)	% Elongation	Density (g/mm^3)
As-Cast	91	136	4.5	4.484
After Homogenization	95	145	4.1	4.484

homogenization treatment. Due to the furnace cooling during the treatment, a minor improvement was obtained. The mechanical properties can still increase by varying other heat treatment parameters.

3.3 Friction and Wear Behaviour

The volume loss is one of the significant parameters of the wear behaviour. So the weight loss is converted into volume loss. The volume loss was gradually increasing with respect to the sliding distance. With the volume loss, the wear rate is calculated and the coefficient of friction is identified using the Winducom software. The adhesive wear test was conducted in the Pin on Disc wear tester. The wear rate was gradually increasing with respect to applied load because the interface temperature between the pin and the disc increases with increase in load [14] conducted at different speeds as revealed Fig. 3. As the speed increases, the wear rate also increases gradually. Figure 6a–c represents the SEM micrographs of the minimum wear condition. The minimum wear rate was observed at minimum load and only scratches were visible on the surface, but further increase in load makes

Fig. 3 Load versus wear rate

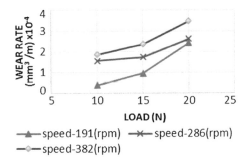

Fig. 4 Load versus specific
wear rate

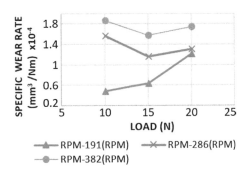

the scratches turns to the grooves. At high load the scratches and grooves were deep and peel of material was seen because of the maximum wear rate. The minimum wear rate was observed due to the smearing and embedding of oxidized Zn–Al alloy. The Zn–22.3Al alloy shows better wear resistance from the observed sliding distance and load conditions. The specific wear rate was calculated under varying load conditions and identified that when speed increases specific wear rate decreases initially up to 15 N load and then increases when the load was further increased as shown in Fig. 4. The friction coefficient as revealed in Fig. 5 has a sharp increase initially and gradually varies with respect to load and speed.

Fig. 5 Load versus
coefficient of friction

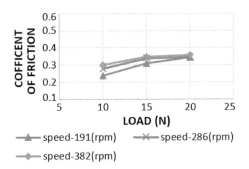

(a)   (b)   (c)

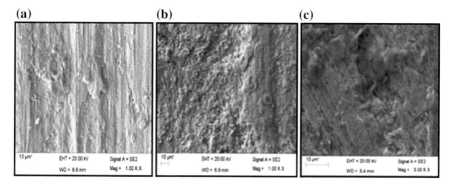

Fig. 6 SEM of wear specimen at load **a** 10 N, **b** 15 N, **c** 20 N

3.4 EDX Analysis

The EDX are analysed for the wear specimen. It was found that iron and carbon content was present in the wear sample. The diffusion of disc material into the specimen material changes the alloy composition which leads to variation in wear rate. Traces of oxygen content were also found due to the atmospheric contamination as revealed in Fig. 7.

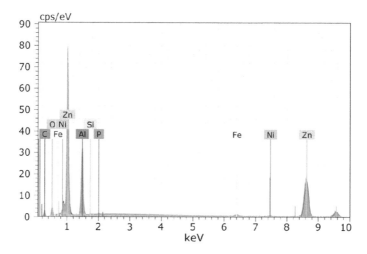

Fig. 7 EDX analysis of the Zn–22.3Al

4 Conclusions

- In the newly developed monotectoid based Zn–22.3Al alloy, the dendritic microstructure disappears after the homogenization treatment and fine grained microstructure was obtained.
- The mechanical properties like Micro-hardness, Ultimate Tensile Strength was improved and Percentage Elongation was reduced after the homogenization treatment.
- The wear rate of the Zn–22.3Al alloy increases with increase in load and speed. The adhesive and smearing are the wear mechanisms which were observed for Zn–22.3Al alloy. The specific wear rate decreases initially up to 15 N load and then increases when the load is further increased. The friction coefficient varies with an increase in load and speed.
- EDX analysis shows the diffusion of disc materials like carbon and iron in the specimen material which changes the alloy composition. Oxygen content was found due to the atmospheric contamination.
- The SEM analysis shows the surface morphology of the wear specimens conducted at various loads respectively.

References

1. Jian L, Laufer EE, Masounave J (1993) Wear in Zn–Al–Si alloys. Wear 165:51–56
2. Murphy S, Savaskan T (1984) Comparative wear behaviour of Zn–Al-based alloys in an automotive engine application. Wear 98:151–161
3. Ilangovan S, Arul S, Shanmugasundaram A (2016) Effect of Zn and Cu content on microstructure, hardness and tribological properties of cast Al–Zn–Cu alloys. Int J Eng Res Africa. 27:1–10
4. Raj SRK, Ilangovan S, Arul S, Shanmugasundaram A (2015) Effect of variation in Al/Si content on mechanical properties of Zn–Al–Si alloys. Int J Appl Eng Res. 10:2723–2731
5. Saravanan R, Sellamuth R (2014) Determination of the effect of Si content on microstructure hardness and wear rate of surface-refined Al–Si alloys. Procedia Eng 97:1348–1354
6. Lee PP, Savaskan T, Laufer E (1987) Wear resistance and microstructure of Zn–A–Si and Zn–Al–Cu alloys. Wear
7. Savaskan T, Murphy S (1987) Mechanical properties and lubricated wear of Zn–25Al-based alloys. Wear
8. Savaşkan T, Pürçek G, Murphy S (2002) Sliding wear of cast zinc-based alloy bearings under static and dynamic loading conditions. Wear 252:693–703
9. Prasad BK (2003) Influence of some material and experimental parameters on the sliding wear behaviour of a zinc-based alloy, its composite and a bronze. Wear 254:35–46
10. Prasad BK, Patwardhan AK, Yegneswaran AH (1996) Dry sliding wear characteristics of some zinc-aluminium alloys: a comparative study with a conventional bearing bronze at a slow speed. Wear
11. Prasad BK (1998) Tensile properties of some zinc-based alloys comprising 27.5% Al: effects of alloy microstructure, composition and test conditions. Mater Sci Eng A245:257–266

12. Modi OP, Yadav RP, Prasad BK, Jha AK, Dasgupta R, Dixit G (1998) Effects of swaging and aluminium content on the microstructure and mechanical and sliding wear properties of Zinc based alloys. Mater Trans 39:582–586
13. Prasad BK, Patwardhan AK, Yegneswaran AH (1997) Dry sliding wear response of a modified Zinc-based alloy. Mater Trans 38:197–204
14. Radhika N, Subramanian R, Venkat Prasat S, Anandavel B (2012) Dry sliding wear behaviour of aluminium/alumina/graphite hybrid metal matrix composites. Ind Lubr Tribol 64:359–366

Evaluation of Mechanical and Thermal Behaviour of Particle-Reinforced Metal Matrix Composite Using Representative Volume Element Approach

P. Vignesh, R. Krishna Kumar and M. Ramu

Abstract In this present work, material properties (Young's modulus and thermal conductivity) of Silicon Carbide (SiC) particle-reinforced Aluminium 1100 (Al) matrix composite have been evaluated using Representative Volume Element (RVE) technique. A three-dimensional RVE model is generated by finite element-based homogenization method for Al–SiC composite. SiC particles are randomly dispersed in various volume portions in the range of 5–20% into the matrix. Examinations are done to ponder the impact of spherical particle geometry on the material properties for all volume fractions using numerical analysis. The numerical results from the RVE approach are compared with the values obtained from the experimental results, rule of mixtures (ROM) and Halpin–Tsai equations and found that the results of RVE approach are in good agreement with the other approaches.

Keywords Particle-reinforced composites · FEM · Representative volume element · Mechanical and thermal properties

1 Introduction

A composite material is produced using a synergistic mix of at least two constituents that vary in physical form and chemical composition such that the subsequent material has prevalent properties than conventional materials [1]. In this current industrial scenario, the need for lightweight and high-performance material has increased. Therefore, MMCs have emerged as the strongest contenders against their monolithic alloys and are being used in aerospace, automotive and marine industries [2]. Among different types of MMCs, particle-reinforced metal matrix

P. Vignesh (✉) · R. Krishna Kumar · M. Ramu
Department of Mechanical Engineering, Amrita School of Engineering,
Coimbatore, Amrita Vishwa Vidyapeetham, Coimbatore, India
e-mail: vigneshp508@gmail.com

© Springer Nature Singapore Pte Ltd. 2019
A. K. Lakshminarayanan et al. (eds.), *Advances in Materials and Metallurgy*,
Lecture Notes in Mechanical Engineering,
https://doi.org/10.1007/978-981-13-1780-4_40

composites have turned out to be extremely popular because they are more affordable, and they have moderately isotropic properties than persistent fibre–matrix composites [3].

Generally, the properties of composites are evaluated by experimental methods which are very difficult and tedious process. Rule of mixtures (ROM) is one of the widely used analytical relations for deciding the properties of a composite by considering the volume content and the material property of each constituent. However, ROM does not consider the particle shape, size, orientation and interfacial bonding which leads to the unsatisfactory solution [4]. Another analytical formulation developed by Halpin and Tsai is a self-consistent method which exactly correlates with the elastic constants confirmed by experimental measurements. HT equations consider volume fraction and particle geometry in calculating the elastic modulus [5].

On the other hand, finite element-based homogenization method is being commonly used in evaluating the material properties and behaviour by considering a RVE or unit cell of the whole composite [6]. The characterization of RVE as a specimen of heterogeneous material was defined by Rodney hill, which is typically the average of the whole mixture and contains enough number of inclusions of the evidence material properties to be autonomous of the surface estimations of displacement and traction, so far as the values are infinitesimally uniform [7]. A 3D RVE model was recognized by Srivastava et al. [8] to explore the effect of volume fraction, molecule dissemination and perspective proportion utilizing finite element-built homogenization system should foresee the mechanical conduct technique for silicon carbide fortified aluminium MMC to separate sizes. Examination on the deformation behaviour of arbitrarily situated stubbles of aluminoborate strengthened magnesium matrix composites was directed by Lee et al. [9] utilizing limited component investigation and arbitrary successive assimilation to foresee Young's modulus and stress–strain response on the 3D RVE model of the composite for various RVE lengths.

The objective of this research work is to predict the Young's modulus and thermal conductivity of silicon carbide-reinforced aluminium (1100) matrix composite by RVE approach. A 3D cubical RVE of the composite is generated using a recent nonlinear multiscale material modelling software DIGIMAT. ANSYS is used as a background FE solver to run the analysis of the generated RVE. Spherical silicon carbide particles are randomly dispersed inside the matrix for volume fractions ranging from 5 to 20%. The numerical simulation results from RVE approach are compared with experimental results, ROM and Halpin–Tsai equations.

2 Methodology

2.1 Experimental Work

The composite was fabricated by taking aluminium (1100) alloy in the matrix part which has a chemical composition of (Al—99%, Fe—0.5%, Si—0.35%,

Cu—0.1%, Mn—0.05%) from the spectroscopy test and silicon carbide in the reinforcement phase. As Al (1100) belongs to the commercially pure category, it is broadly utilized in chemical and food processing ventures for food packaging foils and trays, chemical processing equipment, and furthermore in sheet metal work. Silicon carbide particles are mixed in four volume fractions 5, 10, 15 and 20% with aluminium (1100) and fabricated by squeeze casting method. Fine silicon carbide powder particles in the range below 10 μm available in the commercial metal marts were used. Four cylindrical specimens of height 140 mm and diameter 50 mm of the composite are produced. Squeeze cast specimens were machined to produce tensile test specimen according to ASTM E8 M-04 standard dimensions. The Young's modulus was calculated from the stress–strain plot obtained from the tensile test data. Thermal conductivity was evaluated by the flash method which is popularly known as laser flash analysis (LFA). The experiment was conducted using LFA467 Hyperflash apparatus from NETZSCH technologies for a standard specimen of diameter 25 mm and thickness 4 mm of the composite. Thermal diffusivity and specific heat of the sample was obtained as outputs from which the thermal conductivity is measured.

2.2 Rule of Mixtures (ROM)

The rule of mixtures (ROM) is a simple relation between the properties of composites and those of its constituents. In this method, the parameters considered are the volume portion of each constituent in the composite and the properties like Young's modulus and thermal conductivity for a respective constituent. The properties of the composites achieved should be in the range characterized by ROM [10]. Equations (1) and (2) shows the relation to calculate upper and lower bound values of Young's modulus of the composite. Similarly, Eqs. (3) and (4) are used to calculate upper and lower limit values of thermal conductivity of the composite.

For Young's modulus,

$$\text{Upper bound is represented by} : E_c(u) = E_m V_m + E_p V_p \tag{1}$$

$$\text{Lower bound is represented by} : E_c(l) = E_m E_p / V_m E_p + V_p E_m \tag{2}$$

For thermal conductivity,

$$\text{Upper bound is represented by} : k_c(u) = k_m V_m + k_p V_p \tag{3}$$

$$\text{Lower bound is represented by} : k_c(l) = k_m k_p / V_m k_p + V_p k_m \tag{4}$$

where

E_c, k_c Young's modulus and thermal conductivity of composite (N/mm^2, W/mK),

E_m, k_m Young's modulus and thermal conductivity of matrix (N/mm^2, W/mK),

E_p, k_p Young's modulus and thermal conductivity of reinforcement (N/mm^2, W/mK), and

V_m, V_p Volume fraction of matrix and reinforcements in the composite (%).

2.3 Halpin–Tsai (HT) Equations

It is a semi-empirical model for evaluation of elastic constants of the composite material by taking into account the geometry and orientation of the filler and adaptable properties of filler and matrix. According to Halpin and Tsai, the elastic modulus of a composite can be determined from Eq. (5) and geometry factor η is evaluated by Eq. (6). The reinforcing factor ζ is used to describe the influence of the geometry of the reinforcing phase on the elastic modulus. In this research, the reinforcements are considered as spherical particles and the formula for calculating ζ is given in Eq. (7) [11].

$$E_c = E_m \left(\frac{1 + \zeta \eta f}{1 - \eta f} \right) \tag{5}$$

$$\eta = \frac{\left(\frac{E_f}{E_m} \right) - 1}{\left(\frac{E_f}{E_m} \right) + \zeta} \tag{6}$$

$$\zeta = 2 + 40 f^{10} \tag{7}$$

2.4 RVE Approach

The numerical simulation is carried out using DIGIMAT-FE which is the finite element-based homogenization module from DIGIMAT. It is used to generate the stochastic RVEs that gives a measurement which yield a representative value of the whole element and ANSYS software is used as FE solver and analysis is carried out in five steps as given below:

2.4.1 Defining Material Properties

The material properties for matrix and reinforcements given in Table 1 were used for the entire analysis. An elastic model with isotropic symmetry is considered for mechanical analysis and a Fourier model with isotropic symmetry is chosen for evaluating thermal conductivity for both matrix and reinforcement.

2.4.2 Microstructure Information

Aluminium is defined in the matrix phase and silicon carbide particles in the inclusion phase whose interface behaviour is assumed as perfectly bonded. Silicon carbide is taken in terms of volume fraction for four proportions 5, 10, 15 and 20%. The silicon carbide particles are randomly dispersed inside the matrix with an average size of 5–10 μm and aspect ratio 1. The aluminium matrix is added with enough number of inclusions depending upon the volume fractions, and the silicon carbide particles' geometry is considered as spherical in nature.

2.4.3 RVE Generation and Meshing

A 3D finite element model which is the RVE of the whole composite is generated based on the material input and microstructure information. The size of the RVE is defined as per the formula given in Eq. (8) and the maximum RVE size is taken as $20 \times 20 \times 20 \ \mu m^3$ for 20% volume fraction of silicon carbide.

$$v_f = \frac{\sum_{j=1}^{n} v_j}{L^3} \tag{8}$$

where

v_f volume fraction of reinforcement (%),
v_j volume of the inclusion (μm^3), and
L^3 size of the RVE (μm^3).

The generated RVE is meshed using inbuilt DIGIMAT-FE preprocessor. The RVE is discretized into conforming tetrahedral elements. The number of elements varies from 31,216 to 55,484 for volume fractions ranging from 5 to 20%.

Table 1 Material properties of the constituents

Material	Young's modulus (GPa)	Thermal conductivity (W/mK)
Aluminium 1100	68.9	218
Silicon carbide	410	120

2.4.4 Applying Boundary Conditions

Periodic boundary conditions are imposed to ensure that the flux of the field variable displacement and the temperature is periodic with respect to all the faces of the element. This is authorized through the substantial arrangement of conditions relating the degrees of freedom of the nodes lying on one face with those of the comparing nodes lying on the contrary face on a 2-by-2 premise. Mechanical and thermal loadings are applied for evaluating Young's modulus and thermal conductivity, respectively. For example, maximum uniaxial peak strain ε_{11} is applied in the case of mechanical loading as shown in Fig. 1.

2.4.5 RVE Analysis

Once the boundary conditions and loading are applied, the RVE undergoes a postprocessing which is done by ANSYS in the background. For evaluating Young's modulus, the RVE is subjected to static structural analysis and for determining thermal conductivity, thermal analysis for the generated RVE is performed.

3 Results and Discussion

3.1 Experimental Results

The tensile test conducted on the squeeze cast Al–SiC composite shows an enhancement in the mechanical properties than conventional material. The Young's modulus and thermal conductivity obtained from experimental tensile test and flash method of the composite for all volume fractions is given in Table 2.

Fig. 1 Periodic boundary condition for mechanical loading

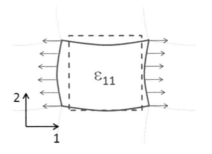

Table 2 Young's modulus and thermal conductivity from experimental tests

Volume fraction (%)	Young's modulus (GPa)	Thermal conductivity (W/mK)
5	74	209
10	80	204
15	87	198
20	95	194

3.2 Rule of Mixtures (ROM) Results

The upper bound and lower bound limit of Young's modulus and thermal conductivity of the composite found using ROM relation explained in Sect. 2.2 are given in Table 3.

3.3 Halpin–Tsai Model Results

By considering the geometry factor and reinforcing factor ζ, the values of elastic constants of the Al–SiC composite obtained from the Halpin–Tsai equations are given in Table 4. As the silicon carbide particles are in spherical shape, the value of ζ calculated from Eq. (7) is approximately 2 for all volume fractions of SiC.

Table 3 Young's modulus and thermal conductivity from rule of mixtures

Volume fraction (%)	Young's modulus (GPa)		Thermal conductivity (W/mK)	
	Lower limit	Upper limit	Lower limit	Upper limit
5	71.89	85.95	209.44	213.10
10	75.15	103.01	201.54	208.20
15	78.72	120.06	194.20	203.30
20	82.65	137.12	187.39	198.40

Table 4 Young's modulus from Halpin–Tsai equations

Volume fraction (%)	Young's modulus (GPa)
5	75.54
10	82.62
15	90.19
20	98.39

3.4   Results from RVE Approach

The numerical simulation carried out for a three-dimensional cubical RVE composed of Al–SiC composite is generated using DIGIMAT-FE. The same 3D cubical
RVE is used for both mechanical and thermal analyses. The RVEs for different
volume fractions are shown in Fig. 2 in which red colour spherical particles rep-
resent the randomly distributed silicon carbide inside the aluminium matrix. The
average number of inclusions of silicon carbide varies from 5 to 45 according to the
volume fraction of SiC. The meshed model of the generated RVE is shown in
Fig. 3. Static structural analysis and steady-state thermal analysis is performed with
the meshed models of RVE using ANSYS in the background solver. Equivalent
stress distribution of the RVEs for all volume fractions is shown in Fig. 4, and the
total temperature distributions of the RVEs for all volume fractions are shown in
Fig. 5. Postprocessing of the results is carried out in DIGIMAT-FE with the
automatic evaluation of properties for the composite, and the values of effective
Young's modulus and thermal conductivity obtained from RVE approach are
tabulated in Table 5. The Young's modulus and thermal conductivity for different
volume fractions obtained from all the approaches are compared with results from
RVE approach which is shown in Fig. 6.

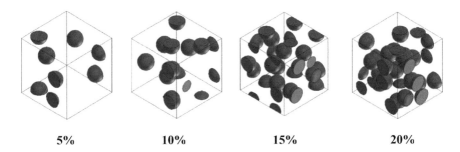

| 5% | 10% | 15% | 20% |

Fig. 2 RVEs for different volume fractions of silicon carbide

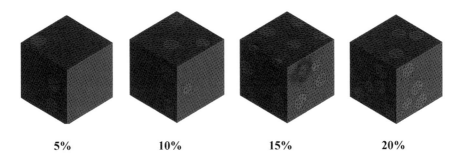

| 5% | 10% | 15% | 20% |

Fig. 3 Meshed models of RVE for different volume fractions of silicon carbide

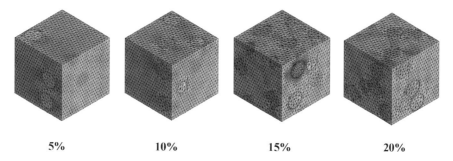

| 5% | 10% | 15% | 20% |

Fig. 4 Equivalent stress distributions of the RVE for 5, 10, 15 and 20% volume fractions

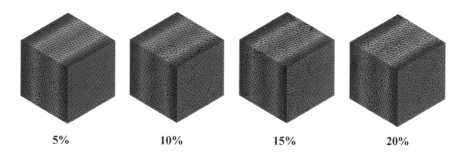

| 5% | 10% | 15% | 20% |

Fig. 5 Total temperature distributions of the RVE for 5, 10, 15 and 20% volume fractions

Table 5 Effective Young's modulus and thermal conductivity obtained from RVE approach

Volume fraction (%)	Young's modulus (GPa)	Thermal conductivity (W/mK)
5	78.64	212.68
10	84.39	207.58
15	92.61	202.44
20	99.52	199.37

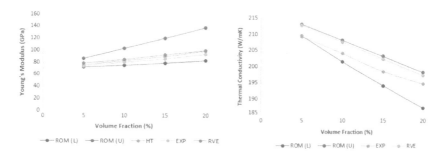

Fig. 6 Comparison of material properties evaluated in different approaches

It is seen from Fig. 6 that Young's modulus of the silicon carbide-reinforced aluminium matrix composite increases on increasing the volume percentage of silicon carbide in the composite. But the thermal conductivity decreases because the increase in volume content of SiC particles may cause poor adhesion of dispersion in the matrix, which decreases the interfacial thermal conductance and the effective thermal conductivity of the composite. It is also due to higher magnitude of thermal conductivity in the matrix than reinforcement part.

4 Conclusions

The Young's modulus and thermal conductivity obtained from the finite element analysis of the generated RVE for different volume fractions are compared with the experimental results, ROM and HT equations. The RVE approach results have a deviation of $\pm 5\%$ from the experimental values since some of the external factors like temperature, pressure, humidity, skilled labour and type of manufacturing method cannot be controlled practically. On comparing with ROM, Young's modulus and thermal conductivity determined from numerical simulation lie within the upper and lower limits for all volume fractions. The elastic modulus obtained from the Halpin–Tsai equations also stays within $\pm 2\%$ tolerance limit with the RVE technique results. Therefore, the results obtained from the RVE approach are in good correlation with other methods which proves that RVE method is a suitable technique for evaluating the properties of new materials.

References

1. Radhika N, Raghu R (2015) Mechanical and tribological properties of functionally graded aluminium/zirconia metal matrix composite synthesized by centrifugal casting. Int J Mater Res 106:1174–1181
2. Raviteja T, Radhika N, Raghu R (2014) Fabrication and mechanical properties of stir cast Al–Si$_{12}$Cu/B$_4$C composites. Int J Res Eng Technol 3(7):343–346
3. Somasundara, Vinoth, K (2013) Studies on mechanical and tribological behaviour of particulate aluminium metal matrix composites. Ph.D. thesis, Anna University, Chennai, India
4. Pal B, Haseebuddin MR (2012) Analytical estimation of elastic properties of polypropylene fiber matrix composite by finite element analysis. AMPC 2:23–30
5. Alfonso I, Figueroa IA, Rodriguez-Iglesias V, Patino-Carachure C, Medina-Flores A, Bejar L, Perez L (2016) Estimation of elastic moduli of particulate-reinforced composites using finite element and modified Halpin-Tsai models. J Braz Soc Mech Sci Eng 38:1317–1324
6. Tian W, Qi L, Zhou J, Liang J, Ma Y (2015) Representative volume element for composites reinforced by spatially randomly distributed discontinuous fibers and its applications. Compos Struct 131:366–373
7. Hill R (1963) Elastic properties of reinforced solids: some theoretical principles. J Mech Phys Solids 11:357–372

8. Srivastava VK, Gabbert U, Berger H, Singh S (2011) Analysis of particles loaded fiber composites for the evaluation of effective material properties with the variation of shape and size. Int J Eng Sci Technol 3(1):52–68
9. Lee WJ, Son JH, Park IM, Oak JJ, Kimura H, Park YH (2010) Analysis of 3D random $Al_{18}B_4O_{33}$ Whisker reinforced Mg composite using FEM and random sequential absorption. Mater Trans 51(6):1089–1093
10. Callister WD Jr. (2007) Materials science and engineering, 7th edn. Wiley, Hoboken. ISBN: 10 0-471-73696-1
11. Michigan Tech. http://www.mse.mtu.edu/~drjohn/my4150/ht/ht.html

Understanding the Effect of Tool Rotational Speed on Microstructure and Mechanical Properties of Friction Stir Processed ZE41 Grade Magnesium Alloy

Radhika Koganti, A. K. Lakshminarayanan and T. Ramprabhu

Abstract Friction stir processing (FSP) is a novel metal processing technique which is used for improving the material properties locally through significant grain refinement and homogenization. In FSP, processing parameter, namely, tool rotational speed plays an important role in producing a defect-free processed zone and enhancing the properties. In this work, FSP of ZE41 magnesium alloy was conducted at different tool rotational speeds (450, 650, 850, 1050, and 1250 rpm) with constant tool traversing speed of 50 mm/min. It is observed that FSP of magnesium alloy at rotational speed of 650 rpm produced superior mechanical properties as compared to other rotational speeds. This is mainly due to the optimum heat input conditions, grain refinement, and favorable distribution of second phase particles throughout the processed zone.

Keywords Friction stir processing · Magnesium alloy · Tool rotational speed · Microstructure · Mechanical properties

1 Introduction

Magnesium alloys are invariably gaining importance as lightest structural material for functional parts in automotive and aerospace industries due to their low density, high specific gravity, and recyclability [1]. In recent times, new magnesium alloys have been evolved with rare earth (RE) additions in order to improve their properties for particular applications [2]. ZE41 alloy is a rare earth cast magnesium

R. Koganti (✉) · A. K. Lakshminarayanan
Department of Mechanical Engineering, SSN College of Engineering,
Chennai 603110, India
e-mail: k2uradhika@gmail.com

T. Ramprabhu
Manufacturing and Materials, Defence Research & Development
Organization, Bangalore, India

© Springer Nature Singapore Pte Ltd. 2019
A. K. Lakshminarayanan et al. (eds.), *Advances in Materials and Metallurgy*,
Lecture Notes in Mechanical Engineering,
https://doi.org/10.1007/978-981-13-1780-4_41

alloy with enhanced mechanical properties compared to conventional magnesium alloys. Controlled additions of zinc and zirconium in magnesium alloys result in increased hardness and castability. This is due to the formation of rare earth precipitates within grains and at the grain boundaries [3]. However, magnesium alloys generally exhibit low ductility and formability at room temperature due to their hexagonal close-packed crystal structure and limited number of independent slip systems [4]. It is well established that grain refinement can significantly enhance the strength and ductility of metallic alloys [5]. Various techniques such as grain refiners, rapid solidification, recrystallization, and severe plastic deformation (SPD) [6, 7] have been identified as effective in modifying the grain size. Friction stir processing (FSP), which is based on the principle of friction stir welding (FSW) is an emerging solid-state processing technique for microstructure modification. Several studies have been carried out to understand the effect of FSP on microstructure and mechanical properties of conventional Mg alloys such as AZ91 [8], AZ31 [9], and Mg–RE alloys. Feng et al. reported that FSP on AZ91 casting has resulted in dissolution of coarse eutectic β–$Mg_{17}Al_{12}$ and significant grain refinement of 15 μm. Also, it was concluded that the tensile properties particularly ductility of the FSPed alloy have been significantly improved [8]. In similar study, Darras et al. have analyzed the effect of friction stir processing on the microstructure of AZ31 and reported that FSP resulted in more equiaxed and homogenized grain structure. The author also reported that the rotational speed of the tool has influenced the hardness of the material. This is due to the increase in the temperature with rotational speed which usually softens the grain growth [10]. Ramesh babu et al. investigated the effect of tool shoulder diameter on the friction stir processing of AZ31B alloy. Tool shoulder diameters of 18 and 24 mm were used for different plate thicknesses. The results concluded that FSP has refined and homogenized the grain structure irrespective of tool shoulder diameter. The results also indicated that various factors like tool axial force, tool rotational speed, and tool traversing speed play important role in eliminating the defects in the processed zone [11]. In past few years, many researchers have reported the effect of FSP on AZ series of magnesium alloys. However, no study is reported to evaluate the effect of tool rotational speed on microstructure and mechanical properties of friction stir processed ZE41 alloy.

2 Experimental Work

Commercially available as-cast ZE41 magnesium alloy was used for this investigation. The chemical composition and the mechanical properties of alloy are given in Tables 1 and 2. Plates of dimensions 100 mm × 80 mm × 6 mm were machined from the base metal specimens by wire electrical discharge machining (EDM).

Table 1 Chemical composition (wt%) of ZE41 magnesium alloy

Material	Zn	Ce	Zr	Cu	Al	Mn	Fe	Si	Ni	Mg
ZE41	3.85	1.27	0.53	0.002	0.006	0.008	0.004	0.003	0.002	Balance

Table 2 Mechanical properties of as-cast ZE41 magnesium alloy (measured)

Material	Grain size (μm)	Microhardness (HV)	Tensile strength (Mpa)	Yield strength (Mpa)	Elongation (%)
ZE41	60	56	210	150	5

FSP was performed in a vertical milling machine which allowed the control of tool rotational and traverse speeds. An HSS tool with tapered pin (6 mm minor dia and 4 mm major dia) and a shoulder diameter of 16 mm was adopted. The FSP tool was tilted by 2.5° from the workpiece for effective stirring action and to prevent the formation of flash, cavities. Friction stir processing was conducted on the machined plates (Fig. 1) at five different tool rotational speeds (450, 650, 850, 1050, and 1250 rpm), whereas the tool traversing speed is kept constant at 50 mm/min. FSP was carried out with constant plunge depth mode and along the longitudinal direction of the specimen. The principal directions of FSP are processed direction (PD), transverse direction (TD), and normal direction (ND).

The microstructures of as-cast, friction stir processed samples were examined under the optical microscope. FSP processed samples were taken along the normal direction of the processed area. All the samples were grinded, polished, and etched with a solution consisting of 4.2 g picric acid, 10 ml acetic acid, 10 ml distilled water, and 70 ml ethanol. The average grain sizes of specimens were measured from micrographs using mean linear intercept method. The hardness of the processed samples was measured at the stir zone using Vickers hardness tester (Vickers EN ISO 6507). A load of 500 gf with 20 s loading cycle was applied. Each tested area, an average of three readings were considered. FSPed specimens were cut in the processed direction and then machined to the micro-sized tensile specimens with gauge length of 10 mm and gauge width of 3 mm. Tensile tests were performed on universal testing machine with a strain rate of 10^{-4} s^{-1}. The tests were repeated with three samples to evaluate the average property value.

Fig. 1 Friction stir processed plate at different speeds

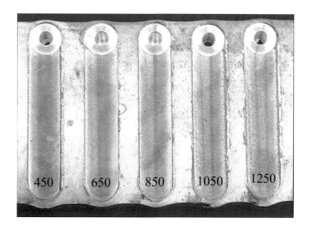

3   Results and Discussion

3.1   Microstructure

Figure 2 shows the typical optical image of microstructure of as-cast ZE41 magnesium alloy. The microstructure consists of α-Mg matrix with eutectically distributed precipitates (T-Phase: Mg_7Zn_3 RE) at the grain boundaries. Also, a small amount of fine particles were identified within the localized matrix. It is evident that these are zirconium-rich particles (Zr_4Zn_3 or Zn_2Zr_3) with an average grain size of 4 μm [12]. The Mg–Zr binary diagram shows the existence of α-Zr particles with α-Mg in Mg matrix below 653.56 °C [13]. Hence, these are the particles remained undissolved in base metal (BM). The grain boundaries are outlined by the distribution of secondary phase particles since they contain higher atomic number elements with brighter contrast. Neil et al. reported similar observation that the SEM image of ZE41 alloy shows the T-phase with white-colored and a light-colored zone (Zr-rich particles) within the Mg matrix [14]. Higher magnifications of SEM revealed that the Zr-rich particles are protruded into the matrix.

The macrostructural and microstructural images of friction stir processed samples at five different tool rotational speeds, and their corresponding mechanical properties were presented in Table 3. The macrostructures of all the samples show a basin-shaped nugget (Fig. 3) with three distinct regions, namely, stir zone (SZ), transition zone (TZ), and crown zone (CZ) which are evident in Mg alloys [15]. FSP has produced more homogenous, equiaxed fine grains in SZ compared to CZ and TZ. Complete dynamic recrystallization (DRX) of grains happens in SZ due to severe plastic deformation and frictional heat input by the tool pin. The low stacking fault energy of Mg alloys and their high gain boundary diffusion rate becomes favorable for dynamic recrystallization. The rain sizes are increased in the CZ due to the variation in frictional heat input. TZ is the region close to SZ where

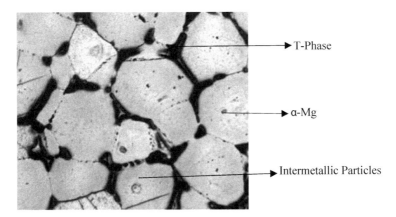

Fig. 2   Optical micrograph of ZE41 alloy

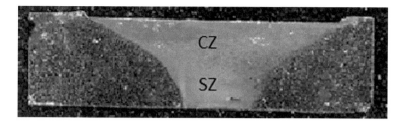

Fig. 3 Macrostructure of FSPed sample at 450 rpm speed

part of grains are subjected to continuous dynamic recrystallization, and original grain boundaries become bulging. TZ shows a clear transition of microstructure from coarser grains of BM to finer grains in SZ.

The microstructure of FSP sample with rotational speed of 450 rpm shows coarser grain bands of average size 37 μm and fine grains in a range of 4–6 μm. The grain size has reduced compared to as-cast alloy (60 μm). The grain boundary precipitates are not completely dissolved. There is a longitudinal increase in the grain boundaries due to low temperature at the SZ. As the speed increases to 650 rpm, the microstructure clearly shows fine, equiaxial grains throughout the SZ. Large amount of eutectially distributed secondary phase particles along the grain boundaries are completely dissolved into the matrix. Higher rotational speed intensifies the frictional heat and stirring action resulting in thorough mixing of material. At rotational speed of 850 rpm, there is an increase in the average grain size compared to that of 650 rpm. Narrow coarser grain bands aligned in discontinuous manner are visible in entire SZ. At higher heat inputs, the material softens and flows along the direction of processing even before recrystallization. Moreover, the secondary phase particles are broken up and partly dissolved into α-matrix. A microstructure with bimodal band, i.e., Mixture of coarser grains of average size 12 μm and finer grains of average size 4–6 μm, is observed clearly when the speed increases to 1050 rpm. With increase in the rotational speed, the grain size decreases first, and then increases. This is due to the increase in the peak temperature of SZ.

3.2 Mechanical Properties

Tensile properties such as ultimate tensile strength (UTS), yield strength (YS), % elongation, and hardness values of friction stir processed samples at different speeds are presented in Table 4. The base metal shows a UTS of 210 Mpa and 5% elongation. FSPed sample with 450 rpm speed shows an increase in tensile strength and elongation. During FSP, higher rotation speed increases the temperature of the metal, resulting in higher heat input, more stirring action, and grain refinement. Segmentation of coarser grains and the partial dissolution of secondary phases alter

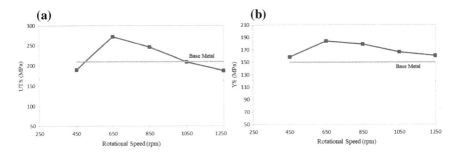

Fig. 4 Variation of **a** UTS and **b** YS with rotational speed

the fracture mode of BM from brittle to dimple fracture, which results in an increase in the ductility of the sample.

At 650 rpm, the sample with uniform, homogenized, and refined grain structure attained the highest strengths and elongation. Compared to BM, the average grain size of the sample has decreased to 1.6–2.8 µm. Complete dissolution of coarser grain boundary networks increases the precipitation strengthening of samples. Also, FSP eliminates casting porosities which will effect strength and ductility of sample. Figure 4 shows the variation of UTS and YS with increase in rotation speed. When the speed increases above 650 rpm, tensile strength of samples reduced. At higher speeds, SZ reaches maximum temperature resulting in higher heat input conditions and excessive grain growth. Therefore, samples exhibit lower tensile strengths and elongation. Variation of elongation with increase in rotational speed is represented in Fig. 5b. The bimodal structure at 1050 rpm results in softening of coarser grains which dominates the strain hardening provided by fine grain structure of SZ. This deteriorates the ductility of the sample.

The results from the Vickers hardness test clearly show that the hardness of the processed material is sensitive to rotational speed. Figure 5a shows the effect of rotational speed on hardness where it is observed that hardness increases initially with speed and then decreases with increase in speed. Higher hardness value (61 Hv) than the base metal (58 Hv) is attained at an optimum rotation speed of 650 rpm. With increase in rotational speed, the grain size in SZ decreased. It is evident that at higher speeds, increased temperature leads to softening due to grain growth. This is in agreement with Hall–Petch behavior of Mg alloys [10], where hardness is inversely proportional to grain size.

Table 3 Macrostructures and Microstructures of FSPed ZE41 alloy at different rotational speeds

S. No	Rotational speed (rpm)	Macrostructure	Microstructure	Grain size (μm)
1	450			Coarser grains of average size 37 μm and fine grains of 4–6 μm
2	650			
			Finer grains of 1.6–2.8 μm	
3	850			
			Average grain size of 8–10 μm	
4	1050			
			Coarser grains of average size 12 μm fine grains of 3–6 μm	
5	1250			
			Grain size of 5–7 μm	

Table 4 Mechanical properties of FSPed ZE41 alloy at different rotational speeds

S. No	Rotational speed (rpm)	Tensile properties			Hardness (HV)
		UTS (MPa)	YS (MPa)	Elongation (%)	
1	450	190	158	5.28	58
2	650	272	184	8.32	63
3	850	246	179	6.05	61
4	1050	209	166	4.91	54
5	1250	187	160	4.55	52

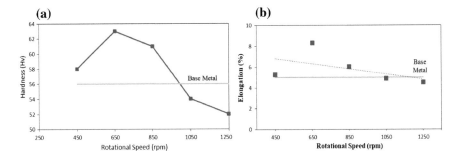

Fig. 5 Variation of **a** Hardness and **b** Elongation with rotational speed

4 Conclusions

- FSP on as-cast ZE41 alloy produced more uniform and fine grain stir zone microstructure at a tool rotation speed of 650 rpm compared to other speeds.
- FSP at 650 rpm caused significant breakup and complete dissolution of coarser T-phase network and remarkable grain refinement (1.6–2.8 μm), thereby improving the tensile properties.

References

1. Mordike BL, Ebert T (2001) Magnesium-Properties—applications—potential. Mater Sci Eng A 302:37–45
2. Rokhil LL (2003) Magnesium alloys containing rare earth metals. Taylor and Francis, London
3. Kannan MB, Dietzel W, Blawer T, Atrens A, Lyon C (2008) Stress corrosion cracking of rare earth containing magnesium alloys ZE41, QE22, and Elekron 21 (EV31A) compared with AZ80. Mater Sci Eng A 480:529–539
4. Kim WJ, Hong SI, Kim YS, Min SH (2003) Texture development and its effect on mechanical properties of an AZ61 Mg alloy fabricated by equal channel angular pressing. Acta Mater 51:3293–3307

5. Yuan W, Panigrahi SK, Mishra RS (2013) Achieving high strength and high ductility in friction stir processed cast magnesium alloy. Metallurgical and Materials Transactions A 44:3675–3684
6. Lin XB, Chen RS, Han EH (2008) High temperature deformations of Mg–Y–Nd alloys fabricated by different routes. Mater Sci Eng A 497:326–332
7. Tang LC, Liu CM, Chen ZY, Ji DW, Xiao HC (2013) Microstructures and tensile properties of Mg–Gd–Y–Zr alloy during multidirectional forging at 773 K. Mater Des 50:587–596
8. Feng AH, Ma ZY (2007) Enhanced mechanical properties of Mg–Al–Zn cast alloy via friction stir processing. Scripta Mater 56:397–400
9. Kang JD, Wilkinson DS, Mishra RK, Yuan W, Mishra RS (2013) Effect of inhomogeneous deformation on anisotropy of AZ31 magnesium sheet. Mater Sci Eng A 567:101–109
10. Darras BM, Khraisheh MK, Abu-Farah FK, Omar MA (2007) Friction stir processing of commercial AZ31 magnesium alloy. J Mater Process Technol 191:77–81
11. Babu SR, Pavithran S, Nithin M, Parameshwaran B (2014) Effect of tool shoulder diameter during friction stir processing of AZ31B alloy sheets of various thicknesses. Procedia Eng 97:800–809
12. Sanchez C, Nussbaum G, Azavant P, Octor H (1998) Structural material properties, microstructure and processing. Mater Sci Eng A 252:269
13. Genghua C, Datong Z, Wen Z, Cheng Q (2015) Microstructure evolution and mechanical properties of Mg–Nd–Y alloy in different friction stir processing conditions. J Alloy Compd 636:12–19
14. Neil WC, Forsyth M, Howlett PC, Hucthinson CR, Hinton BRW (2009) Corrosion of magnesium alloy ZE41—the role of microstructural features. Corros Sci 51:387–394
15. Renlong X, Xuan Z, Zhe L, Dejia L, Risheng Q, Zeyao L, Qing L (2016) Microstructure and texture evolution of an Mg–Gd–Y–Nd–Zr alloy during friction stir processing. J Alloy Compd 659:51–59

Effect of Bamboo Fiber on Mechanical Properties of Fly Ash with Polypropylene Composites

T. Venkateswara Rao, M. Somaiah Chowdary,
Ch. Siva Sanakara Babu and Ch. Mohan Sumanth

Abstract In research, there are many advanced methods in material science but attention should be kept on some of the areas. One such area is natural-reinforced composites. In this work, the natural fiber is bamboo fiber along with filler material. Fly ash has been taken into consideration. Extraction of fibers from jute, coir, and bamboo belongs to this category. To enhance the properties of polymers, various filler materials such as silica, fly ash, etc. are used along with natural fiber-reinforced composite. In this project, bamboo fiber hybrid composites are fabricated with polypropylene along with 10% wt concentration of fly ash. Concentration of fiber was varied up to 25% by weight, and the test specimens were prepared by injection molding. Bending properties were tested using tensometer. Bending strength and bending moment were increased with the increment of weight % of fiber in the composition.

Keywords Natural composite · Bamboo fiber · Polypropylene(PP)
Fly ash · Injection molding · Bending properties

1 Introduction

The properties of products which are manufactured by conventional materials such as metal alloys, polymeric materials, and ceramics are not meet the fast developing society. So there is a necessity to increase the properties of products with unusual combinations of composite materials.

T. Venkateswara Rao (✉) · Ch.Siva Sanakara Babu
Department of Mechanical Engineering, Lakireddy Bali Reddy College
of Engineering, Mylavaram, Andhra Pradesh 521230, India
e-mail: tvrao722@lbrce.ac.in

Ch.Siva Sanakara Babu
e-mail: sivasankarchinka@lbrce.ac.in

M. Somaiah Chowdary · Ch.Mohan Sumanth
Department of Mechanical Engineering, PVP Siddhartha Institute
of Technology, Kanuru 52007, India

© Springer Nature Singapore Pte Ltd. 2019
A. K. Lakshminarayanan et al. (eds.), *Advances in Materials and Metallurgy*,
Lecture Notes in Mechanical Engineering,
https://doi.org/10.1007/978-981-13-1780-4_42

For the recent technologies in aerospace field, the major properties affect the service life of aerostructures, which are high strength/weight ratio and high stiffness/weight ratios. Many researches are focussed to improve the properties of the materials like strength, stiffness, and lightweight by adding different combinations of materials. Khan and Yousif [1] determined tensile strengths of bamboo fiber-reinforced composites with various NaOH-treated fibers through tensile test and observed surface phenomenon using SEM. Finally, they obtained maximum tensile strength (234 MPa) of bamboo fiber-reinforced epoxy composite with 6% of NaOH-treated fibers. Da Costa Correia and Santos [2] explained the mechanical properties of composites which are improved by considering the composites as porosity refining due to matrix densification after aging. Composite material has the more flexibility that can be used in many applications in different areas. By varying the composition, properties can be changed, thus making the composites [3].

The increasing population needs more and more construction materials. Wood and metals are the construction materials, which have been extensively consumed in building construction, vehicle body, furniture, etc. The growth rate of material resources is not in pace with that of population. To meet the deficiency, man has to find suitable substitutes [4]. The materials from natural resources under-explored and unused so far, if put to effective. Utility will solve the problem arising out of the inadequacy of conventional materials [5]. Over the last four decades, composites of synthetic fibers have been developed adopting the knowledge of naturally available vegetative organs of bamboo, palm trees, sisal, etc. [6]. It is concerned with the reduction of weight of space vehicles and aircraft for optimum payload [7]. The constituents of these exotic composites are of high cost. They are cost-effective only in high-tech fields. They do not meet the needs of common man [8].

2 Experimental Work

Industries widely used injection molding process for the fabrication of various plastic parts like complexity in shapes, sizes, and different applications. The required component for injection molding process is raw plastic material, a mold, and injection molding machine.

Pellets are the raw plastic material forms, which are sent to the injection molding machine. The material was melted by heat and pressure supplied to machine by external setup in machine. Melted plastic material is transferred to mold quickly, which created the pressure pack, and it holds the material. Melted plastic material is cooled and then solidified into the final part in mold itself. The injection time is complicated to calculate exactly due to the complexity and velocity difference of the molten plastic material from machine to the mold. Injection power, injection pressure, and shot volume are the major parameters to estimate the injection time.

Materials used in this work are

Polypropylene, fly ash, and bamboo fiber.

A. ***Polypropylene polymer (PP)*** is one of the polymers of a thermoplastic material, a wide variety of applications such as plastic parts, packaging, stationery, automotive components, textiles, etc.
B. ***Fly ash*** is a byproduct from burning pulverized coal in electric power generating plants.
C. ***Bamboo*** is used as fiber with the aim of using them as reinforcement in composite materials as an alternative to replace glass fibers alternative [9–11].

3 Experimental Procedure

Compositions:

	Plain Polypropylene
Polypropylene + 10 wt% Fly ash	Polypropylene + 10 wt% Fly ash + 10 wt% Fibre
Polypropylene + 10 wt% Fly ash + 20 wt% Fibre	Polypropylene + 10 wt% Fly ash + 30 wt% Fibre
Polypropylene + 10 wt% Fly ash + 40 wt% Fibre	Polypropylene + 10 wt% Fly ash + 50 wt% Fibre

These compositions were mixed and added according to their weight percentages and are injected into the injection molding machine without inserting the die, and the exerted irregular materials were collected according to their composition because irregular composition occurs in the final composite, in order to deplete that problem die was not inserted (Figs. 1, 2, and 3).

Fig. 1 Composition mixture

Fig. 2 Complex-shape specimens

Fig. 3 Grinding machine

These complex-shaped compositions were grinded in grinding machine, and they were made into fine pieces except plain polypropylene which can be directly prepared without making of any irregular shapes (Figs. 4 and 5).

These were again injected in injection molding machine, introducing mold in its place and the process continues. For measurement of flexural strength and flexural modulus, by taking standard test method ASTM D638–89, flexural parameter of the composites is valued. The specimens having a dimension of length 98 mm, width 10 mm, and thickness 10 mm were prepared. Five sample specimens were tested for each weight % of fiber. The specimens are tested at a crosshead speed of 2 mm/min, using an electronic tensometer (Model METM 2000 ER-1).

4 Results and Discussion

Measurement of flexural strength and flexural moment:

Flexural strength and flexural modulus of the specimens varying with weight % of fiber composites are tested by using tensometer (Figs. 6, 7, 8, and 9).

Fig. 4 Fine particles formed by grinding

Fig. 5 Final specimens of some compositions

Fig. 6 % of fiber versus
bending strengths

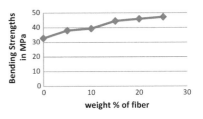

Fig. 7 % of fiber versus bending moment

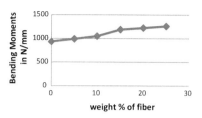

Fig. 8 % of fiber versus elongation

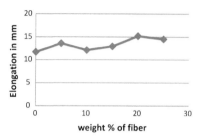

Fig. 9 % of fiber versus load

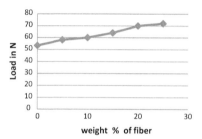

$$\text{Calculation of flexural strength} : \sigma = \frac{3FL}{2bd^2}$$

Discussion on Results: The flexural properties show a gradual increment in load, bending strength, and bending moment because of its high strength-to-weight ratio of bamboo fiber and due to lower specific gravity and higher shear strength of fly ash (Tables 1, 2, 3, 4, 5, 6, and 7).

Table 1 Plain polypropylene (PP)

Load (N)	Elongation (mm)	Bending stress (MPa)	Bending moment (N/mm)
40	9.2	26.25	699.96
50	12.6	32.81	874.95
70	13.2	45.94	1225
Avg-53.33	Avg-11.67	Avg-32.81	Avg-933.3

Table 2 90%PP + 10% fly ash

Load (N)	Elongation (mm)	Bending stress (MPa)	Bending moment (N/mm)
50	11.4	32.81	874.95
60	14.7	39.37	1049.93
70	15.9	45.94	1225
Avg-58	Avg-13.56	Avg-38.06	Avg-991.60

Table 3 10% fly ash + 5% fiber + 85%PP

Load (N)	Elongation (mm)	Bending stress (MPa)	Bending moment (N/mm)
40	10	26.25	699.96
50	12.3	32.81	874.95
70	11.9	45.94	1225
Avg-60	Avg-12.08	Avg-39.38	Avg-1049.98

Table 4 10% fly ash + 10% fiber + 80%PP

Load (N)	Elongation (mm)	Bending stress (MPa)	Bending moment (N/mm)
60	11.0	39.37	1049.93
70	13.5	45.94	1225
80	14.7	52.04	1399.91
Avg-64	Avg-12.92	Avg-44.53	Avg-1189.96

Table 5 10% fly ash + 15% fiber + 75%PP

Load (N)	Elongation (mm)	Bending stress (MPa)	Bending moment (N/mm)
50	12.3	32.81	874.95
70	15.1	45.94	1225
80	15.9	52.04	1399.91
100	18.3	65.62	1749.89
Avg-70	Avg-15.18	Avg-45.94	Avg-1224.94

Table 6 10% fly ash + 20% fiber + 70%PP

Load (N)	Elongation (mm)	Bending stress (MPa)	Bending moment (N/mm)
50	10.7	32.81	874.95
60	14.6	39.37	1049.93
90	15.6	59.06	1574.90
100	18.9	65.63	1749.89
Avg-72	Avg-14.5	Avg-47.25	Avg-1259.92

Table 7 10% fly ash + 25% fiber + 65%PP

Load (N)	Elongation (mm)	Bending stress (MPa)	Bending moment (N/mm)
40	13.2	26.25	699.95
50	14.1	32.82	874.95
60	14.4	39.38	1049.93
Avg-50	Avg-13.9	Avg-32.81	Avg-874.94

5 Conclusions

From the previous study, it was known that polypropylene with fly ash composition and the maximum values are attained with 10% fly ash + Polypropylene. So in our hybrid composite, we fixed the 10 wt% of fly ash as constant and differed the weight composition of fiber and polypropylene.

Mechanical Properties of bamboo fiber-reinforced hybrid composites (Polypropylene + Fly Ash + Bamboo Fiber) are evaluated and the final conclusions are identified.

- Flexural strength increases with the increment of weight percentage of fiber according to the compositions mentioned above and it shows maximum at 25% of fiber in the composition.
- Flexural moment also increases with the increment of weight percentage of fiber according to the compositions mentioned above and it shows maximum at 25% of fiber in the composition.
- Load also increases with the increment of weight percentage of fiber according to the compositions mentioned above and it shows maximum at 25% of fiber in the composition.
- But elongation fluctuates by showing an increment and decrement with the addition of fiber weight percentage.
- It was found that the fiber percentage cannot be increased more than 25% because of the reduction of quantity of polypropylene in the composition; the molds cannot be formed in injection molding machine.

References

1. Khan Z, Yousif BF, Islam MM (2017) Fracture behaviour of bamboo fiber reinforced epoxy composites. Compos B 116:186–199. https://doi.org/10.1016/j.compositesb.2017.02.015
2. da Costa Correia V, Santos SF et al (2014) Potential of bamboo organosolv pulp as a reinforcing element in fiber–cement materials. Constr Build Mater 72:65–71. https://doi.org/10.1016/j.compositesb.2017.02.015
3. da Costa Correia V, Santos SF et al (2014) Natural fibres and polymer composites. Int J Mod Eng Res (IJMER) 12(2):117–136
4. Rassiah K, Megat Ahmad MMH (2013) A review on mechanical properties of bamboo fiber reinforced polymer composite. Aust J Basic Appl Sci 7(8):247–253 ISSN 1991-8178
5. Thwe et al (2003) Comparing tensile strength of bamboo fiber reinforced polypropylene composite BFRP and bamboo- glass fiber reinforced polypropylene hybrid composite BGRP. Aust J Basic Appl Sci 7(8):247–253 ISSN 1991-8178
6. Environmental effects on bamboo-glass/ polypropylene hybrid composites MOE MOE THWE∗, KIN LIAO, Schools of Materials Engineering and ‡ Mechanical & Production Engineering, Nanyang Technological University, Singapore E-mail: askliao@ntu.edu.sg
7. Okubo et al (2004) Using steam explosion technique to extract bamboo fibers. Aust J Basic Appl Sci 7(8):247–253 ISSN 1991-8178
8. Holbery J, Houston D (2006) Natural-fiber-reinforced polymer composites in automotive applications. Springer, Berlin
9. Senapati AK (2014) an extensive literature review on the usage of fly ash as a reinforcing agent for different matrices. Int J Innov Sci Mod Eng (IJISME) 2(3):4–9 ISSN: 2319-6386
10. Venkateswara T, Viswanadh KV, Somaiah Chowdary M (2017) Evaluation of flexural properties of bamboo fiber reinforced hybrid composites. Int J Adv Eng Res Devel (Ijaerd) 4(6):14
11. Venkateswara Rao T, Venkata Rao K, Lakshmi kanth Ch (2014) Mechanical properties of Bamboo Fiber filled with Fly ash filler reinforced hybrid composites. Int J Eng Res Technol (IJERT) 3(9):12

Corrosion and Magnetic Characterization of Electroplated NiFe and NiFeW Soft Magnetic Thin Films for MEMS Applications

S. Venkateshwaran, E. Selvakumar, P. Senthamil selvan,
M. Selvambikai, R. Kannan and A. S. Pradeep

Abstract Electroplating of nanocrystalline NiFe and NiFeW thin films were successfully carried out on the copper substrate from ammonium citrate bath at a current density of 1 mA/dm^2 and controlled pH of 7 with constant bath temperature. The magnetic and corrosion properties were studied using VSM and electrochemical techniques. The results of vibrating sample magnetometer for NiFeW nanocrystalline thin films reveal its soft magnetic properties such as coercivity and saturation magnetization. From the electrochemical studies, corrosion resistance and corrosion inhibition efficiency were calculated. The electrochemical studies of all the coated films reveal that the NiFeW thin film exhibits the enhanced corrosion resistance as compared with NiFe thin film which in turn enhances the soft magnetic nature of NiFeW thin films. Thus, the electroplated NiFeW thin films can be used for Microelectromechanical System (MEMS) and Nanoelectromechanical System (NEMS) applications due to their excellent magnetic- and corrosion-resistant behaviour.

S. Venkateshwaran · E. Selvakumar · P. Senthamil selvan
Department of Mechanical Engineering, Kumaraguru College of Technology,
Coimbatore, Tamilnadu 641049, India
e-mail: venkateshwaran.15me@kct.ac.in

E. Selvakumar
e-mail: selva.15me@kct.ac.in

P. Senthamil selvan
e-mail: senthamil.15me@kct.ac.in

M. Selvambikai · R. Kannan (✉)
Department of Science and Humanities, Physics division, Kumaraguru College
of Technology, Coimbatore, Tamilnadu 641049, India
e-mail: kannan.r.sci@kct.ac.in

M. Selvambikai
e-mail: selvambikai.m.sci@kct.ac.in

A. S. Pradeep
Kumaraguru Center for Industrial Research and Innovation (KC.IRI),
Coimbatore, Tamilnadu 641049, India
e-mail: pradeep.as.kciri@kct.ac.in

© Springer Nature Singapore Pte Ltd. 2019
A. K. Lakshminarayanan et al. (eds.), *Advances in Materials and Metallurgy*,
Lecture Notes in Mechanical Engineering,
https://doi.org/10.1007/978-981-13-1780-4_43

Keywords NiFe · Crystalline · Magnetization · Coercivity · Corrosion and MEMS

1 Introduction

Electrodeposited nanocrystalline magnetic thin films are currently emerging as trending materials due to their immense applications in the areas of sensors, power electronics, actuators and core material for writing elements in recording heads which includes high-density recording media. Soft magnetic thin films with high saturation magnetization have been used in modern nonvolatile magnetic memory, magnetic recording heads, high-frequency plasma inductors [1–4] and Microelectromechanical System (MEMS) due to its good adhesion, lower stress and high corrosive resistance with excellent magnetic properties. Nickel–iron-based alloys are the most important alloy among the known soft magnetic material for magnetic storage applications because of their highest magnetic flux density, lower magnetostriction and lower coercivity [5–7]. Electrodeposition and sputtering are most commonly used techniques for fabricating the NiFe-based thin films. In the present research work, the NiFe and NiFeW thin films were synthesized through electrodeposition method because of its advantages such as large-scale production, high efficiency, minimum wastage of chemicals and easy to control the film growth. The properties of $Ni_{80}Fe_{20}$-based alloys can also be enhanced by adding the suitable element to these alloys as third element. In recent days, researchers show interest to analyse the suitable third elements like Mo, Cr, Ta, P, Mn and W. Tungsten (W) is decided as the best third element to NiFe alloys due to their better electrochemical, mechanical properties, good corrosion resistance, highest melting point (3410 °C), lowest coefficient of linear thermal expansion (4.3×10.6 °C), highest tensile strength (410 kg/mm^2) and highest Young's modulus of elasticity (3500 kg/mm^2) [8–11].

To the best of our knowledge, very few works are documented about the properties of NiFeW alloy thin films and further systematic research is needed to explore the full potential of NiFeW alloy thin films. Our present work mainly focuses on the effect of tungsten(W) on the magnetic properties of NiFe alloys. The addition of tungsten to NiFe alloy may enhance the corrosion resistance, hardness, mechanical strength, ductility and magnetic properties. This research paper investigates the effect of tungsten in NiFe thin films coated on copper plate from ammonium citrate bath for MEMS and NEMS applications.

2 Experimental Work

Nanocrystalline NiFe and NiFeW thin films were electroplated on copper substrate using suitable ultra-pure salts at room temperature of 70 °C. The chemical composition and electroplating bath conditions of NiFe and NiFeW thin films are listed

Table 1 Electroplating bath details of NiFe and NiFeW thin films

S. no	Name of the chemical parameters	Data (g/l)
1	Nickel sulphate	60
2	Ferrous sulphate	30
3	Sodium tungstate	10
4	Diammonium citrate	70
5	Citric acid	10
6	Boric acid	10
7	pH value	7
8	Temperature	30 °C
9	Current density	1 mA/dm^2

in Table 1. A copper plate of size 7 cm as length and 1.5 cm as breadth has been used as a cathode and a pure stainless steel of same size act as an anode. Both the electrodes were cleaned by washing with soap and soaking in 10% H_2SO_4 for 3 min. Both of the plates are rinsed with distilled water just before the deposition. Except the area on which the deposition of films was desired, all the other areas on the substrate are masked off using adhesion tape. All the analytical reagent grade chemicals were dissolved in triple distilled water to prepare the electroplating bath.

The pH of the bath was controlled to 7 by adding few drops of ammonia solution. For electrodeposition of both the films, the Cu substrate is dipped into ammonium citrate bath solution and a current density of 1 mA/dm^2 is maintained as constant for a period of 30 min. All the coated NiFe and NiFeW thin films were subjected to various characterization techniques such as electrochemical studies and vibrating sample magnetometer in order to reveal the corrosion behaviour and magnetic nature of the coated thin films.

The vibrating sample magnetometer (VSM) was used to measure saturation magnetization and the coercivity of NiFe and NiFeW thin films. When samples of thin film have been placed in a uniform magnetic field, a dipole moment proportional to the product of the sample susceptibility and the applied field was induced in the sample. If the sample was made to undergo sinusoidal motion as well, an electrical signal can be induced in suitably located stationary pick-up coils. This signal, which was at the vibration frequency, was proportional to the magnetic moment, vibration amplitude and vibration frequency.

3 Results and Discussion

3.1 SEM and EDAX Analysis of NiFe and NiFeW Thin Films

The chemical composition and surface morphology of the NiFe and NiFeW thin films were analysed using energy-dispersive X-ray analyser and scanning electron

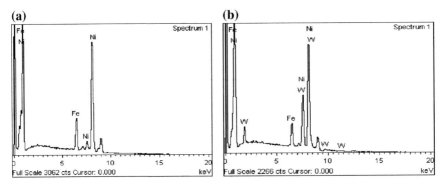

Fig. 1 EDAX spectrographs of **a** NiFe thin film, **b** NiFeW thin film at 70 °C

microscope. The EDAX spectrographs of coated thin films are shown in Fig. 1. The NiFeW thin film has the highest Ni content with lowest Fe content as compared with NiFe thin film. The highest Ni content with moderate W content may enhance the magnetic nature of NiFeW thin films (shown in Table 2).

The SEM pictures of NiFe and NiFeW thin film reveal that the coated thin film has bright and uniform surface morphology with no microvoids and cracks. The SEM micrographs of coated thin films are shown in Fig. 2. The SEM result reveals that the NiFeW thin film coated at bath temperature of 70 °C have bright, uniform surface morphology and well-defined granular size compared with NiFe thin films.

3.2 Corrosion Studies

Corrosion properties of NiFe and NiFeW thin films were studied by electrochemical studies. The results of electrochemical studies are shown in Table 3. The polarization resistance of NiFeW thin film is enhanced to 50% compared to NiFe thin film. The NiFe thin films have 46.5 Ω of charge transfer resistance which is directly proportional to the corrosion resistance and after the addition of W in NiFe thin films, the charge transfer resistance was enhanced to 238 Ω. The corrosion

Table 2 Results of EDAX analysis

Bath identification	Bath temperature (°C)	Ni (wt%)	Fe (wt%)	W (wt%)
NiFe	30	42.07	57.93	–
NiFeW	30	81.83	9.12	9.05
NiFeW	70	82.53	8.98	8.49

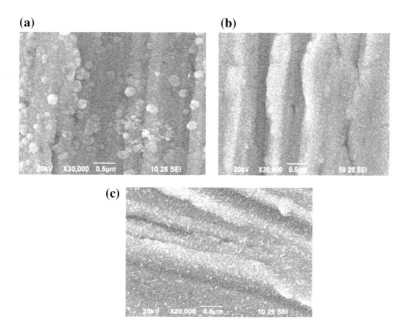

Fig. 2 SEM results for **a** NiFe thin film at 30 °C, **b** NiFe W thin film at 30 °C, **c** NiFe W thin film at 70 °C

Table 3 Electrochemical results of NiFe and NiFeW thin films

Polarization data						Impedance data	
Sample	b_a (V/dec)	b_c (V/dec)	E_{corr} (V)	i_{corr} (A)	Polarization resistance (R_p) (Ω)	Charge transfer resistance (R_{ct}) (Ω)	Corrosion inhibition efficiency (%)
NiFe	0.29931	0.18722	−0.43165	0.000932	53.663	46.5	80.46
NiFeW	0.22164	−2.8232	−0.29139	0.000923	113.11	238	

inhibition efficiency of W-added NiFe thin film is improved to 80.46%. From Figs. 3 and 4, increase in charge transfer resistance and polarization resistance shows that the corrosion resistance behaviour is enhanced.

The enhanced corrosion resistance of NiFeW thin films is mainly due to the presence of tungsten (w) which is the suitable element for NiFe alloys. Tungsten is one of the hardest elements to enhance the corrosion resistance which in turn enhances its magnetic nature due to the unusual properties like tribological, magnetic, electrical and electro-corrosion properties.

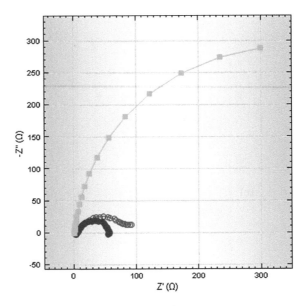

Fig. 3 Impedance curve for NiFe and NiFeW thin films

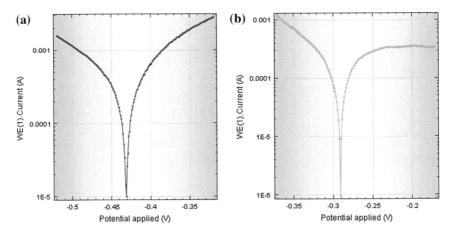

Fig. 4 a Polarization curve of NiFe thin film, **b** Polarization curve of NiFeW thin film at 70 °C

3.3 Magnetic Properties

The magnetic properties of the materials are determined by their crystalline nature, and the important parameters that determine the magnetic properties of soft magnetic materials are saturation magnetization and coercivity [9]. So the hysteresis loop parameters coercivity (H_c), saturation magnetization (M_s) and retentivity(M_r)

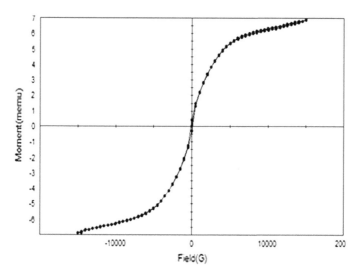

Fig. 5 Magnetic hysteresis loops of NiFeW thin film deposited from ammonium citrate bath at 70 °C

of the electrodeposited thin films were evaluated using VSM data. The higher magnetization density is exhibited by the NiFeW thin film coated at 70 °C. From Fig. 5, the coercivity of NiFeW thin film is calculated as 132.82G. Saturation magnetization and retentivity are calculated as 47886 E-3 emu and 383.72 E-6 emu. Thus, the NiFeW thin film exhibits suitable properties of soft magnetic material which could be used in MEMS devices.

4 Conclusion

Electrodeposition of nanocrystalline NiFe and NiFeW were successfully carried out on copper substrate. The magnetic properties of NiFeW thin film synthesized at 70 °C have been enhanced when compared to NiFe thin film, where ammonium citrate bath is used for both the films. The corrosion resistance of NiFeW thin film is also enhanced due to addition of tungsten.

Thus, the electroplated NiFeW thin film is a soft magnetic material which can be used for Microelectromechanical System (MEMS) and Nanoelectromechanical System (NEMS) applications due to their enhanced magnetic- and corrosion-resistant behaviour.

Acknowledgements The research work was carried out with the support of the **Management of Kumaraguru College of Technology**, Coimbatore and the authors are grateful for their immense support.

References

1. Chechenin EV, Khomenko EV, de Hosson JThM (2007) FCC/BCC competition and enhancement of saturation and magnetization in nanocryatalline Co-Ni-Fe films. JETP Lett 85:212–215
2. Gavrila H, Ionita V (2002) Crystalline and amorphous soft magnetic materials and their applications- status of art and challenges. J Optoelectron Adv Mater 4(2):173–192
3. Kannan R, Ganesan S, Selvakumari TM (2012) Synthesis and characterization of nanocrystallineNiFeWS thin films in diammonium citrate bath. Digest J Nanomater Biostruct 7(3):1039–1050
4. Kannan R, Ganesan S, Selvakumari TM (2013) An investigation on effect of annealing on magnetic properties of NiFeWS electrodeposited coatings in tri sodium citrate bath. Iran J Sci Technol (Sciences) 37(2):181–187
5. Kannan R et al (2018) Mater Res Express 5:046414
6. Kannan R, Ganesan S, Esther P, Joseph Kenady C (2013) Development of high performance magnetic NiFeWS thin film for microlectro mechanical system. Int J Thin Film Sci Tech 3(3):233–244
7. Kannan R, Ganesan S, Selvakumari TM (2012) Effect of annealing temperature on magnetic properties of nano crystalline NiFeWS thin films in diammonium citrate bath. Optoelectron Adv Mater 14(9–10):774–780
8. Kannan R, Ganesan S, Balasubramanian V, Rajeswari M (2014) Development of nano structured Ni rich NiFeWS films for micro electro mechanical system applications. Optoelectron Adv Mater 16(1–2):156–161
9. Kannan R, Kokila S (2015) Synthesis and structural characterization of CoNiW alloy thin films by electrodeposition. Int J Thin Film Sci Tech 4(1):59–62
10. Sundaram K, Dhanasekaran V, Mahalingam T (2011) Structural and magnetic properties of high magnetic moment electroplated CoNiFe thin films. Ionics 17(9):835–842
11. Kannan R, Selvambikai M, JeenaRajathy I, Ananthi S (2017) A study on structural analysis of electroplated nano crystalline nickel based thin films. Rasayan J Chem 10(4):1213–1217

Prediction and Analysis of Microstructure and Mechanical Properties After Equal Channel Angular Pressing of EN 47 Spring Steel

S. Ramesh Kumar, V. Hari Shankar, R. E. Krishna Sangeethaa
and K. V. Sai Tejaswy

Abstract In this paper, ECAP process under room temperature conditions was carried out on EN47 spring steel and its properties were analyzed. Spring steel is corrosion resistant, and it exhibits high tensile strength and has high formability. Because of its desirable properties, it finds application in automotive and industrial suspension applications. The effects on microstructural and mechanical properties are analyzed in this study. Severe plastic deformation causes refinement of the grain structure, and it has been observed after the process. The ECAP was performed at different corner angles $\varphi = 120°$ and $150°$ with the radius of curvature $\psi = 20°$ through route A. The microstructural and mechanical properties were analyzed. After the pass, the values of hardness and tensile strength were measured for the different angles. After processing the sample, more columnar structures were observed in $120°$, and uniformly distributed grains with equiaxed and less columnar structures were observed in $150°$. This is mainly due to the occurrence of grain strengthening effect at the corner angles.

Keywords EN 47 spring steel · ECAP · Microstructure · Tensile strength
Hardness · SEM · Fractography

1 Introduction

Severe plastic deformation techniques are utilized to alter the microstructure of the metals and alloys which include extrusion, high-pressure torsion, and equal channel angle pressing (ECAP). Among these, ECAP is preferred because it has the ability to provide significant grain refinement and homogeneous microstructure to the bulk materials. This technique is accomplished without change in cross section by

S. Ramesh Kumar (✉) · V. Hari Shankar · R. E. Krishna Sangeethaa ·
K. V. Sai Tejaswy
School of Mechanical Engineering, SASTRA Deemed to be University,
Thanjavur, Tamilnadu 613401, India
e-mail: rameshkumar@mech.sastra.edu

© Springer Nature Singapore Pte Ltd. 2019
A. K. Lakshminarayanan et al. (eds.), *Advances in Materials and Metallurgy*,
Lecture Notes in Mechanical Engineering,
https://doi.org/10.1007/978-981-13-1780-4_44

extruding the material around a corner. The channel angles can be varied in this method. Owing to various advantages, grain structure refinement tends to be the most successful via ECAP [1, 2]. Investigation on martensitic steel of heat-resistant type undergoing ECAP has revealed improved mechanical properties including tensile strength and hardness [3]. Microstructure of high-carbon steel consisting of ultrafine grains was changed to equiaxed structure after processing by ECAP and its properties were enhanced [4]. Adopting route Bc, spheroidization occurred and ductile fracture was observed after four passes in high-carbon steel [5]. ECAP processing of low-carbon steel through Bc route has proven to be successful compared to other routes, and also channel angles of the order of 90° were implemented and better grain structures and mechanical properties were obtained [6, 7]. ECAP processing of low-carbon steel also revealed significant improvement in fatigue strength and fracture characteristics [8]. Investigations were made to study the bake-hardening effect of dual-phase steel and have proved to improve the microstructural as well as mechanical properties [9]. Cold rolling and inter-critical annealing at elevated temperatures of plain low-carbon steel have exhibited excellent mechanical properties [10]. Equal channel angular pressing of TA15 titanium alloys has led to grain refinement, and this stands an evidence for the advantage of this technique [11]. ECAP studies of low-carbon steel have revealed that ECAP is a better technique for improving hardness compared to quenching and also it promotes anodic passivation [12]. ECAP-processed low-carbon steels at elevated temperatures have also proven to exhibit better mechanical properties [13]. Influence of temperature on ECAP technique has been explained by experimenting 304L stainless steel and its effects were explained [14].

Our literature studies have revealed lack of investigation of ECAP on EN 47 spring steel and hence we have conducted studies in the same. In this paper, successful attempts were made in studying the microstructural and mechanical properties of ECAP-processed EN 47 spring steel. The process was carried out under ambient conditions via route A at corner angles—120° and 150°. The grain structure was studied using scanning electron microscope images. X-ray diffraction study was carried out to analyze the density of dislocation and crystalline structure.

2 Experimental Work

The equal channel angular pressing was done at room temperature in a 100T Universal Testing Machine. Figure 1 depicts the schematic representation of the setup used in the process. Molybdenum disulphide, a lubricant to reduce friction between surfaces, is applied inside the channels, surface of the specimen, and punch. The ECAP process was carried at route A at both 120° and 150° channel angles, after ECAP-processed samples were tested by tensile test with the ASTM E8 standard using Micro-tensile tester of 2 ton capacity. HITACHI S-300H scanning electron microscopic is used to observe the microstructure at the fractured surface, and optical microscopes were used to identify the distribution of grains and

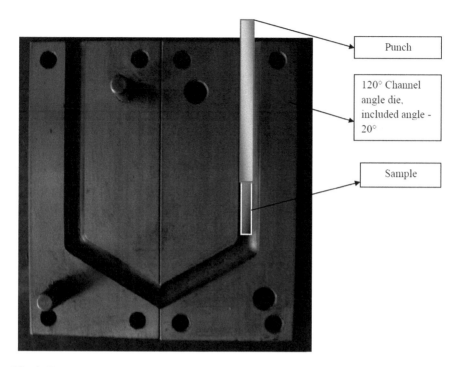

Fig. 1 Representation of sample inside the die

orientation of grains after passes. Three places, namely, near the shear zone, centered off the processed billet, and nearby the dead end of the billet, were identified for microstructural studies. Samples were initially rough polished with emery sheets of grades 320, 400, 600, 800, 1000, 1/0, 2/0, 3/0, and 4/0. For fine polishing, disk polishing with alumina and diamond was done. The microstructures were photographed using Envision 3.0 software and image analyzer.

Figure 2 shows the 3d view and schematic representation of 120° and 150° channel angle die arrangement. Before ECAP-processed sample with 50 mm length and 12 mm diameter as shown in Fig. 3a, processed ECAP billets with different channel angles at route A first pass as shown in Fig. 3b,c

3 Results and Discussion

3.1 Microstructure Analysis

Figures 4, 5, 6, 7, 8, and 9 show the optical microscopic images at different magnifications to identify the distribution of grains near the shear zone, center, and dead and of the 120° and 150° processed billet. Figure 4 shows the formation of

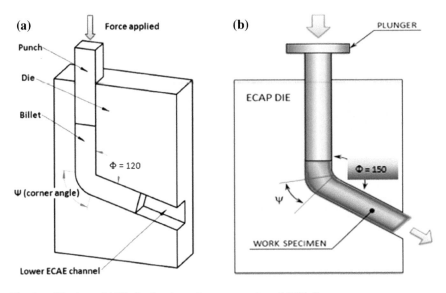

Fig. 2 **a** 3D view of 120° die, **b** schematic representation of 150° die

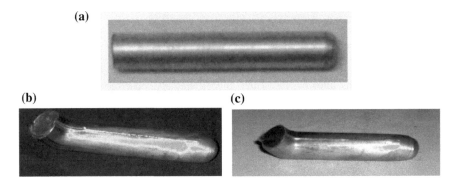

Fig. 3 **a** Before ECAP billet, **b** after ECAP at 120° processed billet, **c** after ECAP at 150° processed billet

epsilon carbide precipitates because of the reduction in the tetragonal nature in martensite, decomposition of retained austenite, spheroidized and coarsened cementite in combination with recrystallized ferrite, loss of tetragonal martensite, and dissolution of the carbides causing formation of plate-like cementite.

Figure 5 shows well-defined pearlitic phase with lamellar packing of ferrite and cementite particles and coarse grains of ferrite and pearlitic colonies. The grains are not aligned or ruptured in a particular fashion. Figure 6 shows an increase in the amount of the pearlite leading to the formation of small lumps of pearlite along with finer ferrite. Relaxing of the martensitic structure is visible in the temper-treated specimen. Quenching of the austenitic phase leads to the growth of lath martensite,

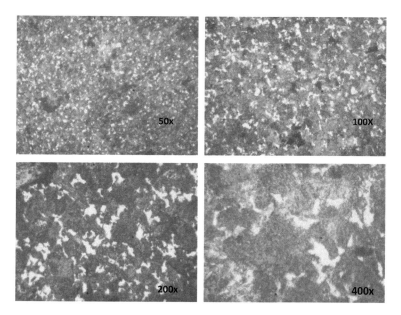

Fig. 4 Microstructure observed from nearby the front end of the billet

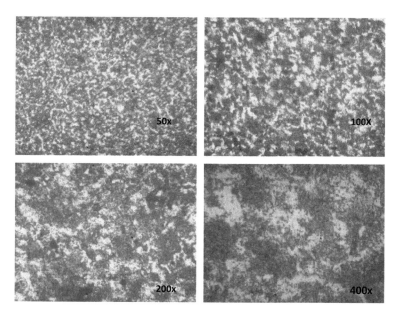

Fig. 5 Microstructure observed from center of the billet

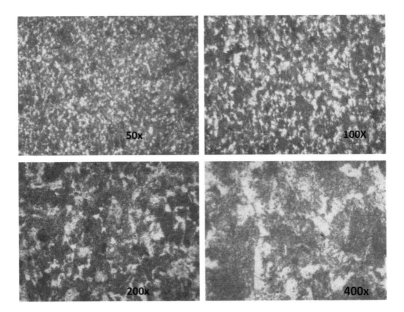

Fig. 6 Microstructure observed from nearby the dead end of the billet

and this is shown in the microstructure of the hardened specimen. Figure 7 depicts a decrease in the percentage of the pearlite and formation of fine grains of ferrite which holds responsibility for the increase in ductility of the specimen while showing only mild increase in hardness. Figure 8 shows the microstructure of the specimen prior to any treatment containing ferrite along the boundaries of pearlite which is acicular in nature. This explains the duplex nature of steel containing both austenite and ferrite. Figure 9 shows the microstructure of the specimen subjected to normalizing treatment. It contains fine yet large grains of ferrite as well as pearlite. Wide distribution of scales on the surface and dispersion of the ferrite was revealed by the subjecting the specimen to age-hardening process.

Figure 10 shows the fracture morphology of 150° ECAPed billet after tensile test. From the SE images, facets like structures appeared with higher the value. Very little amount of cup- and cone-like structures were appeared. Figure 11 shows the fracture morphology of 120° billet after tensile test. From the secondary electron microscopic images, bands of cup and cone-like were observed. And very few facets like structures were appeared near the outer surfaces of the billet. From the tensile test images, ductility is higher in the case of 150°. Vickers microhardness (HV) was measured across the plane of the extruded samples on application of a 300 grams load for a duration of 15 s at 10 different locations on the specimen. The average value is considered. The formation of dead zones due to the frictional effect was the primary reason to avoid the corners of the specimen. For the Vickers test, both the diagonals d_1 and d_2 are measured and the average value is used to compute the Vickers hardness number. The measured base metal hardness was 250 HV,

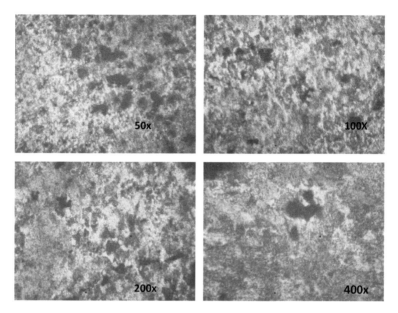

Fig. 7 Microstructure observed from nearby the front end of the billet

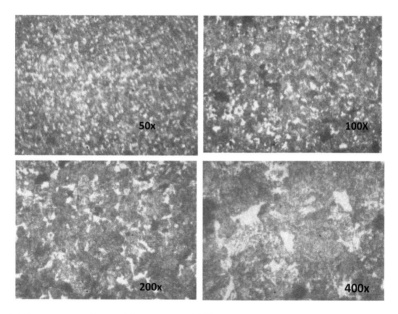

Fig. 8 Microstructure observed from center of billet

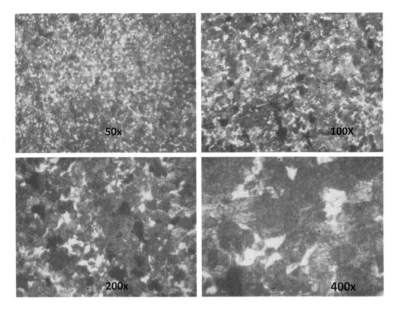

Fig. 9 Microstructure observed from nearby the dead end of the billet

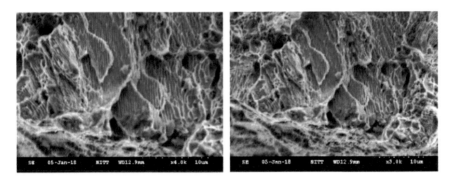

Fig. 10 Fracture Morphology of the 120° ECAPed billet

while it was 296 HV for 120° channel angle and 285 HV for 150° channel angle processed billets.

The tensile test for the specimen was carried out as per the ASTM E8 standards with the help of a tensometer under a constant crosshead speed. Ultimate tensile strengths of 1322 and 1290 N/mm^2 were achieved after tensile test for 120° and

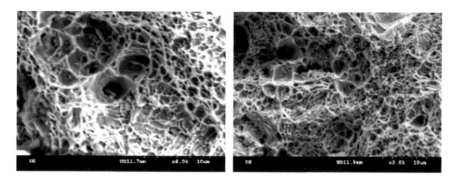

Fig. 11 Fracture Morphology of the 150° ECAPed billet

150° channel angles. To obtain information on the grain refinement and about the lattice distortions caused by the ECAP process, the X-ray diffraction peaks were studied, and hence the crystalline size was calculated for the processed samples. Cross section of the processed samples in a direction perpendicular to the application of the load was used for the X-ray diffraction. A RIGAKU Ultima Model-III X-ray diffractometer using CuKα radiation in the wavelength band of $\lambda = 1.540598A°$ was used to record the peak broadening with 2θ scan 0° to 90°. The indexed peaks were identified, and the width of the peak was measured using the XRDA software. Full width half maximum technique was employed to study the peak broadening of samples after different passes. The calculated theta values were used to determine the crystalline size (Figs. 12, 13; Table 1).

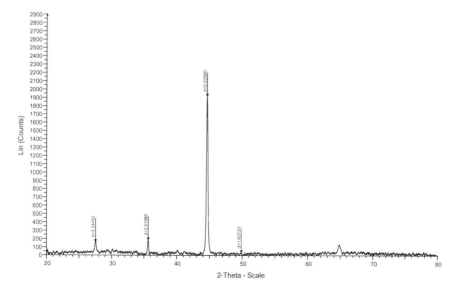

Fig. 12 XRD Peak with d spacing after 120° processed ECAP billet

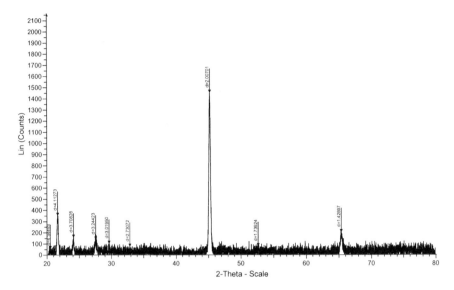

Fig. 13 Fig-13 XRD Peak with d spacing after 150° processed ECAP billet

Table 1 Crystalline size after ECAP of 120° and 150° processed billets

Sample name	Left angle	Observed maximum	d (observed maximum)	Maximum intensity	Average height	Full width half maximum	Mid of chord	Breadth	Gravity C	d of gravity C
	2-Theta °	2-Theta °	Angstrom	Cps	Cps	2-Theta °	2-Theta °	2-Theta °	2-Theta °	Angstrom
120 ECAPed Billet	44.310	44.602	2.02991	1921	1782	0.252	44.605	0.267	44.609	2.02963
150 ECAPed billet	44.900	45.139	2.00701	1452	1186	0.239	45.136	0.246	45.140	2.00697

4 Conclusions

- Equal channel angular pressing process was carried for EN47 steel by means of route A with different channel angles of 120° and 150°.
- Microstructural morphology was observed for the ECAP-processed billet nearby shear zone, center, and dead end. Center zone of the processed billet shows uniform distribution leading to an increase in the mechanical strength.
- X-ray diffraction analysis was performed for 120° and 150° processed billets, and their peaks were identified and its crystalline structure was predicted.
- 120° channel angle processed billets show lesser ductility compared to the 150° ECAP-processed billets.

References

1. Valiev RZ, Langdon TG (2006) Prog Mater Sci 51:881
2. Horita Z, Fujinami T, Nemoto M, Langdon TG (2000) Metall Mater Trans A 31A:691
3. Yang G, Huang CX, Wang C, Zhang LY, Hu C, Zhang ZF, Wu SD (2009) Enhancement of mechanical properties of heat-resistant martensitic steel processed by equal channel angular pressing. Mater Sci Eng, A 515:199–206
4. He T, Xiong Y, Ren F, Guo Z, Volinsky AA (2012) Microstructure of ultra-fine-grained high carbon steel prepared by equal channel angular pressing. Mater Sci Eng, A 535:306–310
5. Xiong Y, He T, Guo Z, He H, Ren F, Volinsky AA (2013) Mechanical properties and fracture characteristics of high carbon steel after equal channel angular pressing. Mater Sci Eng, A 563:163–167
6. Fukuda Y, Oh-ishi K, Horita Z, Langdon TG (2002) Processing of a low-carbon steel by equal-channel angular pressing. Acta Mater 50:1359–1368
7. Shin DH, Kim WJ, Choo WY (1999) Grain refinement of a commercial 0.15% c steel by equal-channel angular pressing, PII S1359-6462(99)00156-6. Scripta Materialia 3:259–262
8. Okayasu M, Sato K, Mizuno M, Hwang DY, Shin DH (2008) Fatigue properties of ultra-fine grained dual phase ferrite/martensite low carbon steel. Int J Fatigue 30:1358–1365
9. Ormsuptave N, Uthaisangsuk V (2017) Modelling of bake-hardening effect for fine grain bainite-aided dual phase steel. Mater Des 118:314–329
10. Phoumiphon N, Othman R, Ismail AB (2016) Improment in Mechanical Properties Plain Low Carbon Steel via Cold Rolling and Intercritical Annealing. Procedia Chem 19:822–827
11. Zhao Y, Guo H, Shi Z, Yao Z, Zhang Y (2011) Microstructure evolution of TA15 titanium alloy subjected to equal channel angular pressing and subsequent annealing at various temperatures. J Mater Process Technol 211:1364–1371
12. Zhang L, Ma A, Jiang J, Jie X (2015) Effect of processing methods on microhardness and acid corrosion behavior of low-carbon steel. Mater Des 65:115–119
13. Eghbali B, Shaban M (2013) Warm deformation of low carbon steel using forward extrusion-equal channel angular pressing technique. J Iron Steel Res Int 20(2):68–71
14. Huang CX, Yang G, Gao YL, Wu SD, Zhang ZF (2008) Influence of processing temperature on the microstructures and tensile properties of 304L stainless steel by ECAP. Mater Sci Eng, A 485:643–650

Microstructural Characteristics and Tensile Properties of Linear Friction-Welded AA7075 Aluminum Alloy Joints

P. Sivaraj, M. Vinoth Kumar and V. Balasubramanian

Abstract Linear Friction Welding (LFW) has been a recently developed solid-state welding process, has capabilities to efficiently join heat treatable Aluminum alloys by limiting the time at the temperature of intermetallic compound formation during the weld cycle. The transverse tensile properties of LFW joints of AA7075 Aluminum Alloy was evaluated and correlated with the microstructural characteristics of the joint. The weld joints exhibited joint efficiency of 75% with the fracture at Thermo-Mechanically Affected Zone. The XRD pattern revealed the absence of major strengthening precipitate $MgZn_2$ in weld nugget and TMAZ of the LFW joint. The absence of strengthening precipitates and coarse grain structure led to the deterioration of the mechanical properties in the TMAZ. The fracture surfaces of the tensile specimens revealed the ductile mode of failure.

1 Introduction

Fusion welding of AA7075 Aluminum alloy is challenging due to hot cracking problems. Solid-state welding processes can produce weld joint way below its melting temperature and at a shorter weld time, which in turn provides an effective solution to reduce the formation of hot cracks [1]. Linear Friction Welding (LFW) is one such recently developed solid-state welding process, having the

P. Sivaraj (✉) · V. Balasubramanian
Department of Manufacturing Engineering, Center for Material Joining
and Research (CEMAJOR), Annamalai University, Annamalai Nagar,
Chidambaram 608002, Tamil Nadu, India
e-mail: cemajorsiva@gmail.com

V. Balasubramanian
e-mail: visvabalu@yahoo.com

M. Vinoth Kumar
Department of Mechanical Engineering, Hindustan Institute of Technology
and Science, Padur, Chennai 603103, Tamil Nadu, India
e-mail: mvinothk@hindustanuniv.ac.in

© Springer Nature Singapore Pte Ltd. 2019
A. K. Lakshminarayanan et al. (eds.), *Advances in Materials and Metallurgy*,
Lecture Notes in Mechanical Engineering,
https://doi.org/10.1007/978-981-13-1780-4_45

capabilities to join workpieces of non-axisymmetrical geometry [2]. LFW is effective in limiting the time at the temperature of intermetallic compound formation during the weld cycle. LFW is accomplished by applying a relative linear motion against the contact faces of the workpiece to plasticize the weld zone using frictional heat and the workpieces were subsequently forged against each other by applying forging pressure [3]. Bhamji et al. investigated the effects of LFW parameters on the strength of AA6082-T6/AZ31 alloy LFW joints and correlated it with the relative fractions of intermetallic phase at the weld line. The relative fractions of intermetallic phase were found to decrease with increase in welding pressures. It is suggested to use welding pressure levels higher than 133 MPa to achieve better joints [4]. Buffa et al. studied the effect of specimen geometries in LFW of AA2011-T8 and AA6082-T6 alloy joints. LFW joints of AA2011 with smaller geometries and AA6082 joints with larger geometries resulted in higher strengths, and this effect is attributed to the flow stress of the materials welded together [5]. Rotundo et al. demonstrated to use of LFW in joining of 2124Al/25 vol.%SiCp metal matrix composites. The weld zone microstructure reveals the uniform distribution of the particles within the weld zone, which accredits the capability of the process to join materials with poor weldability. The test results on LFW joint of composite exhibited promising tensile and fatigue strength of up to 80% of that of the composite [6]. From the literature, it is concluded that LFW is a potential process for joining alloys of limited weldability, and the literature on mechanical properties and microstructural characterization of LFW joints of AA7075 is very scant. Hence, it is necessary to carry out the present investigation to evaluate the microstructural characteristics and tensile properties of LFW AA7075 joints.

2 Experimental Details

AA7075 grade, rolled plates of 6 mm thick received in T651 condition were used as parent metal (PM) in this investigation. The chemical composition of parent metal is presented in Table 1.

The welding was carried out normal to the rolling direction of the parent metal. A series of trial runs were made to attain defect free joint and, subsequently, the weld parameters were optimized to achieve maximum tensile strength. The optimized parameters used to fabricate the LFW joint are Friction stress 27 MPa, Friction time of 30 s, Forging stress 83 MPa, Forging time 3 s, and Frequency 50 Hz. Photographs of the fabricated joint and scheme of specimen extraction are shown in Fig. 1.

Table 1 Chemical composition (wt%) of AA7075-T651 alloy

Zn	Mg	Cu	Fe	Si	Mn	Cr	Ti	Al
6.1	2.9	2.0	0.50	0.4	0.30	0.28	0.20	Bal

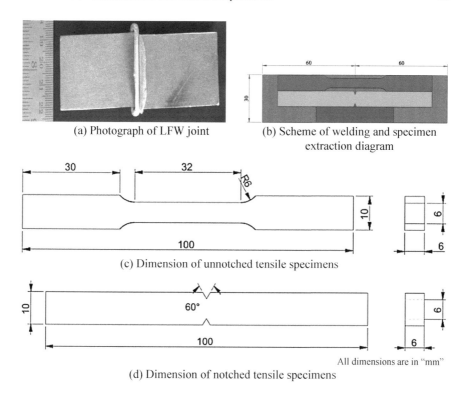

(a) Photograph of LFW joint   (b) Scheme of welding and specimen extraction diagram

(c) Dimension of unnotched tensile specimens

(d) Dimension of notched tensile specimens

All dimensions are in "mm"

Fig. 1 Scheme of specimen extraction and dimensions of specimen

The unnotched and notched tensile specimens were prepared as per ASTM E8 standard and the dimensions are shown in Fig. 1c, d, respectively. Tensile test was carried out using 100 kN electro-mechanical controlled universal testing machine. Hardness measurement was done across the weld centerline using Vickers micro-hardness tester with a load of 50 g and dwell time of 15 s. The sample for microstructural analysis was extracted as shown in Fig. 1b, polished using standard metallographic technique and etched with Keller's reagent (150 ml H_2O, 3 ml HNO_3, and 6 ml HF) for 20 s. The microstructural analysis was done using an optical microscope and scanning electron microscope (SEM) attached with an Energy Dispersive Spectrometer (EDS). The X-ray diffraction (XRD) analysis of the weld joint was carried out using Pan Analytical X-ray diffractor supported by Xpert Pro software.

3 Results

3.1 Microstructure

The optical micrograph of parent metal is shown in Fig. 2a, and it consists of elongated grains along the rolling direction of the AA7075 plates.

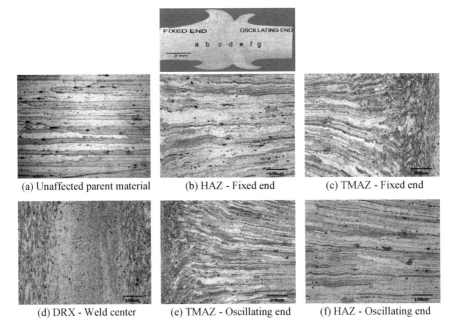

Fig. 2 Optical micrographs of LFW joint

The microstructural features of LFW joints of AA7075 presented in Fig. 2b–f reveals the complex microstructural variation across the joint.

The microstructural regions identified as heat-affected zone (HAZ), thermo-mechanically affected zone (TMAZ), and dynamically recrystallized zone (DRX) are mapped in the macrograph of the weld joint is shown in Fig. 2. The micrograph of the weld center consists of fine grains among the other zones in the weld joint. Grain coarsening together with severe plastic deformation is observed in the TMAZ, while the coarsened grains were observed in the HAZ of the weld joint. The microstructures of fixed and oscillating ends of the joint to their respective TMAZ and HAZ were found similar. The SEM micrograph of weld center is shown in Fig. 3a, with fine grains and undissolved/reprecipitated strengthening precipitates. The EDS composition of weld centre shown in Fig. 3b, closely matches the parent metal composition. The SEM micrograph of TMAZ region shown in Fig. 3c, reveals mechanically deformed grains with precipitates. The spot EDS analysis of the TMAZ precipitates shown in Fig. 3d, confirms the rich presence of alloying elements within the precipitates and the precipitates can be grouped as Mg (Zn_2, AlCu).

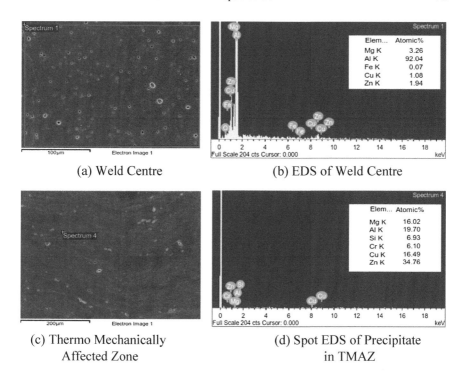

(a) Weld Centre   (b) EDS of Weld Centre

(c) Thermo Mechanically   (d) Spot EDS of Precipitate
Affected Zone   in TMAZ

Fig. 3 Scanning electron micrograph and EDS composition of AA7075 parent metal and TMAZ of LFW joint

3.2 X-Ray Diffraction

The XRD patterns of the parent metal, weld center, and TMAZ are shown in Fig. 4a–c, respectively. The respective PCPDF cards were identified for the peaks.

The XRD pattern of parent metal shown in Fig. 4a, reveals three types of precipitates $MgZn_2$, $Mg_3Cr_2Al_{18}$, and Al_7Cu_2Fe in Aluminum matrix. The XRD pattern of weld center shown in Fig. 4b, reveals the presence of coarse Al_7Cu_2Fe precipitates and dissolution of finer precipitates ($MgZn_2$, $Mg_3Cr_2Al_{18}$) as their respective peaks are missing. However, the XRD pattern of TMAZ shown in Fig. 4c, reveals the disappearance of the finer precipitate $MgZn_2$ from the region, while the precipitate $Mg_3Cr_2Al_{18}$ along with blocky Al_7Cu_2Fe were stable in the region after LFW.

3.3 Tensile Properties

The transverse tensile properties of the parent metal and weld joint are presented in Table 2. The LFW joint exhibited 75% of the PM tensile strength with a reduction

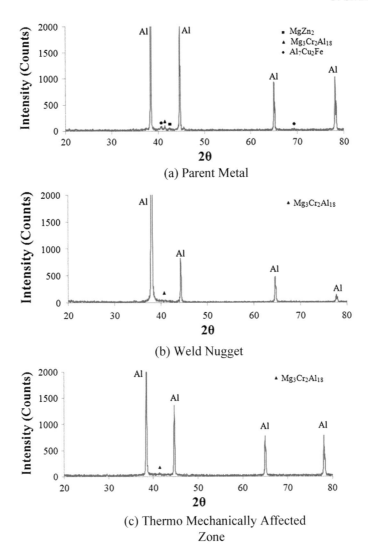

Fig. 4 X-ray diffraction pattern of AA7075 and linear friction weld joint

Table 2 Transverse tensile properties of parent metal and LFW joint

	0.2% Yield strength (MPa)	Ultimate tensile strength (MPa)	Elongation in 50 mm gage length (%)	Notch tensile strength (MPa)	Notch strength ratio (NSR)	Joint efficiency (%)	Fracture location
Parent metal	510	563	16	571	1.01	–	–
LFW joint	360	425	12	410	1.04	75	TMAZ

in elongation of 4%. The Notch Strength Ration (NSR) of both PM and weld joint is greater than unity, which represents both the PM and weld are notch ductile. The fracture in the cross-weld tensile specimen is located in the TMAZ.

3.4   Hardness

The hardness variation across the LFW joint of AA7075 is shown in Fig. 5. The maximum hardness value is recorded in the weld center, while the TMAZ reported the minimum hardness value among the weld regions. A gradual recovery in hardness of the joint is observed from the TMAZ to PM on both sides of the joint.

3.5   Fracture Surfaces

The fracture surfaces of smooth and notch tensile specimens of LFW joints of AA7075 are shown in Fig. 6. The macrograph and low magnification SEM images of the fracture surface of smooth specimen shown in Fig. 6a, b reveals a fracture with a negligible reduction in cross-sectional area of the specimen. The higher magnification image of the fracture surface of the smooth tensile specimen shown in Fig. 6c, reveals dimples of varying sizes. The fracture surface of the notch tensile specimen shown in Fig. 6d–f, reveals a flat featureless region in high magnification image Fig. 6f.

Fig. 5 Hardness distribution across the LFW joint of AA7075

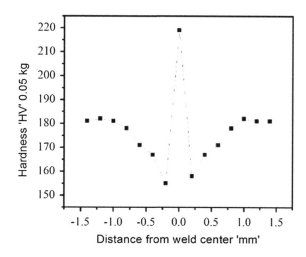

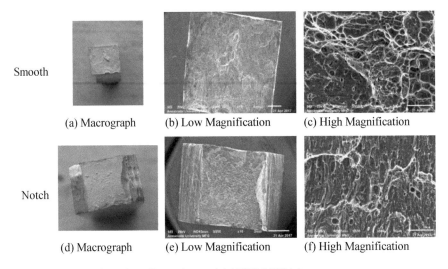

Smooth

(a) Macrograph   (b) Low Magnification   (c) High Magnification

Notch

(d) Macrograph   (e) Low Magnification   (f) High Magnification

Fig. 6 Fracture surface of tensile specimen of AA7075 LFW joints

4 Discussion

The combination of high temperature developed out of friction between the oscillating plates and intense plastic deformation due to the applied forging pressure during LFW resulted in the formation of the dynamic recrystallization (DRX) region at the weld center. The recrystallization experienced at the interface of the LFW joint as any other solid-state welding processes is evidenced by the fine grains in the weld center of the joint (refer Fig. 4a) [7]. The formation of the DRX region in the weld joint is considered as a positive indicator on the structural integrity of the AA7075 joints welded using LFW process and the LFW parameters used. The size and distribution of $MgZn_2$ precipitates plays a major role in deciding the tensile properties and hardness of AA7075 alloy and its joints [8]. The XRD analysis reveals the dissolution of the finer major strengthening precipitate $MgZn_2$ from the weld nugget of the LFW joint (refer Fig. 4b) due to solutionizing effect of the DRX process, while the blocky $Mg_3Cr_2Al_{18}$ precipitate is found to be stable in the nugget after LFW. The DRX process in the nugget during LFW resulted in finer grain size, which is attributed to the highest hardness recorded in the nugget of the joint [1, 9]. The coarse Al_7Cu_2Fe precipitates formed of iron and silicon impurities present in the base metal are dissolved into the matrix during LFW [10, 11]. The thermal cycles encountered by the TMAZ during LFW resulted in grain growth in the region with relatively larger grain within the other regions [12]. The XRD analysis of the TMAZ (refer Fig. 4c), confirms the dissolution of strengthening precipitate $MgZn_2$ in the region, while the $Mg_3Cr_2Al_{18}$ precipitate remains unaltered or may have grown in size. The highly deformed structure of TMAZ, resulted in high dislocation density, which in turn aided the growth of precipitates through diffusion [11]. Hence, the presence of large grains and absence of major strengthening precipitates $MgZn_2$, are attributed to the lowest hardness exhibited by TMAZ. The deteriorated properties of TMAZ can be further evidenced by the fracture of the joint located at TMAZ of the transverse tensile test specimen. The characteristic dimples in fracture surface of LFW joints evidence the adequate level of ductility exhibited by the joint during tensile test.

5   Conclusions

(1) Linear friction-welded AA7075 aluminum alloy joints exhibited a joint efficiency of 75% and the cross-weld tensile specimen failed in the thermo-mechanically affected zone (TMAZ).
(2) The weld nugget region revealed a fine recrystallized micro structure attributed to the dynamic recrystallization (DRX) process at the interface of the joint during Linear Friction Welding.
(3) LFW process resulted in the dissolution of the major strengthening precipitate $MgZn_2$in the weld nugget, while the blocky $Mg_3Cr_2Al_{18}$ precipitate is found to be stable in the nugget and TMAZ.

References

1. Rhodes CG, Mahoney MW, Bingel WH, Spurling RA, Bampton CC (1997) Effects of friction stir welding on microstructure of 7075 aluminum. Scriptamaterialia 36:69–75
2. Shtrikman MM (2010) Linear friction welding. Welding Int 24:563–569
3. Wenya L, Vairis A, Preuss M, Ma T (2016) Linear and rotary friction welding review. Int Mater Rev 61:71–100
4. Bhamji I, Preuss M, Moat RJ, Threadgill PL, Addison AC (2012) Linear friction welding of aluminium to magnesium. Sci Technol Weld Join 17:368–374
5. Buffa G, Cammalleri M, La Commare U, Fratini L (2017) Linear friction welding of dissimilar AA6082 and AA2011 aluminum alloys: microstructural characterization and design guidelines. Mater Form 10:307–315
6. Rotundo F, Ceschini L, Morri A, Jun TS, Korsunsky AM (2010) Mechanical and microstructural characterization of 2124Al/25vol.%SiC p joints obtained by linear friction welding (LFW). Com Part A: Appl Sci Manuf 41:1028–1037
7. Mishra RS, Ma ZY (2005) Friction stir welding and processing. Mater Sci Eng: R: Rep 50:1–78
8. Richard D, Adler PN (1977) Calorimetric studies of 7000 series aluminum alloys: I. Matrix precipitate characterization of 7075. Metall Mater Trans A 8:1177–1183
9. Mahoney MW, Rhodes CG, Flintoff JG, Bingel WH, Spurling RA (1998) Properties of friction-stir-welded 7075 T651 aluminum. Metall Mater Trans A. 29:1955–1964

10. Deshpande NU, Gokhale AM, Denzer DK, John L (1998) Relationship between fracture toughness, fracture path, and microstructure of 7050 aluminum alloy: Part I. Quantitative characterization. Metall Mater Trans A 29:1191–1201
11. Feng AH, Chen DL, Ma ZY (2010) Microstructure and cyclic deformation behavior of a friction-stir-welded 7075 Al alloy. Metall Mater Trans A 41:957–971
12. Kwon YJ, Saito N, Shigematsu I (2002) Friction stir process as a new manufacturing technique of ultrafine grained aluminum alloy. Mater Sci Lett 21:1473–1476

Finite Element Analysis of Polymer Networks with Heterogeneous Random Porous Microstructure

Mehdi Jafari, A. Praveen Kumar and Mohammad Khalili

Abstract An assumption in the micromechanical analysis of polymers is that the constitutive polymeric media is nonporous. Non-porosity of media, however, is merely a simplifying assumption. In this research paper, this assumption is neglected and polymer networks with a different porosity volume fraction are studied. A random morphology description function is used to model the porosity of the network. Nonlinear finite element analyzes are conducted to perform structural analysis of porous polymer networks using an ABAQUS/CAE finite element code. The results reveal that porosity plays a significant role in the mechanical behavior of polymer networks and may increase the maximum von Mises stress drastically.

Keywords Porous media · Porous polymer · Finite element method

1 Introduction

Since the discovery of zeolites and their successful industrial applications, the porous material shave becomes one of the most exciting frontiers in modern science. From the physical point of view, porosity plays a significant role in modifying properties of the materials and has a variety of applications including medical devices and tissue engineering [1, 2]. In the past decade, the field of porous materials has a revolutionized growth history. A number of new porous materials like metal–organic frameworks (MOFs), crystalline covalent organic frameworks (COFs), and amorphous porous organic polymers (POPs) were developed and

M. Jafari · M. Khalili (✉)
Khomeinishahr Branch, Islamic Azad University, Khomeinishahr, Iran
e-mail: dr.mo.khalili@gmail.com

A. Praveen Kumar
Department of Mechanical Engineering, SSN College of Engineering, Chennai, India
e-mail: praveenphd15@gmail.com

© Springer Nature Singapore Pte Ltd. 2019
A. K. Lakshminarayanan et al. (eds.), *Advances in Materials and Metallurgy*,
Lecture Notes in Mechanical Engineering,
https://doi.org/10.1007/978-981-13-1780-4_46

intrigued much attention not only by their high porosity nature (zeolites or activated carbons), but also their capability of incorporating targeted or multiple chemical functionalities into the porous framework by bottom-up or post-synthetic modification approach. They have been recently explored as promising candidates for applications in gas storage and gas separations [3]. These applications of porous polymers make it essential to perform fluid–structure interaction (FSI) analysis in the presence of porous polymers as solid structures. While there are numerous studies conducted on numerical analysis [4], CFD methods and its applications [5–11], investigating the mechanical properties and finite element analysis of porous polymers remains intact. The objective of this paper is to develop and implement a methodology for creating a morphologically realistic heterogeneous random porous microstructures over the entire volume fraction range, and subsequently, analyze their statistical and homogenized material properties attempting to extract valuable insight into the behavior of realistic porous polymer. To that end, the asymptotic expansion homogenization (AEH) method is used in conjunction with multiscale analysis to obtain the stresses at the microscopic and macroscopic levels.

2   Finite Element Model (FEM)

The random morphology description function (RMDF) is implemented to build random microstructure models with different porosity values. Where N is the number of random function, C_i and y^i are random coefficient and random coordinates, respectively.

$$f(y) = \sum_{i=1}^{N} c_i e^{-\left[\dfrac{(y_1 - y_1^{(i)}) + (y_2 - y_2^{(i)})}{w_i^2}\right]}, \quad w_i = \frac{1}{\sqrt{N}} \tag{1}$$

This equation is a summation of two-dimensional Gaussian functions to create realistic random function [12]. Equation (1) with different values of cutoffs can be used to build 2D random porous media with different amounts of porosity. A couple of phenomenal-based models and mechanical statistics-based models [13, 14] have been developed to model mechanical response of the polymer networks. Recently, Tehrani-Sarvestani model [15–17] was developed to model the mechanical behavior and failure of polymer networks. In this study, the Neo-Hookean model is used. For a compressible Neo-Hookean material, the strain energy density function is given by

$$W = C_1(\bar{I}_1 - 3) + D_1(J - 1)^2 \tag{2}$$

where C_1 and D_1 are material constants, $\bar{I}_1 = J^{-2/3} I_1$ is the first invariant of the isochoric part of the right Cauchy–Green deformation tensor.

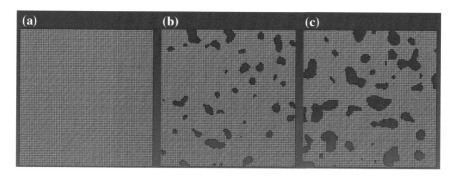

Fig. 1 **a** Nonporous media, **b** porous media with 10% porosity, and **c** porous media with 20% porosity

Table 1 Mechanical property of polymer network [18]	Parameter	Value
	C_1 (MPa)	0.2587
	D_1	1.5828e−3

Figure 1 shows the finite element model of (a) nonporous media, (b) porous media with 10% porosity, and (c) porous media with 20% porosity. Table 1 indicates the mechanical properties of the polymer network used in this simulation. The size of the specimen is 100 mm × 100 mm and six-node-modified quadratic plane stress triangle elements (CPS6 M). These were used for the model in order to reduce mesh density without affecting solution accuracy. Figure 2 shows the schematic boundary condition of the finite element model created using ABAQUS/ CAE. The implemented model is subject to a biaxial deformation control loading.

3 Results and Discussion

In this section, the mechanical behavior of homogenized polymer networks is presented for heterogeneous random porous polymer networks. Von Mises stresses and total deformations are illustrated in Figs. 3 and 4. While the maximum von Mises stress in the nonporous media is found as 1.9 (MPa), the maximum von Mises stress in porous media with $v = 10\%$ and $v = 20\%$ is 21.9(MPa) and 22.6 (MPa), respectively. The stress contours in the porous media indicate that by increasing the porosity of the media, the maximum von Mises stress will increase drastically.

In comparison with nonporous media, by increasing its porosity, the higher volume fraction of the porous media will face the same or lesser amount of the von Mises stress in the nonporous media, i.e., lesser volume fraction of the porous media will withstand higher values of stresses. It can be observed from the stress

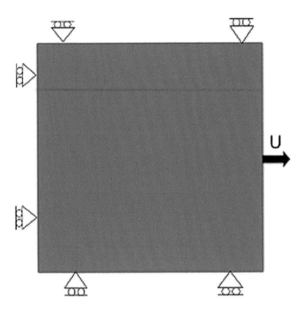

Fig. 2 Schematic boundary condition of finite element model

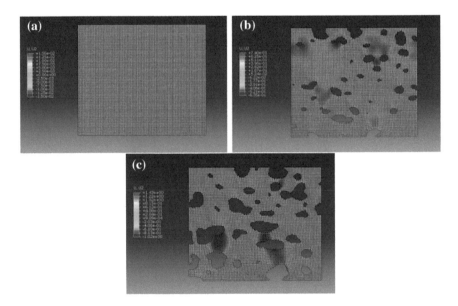

Fig. 3 Deformation perpendicular to the direction of subjected load with (**a**) no porosity, **b** 10% porosity, **c** 20% porosity

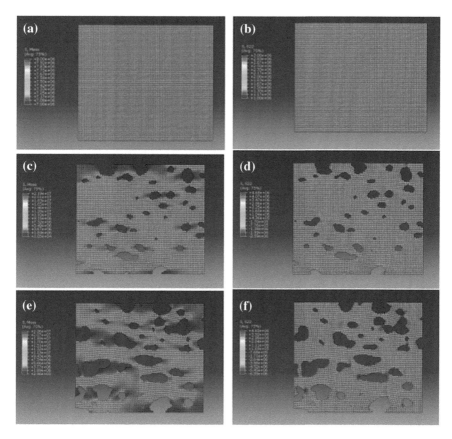

Fig. 4 Von Mises stress of media with (**a**) no porosity, **c** 10% porosity, **e** 20% porosity and stress perpendicular to the direction of the subjected load with (**b**) no porosity, **d** 10% porosity, **f** 20% porosity

contours that the certain regions, i.e., between the holes, will face extremely higher values of stresses due to the geometry of porous media and these regions are more prone to failure. Furthermore; increasing the porosity will increase the volume fraction of the media which is lesser than nonporous media and as a result, a lower volume fraction of the material must withstand the deformation with a higher value of von Mises stress. While there are many ongoing studies in the field of nanoparticles, this study illustrates the importance of polymers with nano-porosity. There are remarks regarding capability and validity of proposed framework which must be mentioned. First, the polymer network is assumed to be polydisperse and the chains length distribution follows a simple exponential distribution. Second, the deformations are assumed to be affine which means that at each instance all the polymer strands are under the same deformation.

4 Conclusions

The overall results of this study reveals that structural analysis using finite element method of porous media is crucial to investigate. In this regard, the finite element method has been successfully used to develop mechanical properties of the heterogeneous random porous polymer network. Initial simulations have shown that increasing the porosity will increase the maximum von Mises stress consequently. This study also shows that in the presence of porous polymer networks, performing structural analysis to predict the stress field and failure is inevitable.

References

1. Rezwan K, Chen Q, Blaker J, Boccaccini AR (2006) Biodegradable and bioactive porous polymer/inorganic composite scaffolds for bone tissue engineering. Biomaterials 27 (18):3413–3431
2. Tehrani M, Moshaei MH, Sarvestani A (2017) Cell mechanotransduction is mediated by receptor diffusion. In: 54th annual technical meeting society of engineering science, no. https://doi.org/10.13140/rg.2.2.34771.89124
3. Lu W, Yuan D, Sculley J, Zhao D, Krishna R, Zhou HC (2011) Sulfonategrafted porous polymer networks for preferential CO_2 adsorption at low pressure. J Am Chem Soc 133 (45):181261–8129
4. Tehrani M, Sarvestani A (2017) Force-driven growth of intercellular junctions. J Theor Bio 421:101–111
5. Asadi S, Mokhtari AA, Suratgar AA, Khodabandeh E, Karimi A (2013) A novel two-phased flow meter design using mems pressure meters array. In: IEEE 3rd international conference on control, instrumentation, and automation (ICCIA), pp 89–94
6. Khodabandeh E, Abbassi A (2018) Performance optimization of water-Al_2O_3 nanofluid flow and heat transfer in trapezoidal cooling microchannel using constructal theory and two phase eulerian-lagrangian approach. Powder Technol 323:103–114
7. Khodabandeh E, Davoodi H, Zargar O, Rozati SA (2016) Evaluation of the performance and structure of an industrial bellows valve using finite element analysis. 2(1)
8. Mashayekhi R, Khodabandeh E, Bahiraei M, Bahrami L, Toghraie D, Akbari OA (2017) Application of a novel conical strip insert to improve the efficacy of water{ag nanofluid for utilization in thermal systems: A twophase simulation. Energy Convers Manage 151:573–586
9. Som D, Li K, Kadel J, Wright J, Modaresahmadi S, Bird JZ, William W (2017) Analysis and testing of a coaxial magnetic gearbox with flux concentration halbach rotors. IEEE Trans Magn 53(11):1–6
10. Li K, Wright J, Modaresahmadi S, Som D, Williams W, Bird JZ (2017) Designing the first stage of a series connected multistage coaxial magnetic gearbox for a wind turbine demonstrator. In: IEEE energy conversion congress and exposition (ECCE), IEEE, pp 1247–1254
11. Ahmadi SM, Ghazavi MR, Sheikhzad M (2015) Dynamic analysis of a rotor supported on ball bearings with waviness and centralizing springs and squeeze film dampers. Int J Eng-Trans C: Aspects 28(9):13–51
12. Vel SS, Goupee AJ (2010) Multiscale thermoelastic analysis of random heterogeneous materials: Part i: Microstructure characterization and homogenization of material properties. Comput Mater Sci 48(1):22–38

13. Arruda EM, Boyce MC (1993) A three-dimensional constitutive model for the large stretch behavior of rubber elastic materials. J Mech Phy Solids 41(2):389–412
14. Tehrani M (2017) Micromechanical analysis of strength of polymer networks with polydisperse structures, Ph.D. thesis, Ohio University
15. Tehrani M, Moshaei MH, Sarvestani A Revisiting the deformationinduced damage in filled elastomers: Effect of network polydispersity, arXiv preprint arXiv:1710.01846
16. Tehrani M, Moshaei MH, Sarvestani AS (2017) Network polydispersity and deformation-induced damage in filled elastomers. Macromol Theory Simul 26:1–9
17. Tehrani M, Sarvestani A (2017) Effect of chain length distribution on mechanical behavior of polymeric networks. Eur Polym J 87:136–146
18. Shahzad M, Kamran A, Siddiqui MZ, Farhan M (2015) Mechanical characterization and FE modelling of a hyperelastic material. Mater Res 18(5):918–924

Numerical and Experimental Evaluation of Hyperelastic Material Parameters

T. Sukumar, B. R. Ramesh Bapu and D. Durga Prasad

Abstract Hyperelastic materials are playing a vital role in hydraulic and pneumatic applications in the automotive industry and acts as a sealing element. Typical hyperelastic sealing elements are O-Rings, gaskets and lip seals, which are used in hydraulic and pneumatic braking systems in passenger car, light commercial and heavy commercial vehicles. If the sealing is not properly done, the efficiency and performance of the system will come down significantly. Since the hyperelastic materials are highly nonlinear, it is very difficult to predict the behaviour like sealing pressure and strain in the classical method. The behaviour of the hyperelastic material can be predicted using a numerical method called finite element analysis. The behaviour is purely based on the type of hyperelastic material model and number of material parameters. This research work will describe the material model and selection of material parameters based on the experimental study. Also, it will describe the sealing behaviour of the gasket which is used in automotive applications.

Keywords Hyperelastic · Material · Sealing element · Nonlinear
Sealing pressure · Strain · Material models · Finite element analysis

Nomenclature

FEA   Finite Element Analysis
IRHD   International Rubber Hardness Degrees
NBR   Nitrile Butadiene Rubber

T. Sukumar (✉)
Department of Mechanical Engineering, JNT University, Anantapur, India
e-mail: thiru_sukumar@yahoo.com

B. R. Ramesh Bapu
Chennai Institute of Technology, Chennai, India

D. Durga Prasad
Department of Mechanical Engineering, JNT University, Anantapur, India

© Springer Nature Singapore Pte Ltd. 2019
A. K. Lakshminarayanan et al. (eds.), *Advances in Materials and Metallurgy*,
Lecture Notes in Mechanical Engineering,
https://doi.org/10.1007/978-981-13-1780-4_47

1 Introduction

Hyperelastic rubber-like materials are extensively used in the automotive application for sealing purpose [1]. The main objective of the sealing is to prevent fluid (air and hydraulic oil) from leaking from high-pressure chamber to lower pressure chamber in the sealing region. The sealing elements are classified into two categories: (1) Static seal (2) Dynamic seal. Static seals are typically subjected to a preload during assembly to reach the sealing mechanism. Preload or axial squeeze occurs when seal height is designed to be higher than the sealing gap. During assembly, the gasket will be subjected to compression and it will deform elastically, produce internal stresses and generate a reaction force on both sides of the gasket top and bottom. When the fluid pressure is applied, the internal stress or sealing pressure will prevent the leakage from the sealing region. When the fluid pressure reaches an internal stress or sealing pressure, it will start to leak; hence, the fluid pressure should be maintained always less than the sealing pressure to ensure the 100% sealing.

One of the critical parameters for face sealing is preload or axial squeeze of the seal. The percentage of the axial squeeze is calculated as follows:

$$\text{Axial Squeeze} = \frac{\text{Initial height of the seal} - \text{Compressoed height of the seal}}{\text{Original height of the seal}} \tag{1}$$

As per industry practice to achieve the maximum squeeze, the seal height is maintained such that, will compress 25% squeeze from its original height. Also considering manufacturing tolerance, the suggested value of axial squeeze is 10–40%. Since the seals are made of hyperelastic material like rubber and highly nonlinear, it is very difficult to predict the sealing behaviour like axial squeeze, sealing pressure and string, etc.

2 Hyperelastic Material Models

Hyperelastic material models are used to predict the behaviour of the rubber materials with the help of material parameters [2]. Mooney–Rivlin Material model is one of the material models which are generated from stress–strain data using uniaxial testing machine. The material parameter or constants are as follows:

1. Mooney–Rivlin—2 parameter material constant.
2. Mooney–Rivlin—3 parameter material constant.
3. Mooney–Rivlin—5 parameter material constant.
4. Mooney–Rivlin—9 parameter material constant.

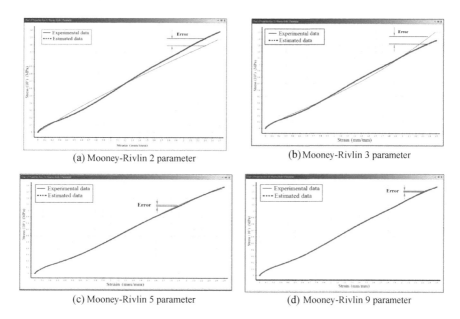

(a) Mooney-Rivlin 2 parameter

(b) Mooney-Rivlin 3 parameter

(c) Mooney-Rivlin 5 parameter

(d) Mooney-Rivlin 9 parameter

Fig. 1 Different types of material constants

The accuracy of the result purely depends on the number of material parameters. Fig. 1a–d shows the comparison between estimated data versus experimental data results.

3 Finite Element Analysis

Finite Element Method is a numerical method and it is used to predict the sealing behaviour using hyperelastic material models and material parameters. In this research, Mooney–Rivlin material model with different material parameters are used to predict the seal behaviour and the results are compared with experimental analysis results.

Mooney–Rivlin introduced a dependence of the strain-energy function on both the first and second invariants. It is one of the first hyperelastic model, and it is highly accurate in predicting the nonlinear behaviour of isotropic rubber-like materials. The strain-energy function for this material model is often seen in literature in the form of

$$W = C_{10}(\bar{I}_1 - 3)C_{01}(\bar{I}_2 - 3) + \frac{1}{d}(J - 1)^2 \tag{2}$$

where

W	Strain Energy Density
$\bar{I}_1$	First deviatoric strain invariant.
$\bar{I}_2$	Second deviatoric strain invariant.
C_{10}, C_{01}	Material Constants (two parameters)

The initial shear modulus

$$\mu = 2\left(C_{10} + C_{01}\right) \tag{3}$$

The initial bulk modulus (K)

$$K = \frac{2}{d} \tag{4}$$

where

$$d = \frac{1 - 2\gamma}{\left(C_{10} + C_{01}\right)} \tag{5}$$

Mooney–Rivlin works well with incompressible elastomers with strain up to 200%. For example, rubber for seals, O-rings, gaiters and an automobile tyre applications.

3.1 Generation of Stress–Strain Data

Dump-bell specimen was prepared for Nitrile butadiene rubber (NBR) material with hardness of 70 IRHD, as per the ISO 37 specification [3]. A uniaxial tensile testing machine is used to generate the stress–strain data as per ASTM 412 D test standard and the strain rate of 500 mm/min. The generated data is input to the finite element analysis software to generate the material constants using Mooney–Rivlin material model [4].

3.2 Finite Element Model

To study the effects of material parameters in gasket, which is used in typical automotive sealing application as shown in Fig. 2 along with its mating parts details. In FEA, it is modelled using 2D axisymmetric element since its symmetry about an axis is shown in Fig. 3. The axisymmetric elements are used to minimize the problem size, solution time and for better solution accuracy and better convergence. In addition to this, to avoid the convergence issue, the gasket and its

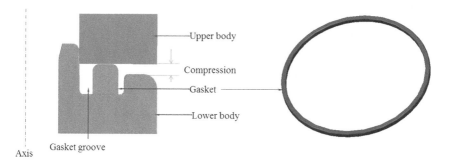

Fig. 2 Gasket and its mating parts details

Fig. 3 Finite element model

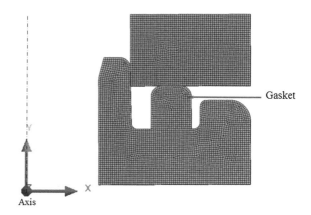

mating parts are modelled using 2D quadrilateral elements instead of triangular elements. The amount of gasket squeeze during assembly is given as a displacement along Y-axis on the top body and the displacement of bottom body is fixed as shown in Fig. 4. Coefficient of friction between the contact surfaces is considered as 0.25.

Nonlinear finite element analysis is carried out with Mooney–Rivlin 2 parameter material constants (C_{10}, C_{01}) and behaviour of the gasket is captured as shown in Fig. 5a–c. When gasket is squeezed (compressed) from free height to assembled height, the height of the gasket reduced and width of the gasket increased Fig. 5a, b. Also the strain will increase from free height to assembled height as shown in Fig. 5c.

Similarly gasket behaviour is studied for considering Mooney–Rivlin 3 parameters, 5 parameters and 9 parameters constants. The analysis results are plotted as graph from the Figs. 6, 7 and 8. It is clearly observed that the number of material parameter constant increases, the strain is decreased and contact pressure is increased [5]. Also, the reaction force or assembly force is increasing with increase in the number of material parameters [6].

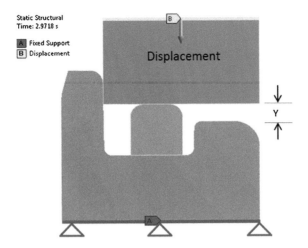

Fig. 4 Boundary condition details

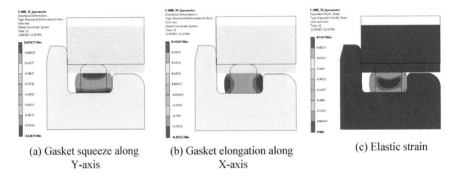

(a) Gasket squeeze along
Y-axis

(b) Gasket elongation along
X-axis

(c) Elastic strain

Fig. 5 Behaviour of the gasket after assembly

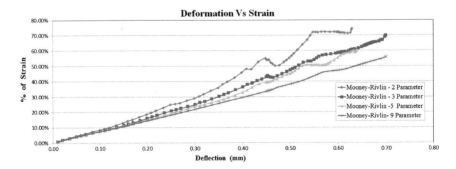

Fig. 6 Strain comparison between different material constants

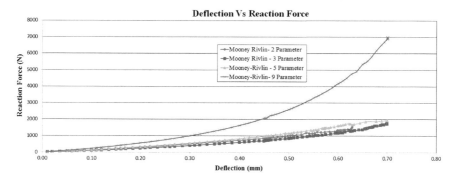

Fig. 7 Reaction force comparison between different material constants

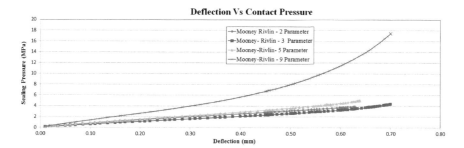

Fig. 8 Contact pressure comparison between different material constants

4 Experimental Verification

To evaluate the finite element analysis results with different material constants, experimental analysis is carried out for the gasket to predict the assembly force. Gasket is developed with hardness of 70 IRHD and assembled in a device with help of compression testing machine as shown in Fig. 9. During assembly, the assembly force is captured at different compression heights and plotted a graph as shown in Fig. 10. The graph shows that the reaction force is increased when the compression is increased.

Gasket assembled in a device

Fig. 9 Experimental arrangements

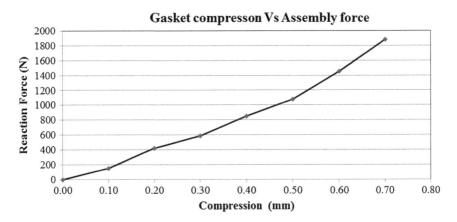

Fig. 10 Gasket assembly force

5 Inferences

The assembly force observed from the experiment at 0.70 mm compression is around 1880 N, and it is comparable with the reaction force of 1950 N obtained from the finite element analysis results of Mooney–Rivlin—5 parameters material constants as plotted in Fig. 12. Whereas, Mooney–Rivlin 9 parameters having high reaction force as shown in Fig. 11 and it is far away from the experimental assembly force because of higher material parameter. Similarly, 2 parameter and 3 parameter are having less reaction force when compared to the experimental results.

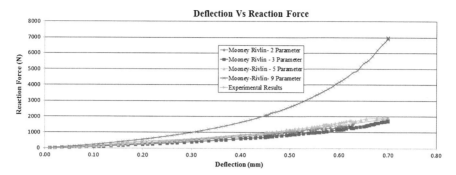

Fig. 11 Reaction force comparison between experimental and finite element analysis

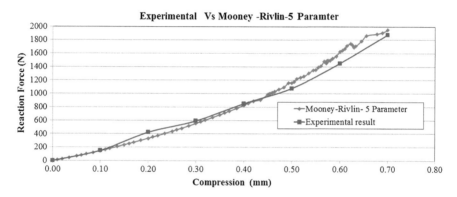

Fig. 12 Reaction force comparison between experimental and Mooney–Rivlin 5 parameters

Based on this inference, Mooney–Rivlin 5 parameters material constant is more appropriate for hyperelastic nonlinear analysis.

6 Conclusion

In this paper, hyperelastic material model and its number of material parameters have been studied. The material parameters are generated with the help of stress, strain data from uniaxial tensile test machine. To understand the effect of a number of material parameters, gasket has been taken for finite element analysis and results are plotted between the material parameters considering elastic strain, reaction force and contact pressure.

Also, to verify the finite element analysis results and conclude the optimum number of material parameters, experimental analysis was carried out. Based on the experimental results, the Mooney–Rivlin 5 parameter material constants results (assembly force) are very close to the experimental analysis results (gasket assembly forcc).

References

1. Boyce MC, Arruda EM (2000) Constitutive models of rubber elasticity: a review. Rubber Chem Technol 73(3):504–523
2. Sussman T, Bathe KJ (2009) A model of incompressible isotropic hyperelastic material behavior using spline interpolations of tension–compression test data. Commun Numer Meth Eng 25:53–63 Published online 1 February 2008 in Wiley InterScience (www.interscience. wiley.com, https://doi.org/10.1002/cnm.1105)
3. Rubber, Vulcanized or Thermo plastics-Determination of Stress-Strain Properties—ISO 37 Fifth Edition-2011
4. Nagdi K (1993) Rubber as an engineering material. Hanser Publisher, Munich, Germany. ISBN 0-06-056388-9
5. Bechir H, Chevalier L, Chaouche MK, Boufala K (2006) Hyperelastic constitutive model for rubber-like materials based on the first Seth strain measures invariant. Eur J Mech A/Solids 25:110–124
6. Markmann G, Verron E (2006) Comparison of hyperelastic models for rubber-like materials. Rubber Chem Technol 79:835–858

Electrospinning of PAN–Hematite Composite Nanofiber

S. J. Vijay, Alexandre Tugirumubano, Sun Ho Go, Lee Ku Kwac and Hong Gun Kim

Abstract PAN-based carbon nanofibers reinforced with nanoparticles are used in various applications including energy, medical, and electronics. Many of the production methods include chemical-based approach to fabricate such fibers. Here, an attempt has been made to fabricate PAN–hematite composite fiber using physical methods. Hematite nanopowders (30–40 nm) are mixed in PAN solution. They are spun into fibers using electrospinning technique. The spun fibers are stabilized and carbonized using a tubular furnace. The fibers are then characterized for their electrical and metallurgical properties. It is found that the nanofibers have hematite reinforced in them at regular interval. The produced PAN-Hematite composite fiber yielded desired electrical and metallurgical properties.

Keywords PAN fibers · Hematite · Composite nanofiber · SEM–EDAX

S. J. Vijay (✉)
Department of Mechanical and Aerospace Engineering, Karunya Institute
of Technology and Sciences, Coimbatore 641114, India
e-mail: vijayjoseph@karunya.edu

A. Tugirumubano · S. H. Go · L. K. Kwac · H. G. Kim (✉)
Institute of Carbon Technology, Jeonju University, Jeonju-si 55069,
Jeollabuk-do, South Korea
e-mail: hkim@jj.ac.kr

A. Tugirumubano
e-mail: alexat123@yahoo.com

S. H. Go
e-mail: royal2588@naver.com

L. K. Kwac
e-mail: kwac29@jj.ac.kr

© Springer Nature Singapore Pte Ltd. 2019
A. K. Lakshminarayanan et al. (eds.), *Advances in Materials and Metallurgy*,
Lecture Notes in Mechanical Engineering,
https://doi.org/10.1007/978-981-13-1780-4_48

1 Introduction

PAN-based carbon nanofibers are finding applications in energy-based industries. Their primary application is in energy storage devices like lithium–ion battery [1]. The property attributes such as controllable fiber diameter, high surface area-to-volume ratio, low density, high pore volume, etc. make it a better contender in many such applications [2–4]. The carbon nanofibers are also finding application in nanoelectronics industries and are used as electromagnetic shields [5]. But the ability of those materials can be improved by reinforcing them with metal nanoparticles [6]. Here, a trial has been made to reinforce Fe_2O_3 nanoparticles so that it increases the magnetic property of the carbon nanofibers. The nanofibers reinforced with nanoparticles are produced through electrospinning process. Further, they are stabilized and carbonized. The metallurgical properties are studied primarily to understand the structure of the nanofibers and the integrity of nanoparticles reinforced in them.

2 Experimental Procedures

Figure 1 shows the methodology followed to produce the nanoparticle-reinforced nanofibers, wherein the nanoparticles are physically mixed in the PAN solution using a mechanical stirrer. It is then spun using an electrospinning machine to desired layer and thickness. The PAN solution was prepared with 11 wt% of polyacrylonitrile (PAN) and 6 wt% Fe_2O_3 nanoparticles of size varying between 30 and 40 nm, in N, N-dimethyl formamide (DMF) solvent [7, 8]. The solution is then mixed using a hot plate stirrer at 70 °C for 6 h. The solution is loaded in an electrospinning machine, and the nanofibers are spun on an aluminum foil fixed to the anode plate of the machine. The voltage was kept at 15 kV, and the injection rate was maintained at 2.1 ml/h, at 27.3 °C. The humidity of the chamber was maintained at 40%. The electrospun nanofibers are then removed from the aluminum foil and stabilized in a tubular furnace for 2 h at 280 °C with atmospheric air. The rate of increase in temperature was maintained at 1 °C/min. Then, the fibers

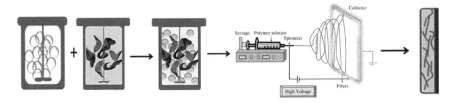

Fig. 1 Methodology to produced nanoparticles reinforced nanofibers

are carbonized at 750 °C for 1 h in nitrogen environment. The rate of increase in temperature was maintained at 5 °C/min. Figures 2, 3, and 4 show electrospun fibers before and after stabilization and carbonization.

3 Results and Discussions

Figure 5 shows the SEM microstructure of the electrospun PAN fibers which were impregnated with Fe_2O_3 nanoparticles. It also shows the EDAX area mapping of the structure which authentically characterizes the presence of carbon and Fe_2O_3 nanoparticles. The SEM image shows that the fibers ranging between 0.44 and 0.8 µm. It also shows that the Fe_2O_3 nanoparticles are embedded well in the PAN

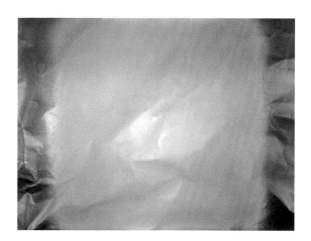

Fig. 2 Electrospun PAN fibers before stabilization

Fig. 3 Stabilized PAN fibers before carbonization

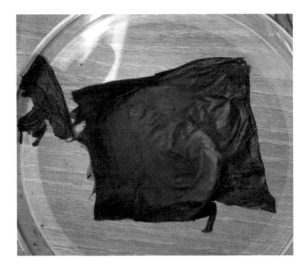

Fig. 4 Carbonized PAN fibers

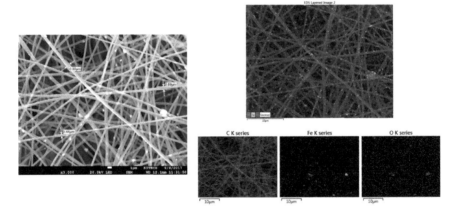

Fig. 5 SEM–EDAX area mapping of electrospun PAN fibers before stabilization

fibers. However, it is also evident that those particles have agglomerated in few locations in the fiber. Nevertheless, they have not disintegrated from the fibers at this stage.

Figure 6 shows the EDAX line mapping of the PAN fibers which also clearly indicates that the nodes which are visible in the fibers are the Fe_2O_3 nanoparticles which were mixed in the PAN solution while preparing the PAN solution. The images prove that the number of Fe_2O_3 nanoparticles nodes is directly proportional to the amount of Fe_2O_3 nanoparticles added during the preparation of the PAN solution.

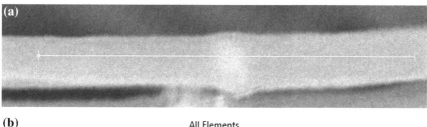

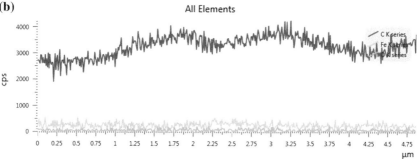

Fig. 6 EDAX line mapping of electrospun PAN fibers before stabilization

Fig. 7 SEM microstructure
of stabilized PAN fibers
before carbonization

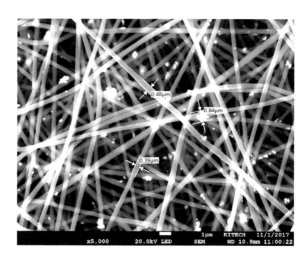

Figure 7shows the SEM image of the PAN fibers after stabilization. The fiber diameter has been stable during the stabilization process. There were no size variations observed during the stabilization. It is also observed that the Fe_2O_3 nanoparticles also stayed intact and have not distorted during this stage. It is found that the particles are prominent in the locations where they are embedded in the fiber. Defects like distortion of fiber or disintegration of particles from the fibers are not visible. However, there are few agglomerations of particles present in the fibers.

Figure 8 shows the EDAX area mapping of the stabilized samples. There are not many variations in the sample, and the compositions have remained as observed in the samples before stabilization.

Figure 9 shows the SEM image of the PAN fibers after carbonization. There are few variations observed, which includes the shrinkage in the carbon fibers. The size of the fiber has reduced and it is found to be around 0.36 μm. It is also observed that the fibers have disintegrated in locations where the particles are agglomerated. It is observed that in many places necking has happened in the locations where the nanoparticles are impregnated but not broken.

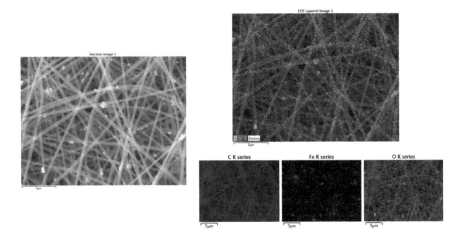

Fig. 8 SEM–EDAX area mapping of stabilized PAN fibers before carbonization

Fig. 9 SEM microstructure of carbonized PAN fibers

Fig. 10 SEM image showing the reinforced Fe_2O_3 nanoparticles in the PAN fibers

Figure 10 clearly visualized that the necking has happened in the locations where the particles are not agglomerated, whereas the fibers are broken in the places where the nanoparticles have agglomerated. This gives clear information that to construct nanofibers reinforced with nanoparticles, the agglomerations of nanoparticles should be avoided to produce nanofibers with better integrity.

4 Conclusions

PAN-based carbon nanofibers reinforced with Fe_2O_3 nanoparticles were successfully fabricated using electrospinning process. It was observed that the carbonized nanofibers exhibited 16 mΩ for the specified process parameters. It can be improved by increasing the carbonization temperature up to 1000 °C. It was found that the EDAX maps agreed well with the compositions that were added during the PAN solution preparation process. The SEM images showed that the Fe_2O_3 nanoparticles were properly embedded in the fibers and it has agglomerated in few locations. The images also proved that the nanoparticles have disintegrated at agglomerations during carbonization process but stayed intact at places where the particles are not agglomerated.

Acknowledgements This research was supported by Basic Science Research Program through the National Research Foundation of Korea (NRF) funded by the Ministry of Education (No. 2016R1A6A1A03012069) and the Korea Government (MSIP) (No. 2017R1A2B4009646).

References

1. Ji L, Zhang X (2009) Electrospun carbon nanofibers containing silicon particles as an energy-storage medium. Carbon 47:3219–3226

2. Agubra VA, Zuniga L, Flores D, Campos H, Villarreal J, Alcoutlabi M (2017) A comparative study on the performance of binary SnO$_2$/NiO/C and Sn/C composite nanofibers as alternative anode materials for lithium ion batteries. Electrochim Acta 224:608–621
3. Wu Q, Tran T, Lu W, Wub J (2014) Electrospun silicon/carbon/titanium oxide composite nanofibers for lithium ion batteries. J Power Sour 258:39–45
4. Xu Z-L, Zhang B, Kimn J-K (2014) Electrospun carbon nanofiber anodes containing monodispersed Si nanoparticles and graphene oxide with exceptional high rate capacities. Nano Energy 6:27–35
5. Mu J, Chen B, Guo Z, Zhang M, Zhang Z, Shao C, Liu Y (2011) Tin oxide (SnO$_2$) nanoparticles/electrospun carbon nanofibers (CNFs) heterostructures: controlled fabrication and high capacitive behavior. J Colloid Interface Sci 356(2011):706–712
6. Xia X, Wang X, Zhou H, Niu X, Xue L, Zhang X, Wei Q (2014) The effects of electrospinning parameters on coaxial Sn/C nanofibers: morphology and lithium storage performance. Electrochim Acta 121:345–351
7. Park S-W, Shim H-W, Kim J-C, Kim D-W (2017) Uniform Si nanoparticle-embedded nitrogen-doped carbon nanofiber electrodes for lithium ion batteries. J Alloy Compd 728:490–496
8. Yu Y, Yang Q, Teng D, Yang X, Ryu S (2010) Reticular Sn nanoparticle-dispersed PAN-based carbon nanofibers for anode material in rechargeable lithium-ion batteries. Electrochem Commun 12:1187–1190

Effect of Al–Si Filler Rods on the Microstructure and Mechanical Properties of TIG-Welded Thick AA5083-F Alloys

R. Aarthi and K. Subbaiah

Abstract The marine aluminum alloy 12 mm thick plates are joined using Tungsten Inert Gas welding technique. The mechanical properties and microstructural features of Tungsten Inert Gas welded joints of the Aluminum Alloy 5083-F with two Al–Si filler rods are investigated. Weldments processed by Al–Si filler rods, such as ER4043 and ER4047 are mechanically softer than the base material AA5083 in F condition. The welded joint tensile strength of ER4043 filler rod-welded joints is 33% higher than that of the ER4047 filler rod-welded joints tensile strength.

Keywords Tungsten inert gas welding · Al–si filler rods · ER4043 and ER4047 · Tensile properties · Microstructure · hardness

1 Introduction

The AA5083 aluminum alloy is widely used in marine, defense, automotive, cryogenics, and construction industries because of its high strength-to-weight ratio, reasonable corrosion resistance, superplasticity, and good weldability [1–4]. The major alloying element in AA5083 is magnesium which provides solid solution strengthening and additional strengthening through its influence on work hardening arising from cold deformation. The amount of magnesium in AA5083-H18 is higher than the equilibrium value of solubility limit of Mg in Al at room temperature. Therefore, the precipitation process can occur in this process. In this regard, Al (Mg, Si) phase is often present in the microstructure as the major constituent. In addition in the presence of Manganese (0.4–1 wt%) results in the precipitation of Al

R. Aarthi · K. Subbaiah (✉)
SSN College of Engineering, Chennai 603110, India
e-mail: subbaiahk@ssn.edu.in

R. Aarthi
e-mail: r.aarthi4@gmail.com

© Springer Nature Singapore Pte Ltd. 2019
A. K. Lakshminarayanan et al. (eds.), *Advances in Materials and Metallurgy*,
Lecture Notes in Mechanical Engineering,
https://doi.org/10.1007/978-981-13-1780-4_49

(Mn, Fe, Cr, and Si) during ingot preheating and high-temperature homogenization treatment [5–8].

Owing to the great prominence of low specific weight, low-temperature toughness, and corrosion resistance in marine environments, thick plates of Al 5083 alloy have been widely applied to liquefied natural gas (LNG) transport and storage tank. For different types of storage tank, the thickness of plate could vary from 35 to 200 mm. Welding is the main joining technology in manufacturing of LNG carrier. So far, arc welding is still the most frequently used welding process, since it shows outstanding advantages such as convenience, cost-effectiveness, simple clamping, and economical equipment [9].

2 Experimental Work

Butt Welding on the AA5083-F alloy 12 mm thick plates were made with two filler rods using alternating current Tungsten Inert Gas Welding machine with a standard 2% Thoriated tungsten electrode. The diameter of the electrode is 2 mm. The argon shielding gas flow rate was 40 l/min. The welding speed was kept at 150 mm/min. The arc welding current used is 230 A.

After welding, samples were cut across the weld for metallographic analysis and tensile tests using EDM machining process. The configuration and size of the transverse tensile specimens were prepared as per ASTM-E8 standard. Prior to the tensile tests, the Vickers hardness profiles across the weld were measured under a load of 1 Kgf for 15 s along the centerlines of the cross section of the tensile specimens using an automatic microhardness tester. The cross sections of the metallographic specimens were polished with alumina suspension, etched with Keller's reagent, and observed by optical microscopy.

3 Results and Discussion

3.1 Optical Microstructure of TIG-Welded AA5083-F with ER4043 Filler Rod

The AA5083-F aluminum alloy base metal microstructure is shown in Fig. 1. The intersection of the heat-affected zone and the weld are shown in Fig. 2. The weld microstructure of the ER4043 filler rod-welded joint is shown in Fig. 3. The weld microstructure consists of Al–Si eutectics and coarse grains. The heat-affected zone and the base metal microstructure are shown in Fig. 4.

Fig. 1 AA5083-F base metal

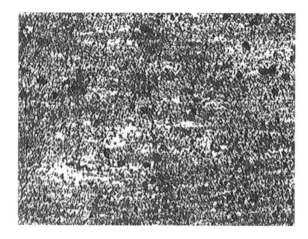

Fig. 2 Heat-affected zone and weld

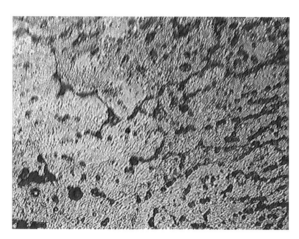

Fig. 3 Weld

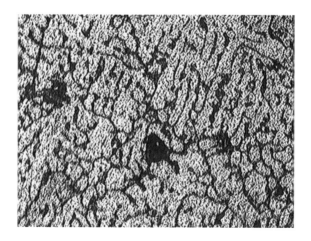

Fig. 4 Heat-affected zone
and base metal

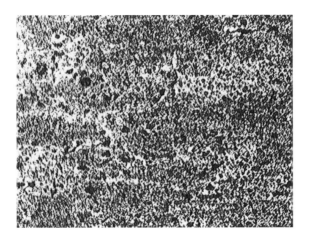

3.2 Optical Microstructure of TIG-Welded AA5083-F with ER4047 Filler Rod

The AA5083-F aluminum alloy base metal microstructure is shown in Fig. 5. The intersection of the base metal and heat-affected zone are shown in Fig. 6. The intersection of the heat-affected zone and the weld are shown in Fig. 7. The weld microstructure of the ER4047 filler rod-welded joint is shown in Fig. 8. The weld microstructure consists of Al–Si eutectoids and coarse dendrites. The weld section with porosity is shown in Fig. 9. The formation of porosity is due to entrapment of hydrogen atoms in molten metal. The shrinkage cavity defect in the weld is shown in Fig. 10. Multi-pass TIG welding is done on AA5053 F and the gap between the adjacent passes are shown in the Fig. 10.

Fig. 5 AA5083-F base metal

Fig. 6 Base metal and heat-affected zone

Fig. 7 Heat-affected zone and weld

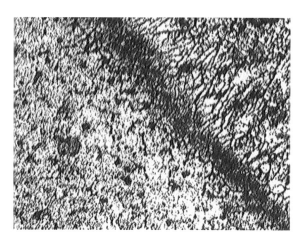

Fig. 8 Weld

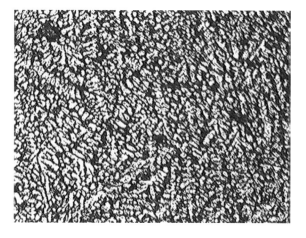

Fig. 9 Porosity in the weld

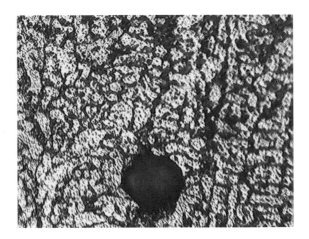

Fig. 10 Defect in the weld

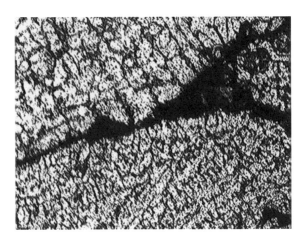

3.3 Tensile Properties

The tensile testing results of the TIG-welded joints of AA5083-F aluminum alloy 12 mm plates with ER4043 and ER4047 filler rods are given in Table 1. The tensile testing results indicate that ER4043 filler rod-welded joints produced better tensile properties than ER4047 filler rod-welded joints. The welded joint tensile strength of ER4043 filler rod-welded joints is 33% higher than that of the ER4047 filler rod-welded joints tensile strength.

Table 1 Tensile Testing of TIG-welded joints of AA5083-F with Al–Si filler rods

Filler rod	UTS (MPa)	El (%)
4043	150.80	4.00
4047	116.28	3.07

Srikrishna and Subbaiah [10] investigated low and high silicon content Al–Si alloy fillers for joining AA5083-H111 and AA6061-T6 aluminum alloys. According to their results, the tensile strength of welds of low silicon content filler rods (ER4043) (136 MPa) has been better when compared to high silicon content filler rod (ER4047) (115 MPa).

3.4 Hardness Curves

The hardness curve of 12 mm thick TIG-welded joints of AA5083-F aluminum alloy due to filler rods ER4043 and ER4047 are shown in Fig. 11. The hardness values of the welded joints at the weld center are less than the base metal value. On comparing the hardness values of the two filler rods used in this experiment, the hardness value of the TIG-welded joints of ER4047 is higher than the TIG-welded joints of ER4043. The loss of hardness in the weld is more in the case of ER4043 filler rod than that of ER4047 filler rod.

3.5 Discussion

The microstructure of the weld cross section of ER4043 filler rod-welded joints contains no visible defects, whereas ER4047 welded joint contains porosity and shrinkage cavity in the cross section. As the ER4047 filler rod-welded joint contains defects, it has produced lesser weld tensile strength (116.28 MPa) compared to ER4043 filler rod-welded joint weld tensile strength (150.80 MPa).

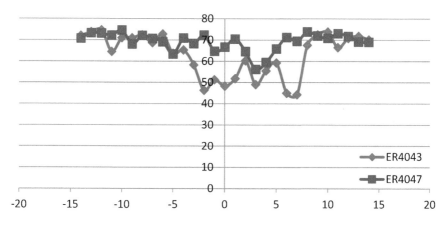

Fig. 11 Hardness curves of thick AA5083-F with ER4043 and ER4047 filler rods

The weld microstructure of TIG-welded joints with ER4043 filler (Fig. 4) was finer compared to the microstructure of TIG-welded joints with ER4047 filler (Fig. 8). The fine grains in the weld with ER4043 filler rod were responsible for higher tensile properties.

4 Conclusions

- The aluminum alloy AA5083-F was welded with two filler metals, viz., ER4043 and ER4047 and the following conclusions were obtained
- The weld microstructure of ER4043 contained relatively larger grains with Al–Si eutectics, whereas ER4047 weld microstructure contains equiaxed grains with Al–Si eutectics. The low Al–Si filler rod (ER4043) welded joints produced 33% of higher tensile property than the high Al–Si (ER4047) welded joints. The high Al–Si filler rod (ER4047) welded joint cross section had minimum drop in hardness when compared to low Al–Si filler rod (ER4043), which shows maximum drop in hardness.
- The ER4043 weld tensile properties were high irrespective of larger grains and porosity and because of lesser Al–Si eutectics formed in the weld. The ER4047 weld tensile properties were less, even though it contains equiaxed grains and porosity and because of huge volume Al–Si eutectics formed in the weld.
- The 12 mm thick AA5083 F plates can be TIG-welded with ER 4043 filler rod is more reliable than ER4047 filler rods.

Acknowledgements The authors are grateful to SSN Trust for their financial support through student internal funding.

References

1. Liu Y, Wang W, Xie J, Sun S, Wang L, Qian Y, Wei Y (2012) Microstructure and mechanical properties of aluminum 5083 Weldments by gas tungsten arc and gas metal arc welding. Mater Sci Eng A 549:7–13
2. Hadadzadeh A, Ghaznavi MM, Kokabi AH (2017) HAZ softening behavior of strain hardened Al-6.7 Mg alloy welded by GMAW and pulsed GMAW processes. Int J Adv Manuf Technol 92:2255–2265
3. Hadadzadeh A, Ghaznavi MM, Kokabi AH (2014) The effect of gas tungsten arc welding and pulsed gas tungsten arc welding processes parameters on heat affected zone softening behavior of strain hardened Al6.7 Mg alloy. J Mater Design 55:335–342
4. Jiang Z, Hua X, Huang L, Wu D, Li F (2017) Effect of multiple thermal cycles on metallurgical and mechanical properties during multi-pass metal arc welding of Al 5083 alloy. Inter J Adv Manuf Technol 93:3799–3811
5. Cam G, Ipekoglu G (2017) Recent developments in joining of aluminum alloys. Inter J Adv Manuf Technol 91:1851–1866

6. Rastkerdar E, Shamanian M, Saatchi A (2013) Taguchi optimization of pulsed current GTA welding parameters for improved corrosion resistance of 5083 aluminum welds. J Mater Eng Perform 22:1149–1160
7. Ishak M, Noordin NFM, Razali ASK, Razali ASK, Romlay FRM (2015) Effect of filler on weld metal structure of AA6061 aluminum alloy by tungsten inert gas welding. Inter J Auto Mech Eng 11:2438–2446
8. Manti R, Dwivedi DK, Agarwal A (2008) Microstructure and hardness of Al-Mg-Si weldments produced by pulse GTA welding. Inter J Adv Manuf Technol 36:263–269
9. Babu NK, Bhikanrao PY, Sivaprasad K (2013) Enhanced mechanical properties of AA5083 GTA weldments with current pulsing and addition of scandium. Mater Sci Forum 765:716–720
10. Srikrishna S, Subbaiah K (2017) Comparative investigation of the effect of Al-Si filler rods on tungsten inert gas welding of AA5083-H111 and AA6061-T6 aluminum alloys. J Chem Pharma Sci 7:207–209